Studies in Computational Intelligence

Volume 841

Series Editor

Janusz Kacprzyk, Polish Academy of Sciences, Warsaw, Poland

The series "Studies in Computational Intelligence" (SCI) publishes new developments and advances in the various areas of computational intelligence—quickly and with a high quality. The intent is to cover the theory, applications, and design methods of computational intelligence, as embedded in the fields of engineering, computer science, physics and life sciences, as well as the methodologies behind them. The series contains monographs, lecture notes and edited volumes in computational intelligence spanning the areas of neural networks, connectionist systems, genetic algorithms, evolutionary computation, artificial intelligence, cellular automata, self-organizing systems, soft computing, fuzzy systems, and hybrid intelligent systems. Of particular value to both the contributors and the readership are the short publication timeframe and the world-wide distribution, which enable both wide and rapid dissemination of research output.

The books of this series are submitted to indexing to Web of Science, EI-Compendex, DBLP, SCOPUS, Google Scholar and Springerlink.

More information about this series at http://www.springer.com/series/7092

Siddhartha Bhattacharyya ·
Debanjan Konar · Jan Platos ·
Chinmoy Kar · Kalpana Sharma
Editors

Hybrid Machine Intelligence for Medical Image Analysis

Springer

Editors
Siddhartha Bhattacharyya
Department of Information Technology
RCC Institute of Information Technology
Beliaghata, West Bengal, India

Jan Platos
Department of Computer Science
VSB-Technical University of Ostrava
Ostrava-Poruba, Czech Republic

Kalpana Sharma
Department of Computer Science
and Engineering
Sikkim Manipal Institute of Technology
Gangtok, Sikkim, India

Debanjan Konar
Department of Computer Science
and Engineering
Sikkim Manipal Institute of Technology
Gangtok, Sikkim, India

Chinmoy Kar
Department of Computer Science
and Engineering
Sikkim Manipal Institute of Technology
Gangtok, Sikkim, India

ISSN 1860-949X ISSN 1860-9503 (electronic)
Studies in Computational Intelligence
ISBN 978-981-13-8929-0 ISBN 978-981-13-8930-6 (eBook)
https://doi.org/10.1007/978-981-13-8930-6

This Springer imprint is published by the registered company Springer Nature Singapore Pte Ltd.
The registered company address is: 152 Beach Road, #21-01/04 Gateway East, Singapore 189721, Singapore

I would like to dedicate this book to my parents Late Ajit Kumar Bhattacharyya and Late Hashi Bhattacharyya, my beloved wife Rashni, and my dearest friends Jaydip, Sandipan, and Swarup, who have always stood by in times of bereavement.
Siddhartha Bhattacharyya

I would like to dedicate this book to my parents Mr. Debidas Konar and Mrs. Smritikana Konar who have encouraged me all the way and to my beloved daughter Sanvi.
Debanjan Konar

I would like to dedicate this book to my wife Daniela and my daughters Emma and Margaret.
Jan Platos

I would like to dedicate this book to my parents, wife, and my little hero.
Chinmoy Kar

I would like to dedicate this book to my family.
Kalpana Sharma

Preface

With the massive influx of multi-modality image data in the field of medical image processing, analysis and understanding the requirement for intelligent paradigms is always prompted for addressing the complex real-life problems of an array of medical imaging applications, including computer-assisted diagnosis, image-guided radiation therapy, landmark detection, imaging genomics, and brain connectomics. The objective of such a computational paradigm is to give rise to fail safe and robust solutions to the emerging problems faced by mankind in the domain of medical science and technology. Typical applications encompass the fields of medical image enhancement, segmentation, classification, object detection, to name a few. Recent years have witnessed the revolution in the fields of medical image analysis and computer vision by employing machine learning and computational intelligent algorithms. However, the applicability of these algorithms remains to be investigated in routine clinical practice owing to the unstructured and considering the volume of the medical image data produced. The traditional medical image processing approaches often need to cope with complex medical image processing problems with a profusion of images like MRI, CT scan, X-ray, and USG in various applications. Owing to the adaptive and complex behavior of intelligent computer vision programs with the advancement and considering the volume of medical image datasets, the combination of medical image analysis and understanding with machine learning algorithms received lots of attention. Imparting intelligence in a machine is the need of the hour. Several intelligent techniques have been in vogue over the years in this direction. Among these techniques, machine learning techniques stand in good stead. However, it is often noted that the classical machine intelligent techniques often fall short in offering a formidable solution in the fields of medical image analysis and pattern recognition. On and above, if the different components of the machine learning are conjoined together, the resultant hybrid machine intelligence is found to be more efficient and robust by design and performance. The expression "machine learning" usually refers to the ability of a computer to learn a specific task from data or experimental observation without explicitly programed. Even though it is commonly considered a synonym of machine intelligence, there is still no commonly accepted definition of machine learning. But generally, machine learning is a set of methodologies and approaches to address complex real-world problems to which mathematical or traditional modeling can be useless for a few reasons: The processes might be too complex for mathematical reasoning, it might contain some uncertainties during the process, or the process might simply be stochastic in nature. Machine intelligence, therefore, uses a combination of four main complementary techniques—the fuzzy logic which enables the computer to understand natural language, artificial neural networks which permits the system to learn experiential data by operating like the biological one, learning theory, and probabilistic methods which helps to deal with uncertainty imprecision. Except those main principles, currently popular approaches include biologically inspired algorithms such as swarm intelligence and artificial immune systems, which can be seen as a part of evolutionary computation, image processing, and artificial intelligence, which tend to be confused with machine intelligence. Although both machine intelligence (MI) and artificial intelligence (AI) seek similar goals, there is a clear distinction between them. Machine intelligence is thus a way of performing like human beings. Indeed, the characteristic of "intelligence" is usually attributed to humans. More recently, many products and items also claim to be "intelligent," an attribute which is directly linked to the reasoning and decision making. We are entering a phase where we are going to see the advances in digitization. As a consequence, the diversified field of applications in medical image processing has been evolved over the years with rapid advancement in digital technologies. An innovative integration of machine intelligence in medical image processing is very likely to have a great benefit to the field, which will contribute to a better understanding of complex and huge volume of unstructured medical images. The number of image processing algorithms that incorporate some learning components is expected to increase, as adaptation is needed. However, an increase in adaptation is often linked to an increase in complexity, and one has to efficiently control any machine intelligent technique to properly adapt it to medical image processing problems. Indeed, processing huge amounts of medical images means being able to process huge quantities of unstructured data often of high dimensions, which is problematic for most machine intelligent techniques. Therefore, interaction with the medical image data and with image priors is necessary to drive model selection strategies.

This volume comprises 11 well-versed chapters focusing on the latest advances in hybrid machine intelligence for medical image analysis.

Content-based medical visual information retrieval (CBMVIR) or content-based medical image retrieval (CBMIR) is an active computer vision and medical imaging research area. This is due to the steady increase in visual data production and the expanding assortment of restorative imaging information. CBMIR helps medical specialists in recovering related medical pictures plus case chronicles to understand the particular patient's infection or damage status and furthermore clarifies the therapeutic result to the patient. Chapter "Content-Based Medical Visual Information Retrieval" briefly surveys the various techniques for retrieval of medical images, and different methodologies adopted in various retrieval techniques are focused.

Image classification has wide applications in many fields including medical imaging. A major aspect of classification is to extract features that can correctly represent important variations in an image. Chapter "Pixel *N*-Grams Representation for Medical Image Classification" describes a novel feature extraction and image representation technique "Pixel N-grams" inspired from "Character N-gram" concept in text categorization. The classification performance of Pixel N-grams is tested on the various datasets including UIUC texture dataset, binary shapes dataset, miniMIAS dataset of mammography, and real-world high-resolution mammography dataset provided by an Australian radiology practice. The results demonstrate promising classification accuracy in addition to reduced computational costs, enabling a new way for mammographic classification on low-resource computers.

Segmentation is considered as one of the challenging and important processes in the field of digital image processing, and there are numerous applications like medical image analysis, satellite data processing, etc., where the digital image processing can be beneficial. Various algorithms have been developed by many researchers to analyze different medical images like MRI, X-rays, etc. Nuclear images is also a promising research topic in medical image analysis. Chapter "Penalized Fuzzy *C*-Means Enabled Hybrid Region Growing in Segmenting Medical Images" provides a hybrid scheme for image segmentation using fuzzy c-means clustering, region growing method and thresholding. The fuzzy c-means (FCM) clustering is used as a pre-processing step. It helps to process the image more accurately in further stages. To eliminate the noise sensitivity property of FCM, the authors have used the PFCM, i.e., penalized FCM clustering algorithm. In this work, the authors have used the region growing segmentation technique coupled with the thresholding to find the appropriate region. In the proposed work, a similarity feature depending on pixel intensity is used. The threshold value can be calculated using different techniques, such as iterative approach, Otsu's technique, local thresholding, manual selection, to determine the optimal threshold. Results are obtained and derived by applying the proposed method on different images obtained from publicly available benchmark datasets of human brain MRI images. Experimental results indicate that PFCM-supported hybrid model is far more superior in segmenting medical images than traditional image segmentation methods.

Diabetic retinopathy (DR) is a prime reason for escapable blindness in the world. As it progresses, the eyesight of the patient starts worsening, which may lead to blindness if not treated in an early stage. Medical image segmentation and analysis techniques are used for this type of detection of abnormality in retina that correlates and define the harshness of DR. Chapter "Classification of Diabetic Retinopathy Based on Segmentation of Medical Images" provides the method of DR medical image segmentation to automatically detect and classify the condition of DR. The chapter also discusses the feature extraction of blood vessels, optic disk, microaneurysm, exudates, and macula. The texture of features (gray-level co-occurrence matrix features, histogram intensity features, moment invariants, and gray-level run-length matrix features) finally classifies the DR images into four classes, viz.

normal, mild, severe, and proliferative. Result obtained with the method is found to be accurate and is able to correctly classify medical images.

Chapter “A New Hybrid Adaptive Cuckoo Search-Squirrel Search Algorithm for Brain MR Image Analysis” presents a new hybrid adaptive cuckoo search-squirrel search (ACS-SS) algorithm for brain (magnetic resonance) MR image analysis. Thresholding is one of the popular methods utilized for brain image segmentation. In this context, the authors present an optimal multi-level thresholding technique for brain MR images using edge magnitude information. The edge magnitude is computed using the gray-level co-occurrence matrix (GLCM) of the brain image slice. The optimum thresholds are found by maximizing the edge magnitude. A new hybrid evolutionary computing technique, namely ACS-SS, is investigated to maximize the edge magnitudes. The proposed scheme is tested with T2-w brain MR images from Harvard medical education database.

X-ray image-based pre-diagnosis modality is the cheapest way of dealing with any bone-related problems. Chapter “Analysis of Human Bone Disorder Using Fuzzy and Possibility Theory” is aimed at identification of the particular region eliminating bones from the flesh and cartilage in the X-ray image where bone disorder may occur. This facilitates the doctors for proper diagnosis. In this chapter, the authors employ possibility theory for analyzing the X-ray images to appraise the possibility of the regions having disorder. The novelty of the work is to facilitate the doctors concentrating only on automatic detection of region of interest (ROI) for correct diagnosis of the patients. The projected technique comprises (i) pre-processing to denoise the X-ray image using fuzzy inference system and (ii) isolation of the bone region from the rest of the X-ray image using Type 1 and Type 2 fuzzy-based edge discovery methods. In order to enable the experts for more precise diagnosis, region of interest (ROI) of the bone has been identified using possibility theory. Finally, experimental results of different bone regions using the proposed approach have been demonstrated and compared with the existing methods showing better performance and diagnosis.

The application of computer-aided algorithms for disease diagnosis and treatment is gaining prominence in the past years, and the role of machine learning algorithms is inevitable. Chapter “Segmentation of Anomalies in Abdomen CT Images by Convolution Neural Network and Classification by Fuzzy Support Vector Machine” focuses on the segmentation of liver and anomalies like tumor and cyst from abdomen CT images using deep learning convolution neural network (DLCNN) and classification of tumor stages by fuzzy support vector machine (FSVM). The segmentation result of DLCNN outperforms the backpropagation neural network, group method data handling neural network, and decision tree algorithm. The fuzzy support vector machine-based tumor classification results were superior when compared with classical SVM. This chapter focuses on the following: (i) machine learning algorithms for classification and segmentation of medical images, (ii) role of DLCNN in medical image segmentation, (iii) role of FSVM in anomalies classification, and (iv) validation of segmentation and classification results by performance metrics. The simulation results are generated in Matlab 2015a and Java and validated on real-time abdomen CT images.

The biofeedback therapies have been popular for long for the treatment of various mental challenges, stress, and chronic headache like TTH-type headache. Stress due to TTH is a very frequent occurrence in all our lives. ATTH symptom experienced for long by an individual with no relief or with increased frequency is termed as “distress.” Weakened cognitive and physiological control is due to TTH-originated stress and results in performance reduction. It can lead to symptoms like headache, gastrointestinal disturbances, elevated blood pressure, chest pain, insomnia, peptic ulcers, sexual dysfunction, skin ailments, etc. Many tools for the research purposes are being created and cross-checked by validation smart techniques, and academy and industries are in a race to get the patents for them so that they can create another mean for moneymaking and business in scientific creations. Chapter “An Optimized EMG and GSR Biofeedback Therapy for Chronic TTH on SF-36 Scores of Different MMBD Modes on Various Medical Symptoms” deals with data related to MMBD, like reliability and scalability, accessibility, heterogeneity, and quality of services for mental health. In the present chapter, the authors have made efforts to check the hypothesis and the efficacy of both biofeedback techniques on possible individual and combined modes.

Computed tomography (CT) has progressively increasing demand in the fields of medicine and industry due to its accuracy and fast image acquisition. CT scanner emanates X-ray to the human being during the image acquisition; it is evident that patient may be at huge risk in inducing cancer on human beings. Despite the radiation risk in CT, the increased attention must be needed during the investigation, and the problem of low-radiation dose is fetching an area of focusing. In wide ranging, it consists of two methods to reduce CT. Initially, the current density of X-ray tube and angle of projection are reduced without modification in radiation dose of each direction. Subsequently, the radiation dose is reduced, the noise increases gradually, and in turn, it will affect the quality of the image. In such case, the reconstruction of the image happens to be an ill-posed problem. Chapter “Proficient Reconstruction Algorithms for Low-Dose X-Ray Tomography” proposes the Landweber iterative algorithm for this purpose and compares the benefits accrued with existing iterative algorithms.

Deep learning performs as a computational tool with various potential utilities in ophthalmology. Retinal infections of the eye need to analyze small retinal vessels, microaneurysms, and exudates in the diagnosis of retinal diseases. Due to the appearance of various noises in the fundus images, the retinal vasculature is complicated to be analyzed for retinal conditions. Chapter “DVAE: Deep Variational Auto-Encoders for Denoising Retinal Fundus Image” focuses on the field of advanced deep learning in which a plethora of architecture is available with the increase in dimension and flexibility of the retinal fundus images. Removal of noise is an essential part to better visibility of noisy fundus, and thus, a deep learning method for degrading retinal fundus image restoration scheme has been suggested in this investigation. A deep convolutional denoising auto-encoder method based on total variational multi-norm loss function minimization with batch normalization approach has been introduced for the restoration of the fundus. The proposed scheme is utilized to restore the perceptible structural details of the fundus as well as to decrease the noise level. Moreover, the speed of the network for target noisy images is faster compared to other models after fine-tuning of the network with the dropout mechanism. The retinal image databases such as DRIVE, STARE, and DIARETDB have been adopted to assess image denoising effects.

Human gait is an important area of study in the fields of physical therapy, medical diagnostics, and biomedical engineering. Using multi-camera system associated with necessary software tools, the 3D human gait analysis is prepared in chapter “Optical Marker- and Vision-Based Human Gait Biomechanical Analysis.” The images captured from each camera are assembled together to develop a 3D model with the complete motion of human gait pattern. The passive optical markers are attached to the different locations of the lower limb of a human subject. The static position and the dynamic movement of the markers are obtained, and a 3D model is initialized. By taking the raw data from each camera, 3D tracking is performed. Also, a force plate placed in the walking platform assists in procuring kinetics data involved in movement and locomotion. Finally, the six joints of the lower limb are tracked, and an inverse kinematics along with inverse dynamics library for the human gait is developed and validated with analytical geometrical results.

The primary purpose of this book is to increase the awareness of medical image processing researchers to the impact of machine learning and computational intelligence algorithms. Thus, the book aims to introduce to the prospective readers the latest trends in machine learning technologies and computational intelligence for intelligent medical image analysis. The book can be used as a useful treatise for the undergraduates and postgraduates in the disciplines of computer science, electrical engineering, and electronics engineering. The editors would like to take this opportunity to render their deepest gratitude to Mr. Aninda Bose, Senior Editor, Hard Sciences, Springer India, for his immense support and cooperation during the project.

Kolkata, India — Siddhartha Bhattacharyya
Gangtok, India — Debanjan Konar
Ostrava, Czech Republic — Jan Platos
Gangtok, India — Chinmoy Kar
Gangtok, India — Kalpana Sharma
April 2019

Contents

About the Editors

Siddhartha Bhattacharyya completed his Ph.D. in Computer Science and Engineering at Jadavpur University, India, in 2008. Currently he is the Principal of RCC Institute of Information Technology, Kolkata, India. In addition, he is a Professor of Computer Application and Dean (Research and Development) of the institute. He served as the Editor-in-Chief of the International Journal of Ambient Computing and Intelligence (IJACI), published by IGI Global. He is the Associate Editor of the International Journal of Pattern Recognition Research, IEEE Access, Evolutionary Intelligence and a member of Applied Soft Computing editorial board. His research interests include soft computing, pattern recognition, hybrid intelligence and quantum computing, and he has published over 230 research articles and patents.

Debanjan Konar is an Assistant Professor at the Department of Computer Science and Engineering at Sikkim Manipal Institute of Technology, India. He is pursuing his Ph.D. at the Indian Institute of Technology, Delhi in Quantum Inspired Soft Computing. His research interests include quantum inspired soft computing, deep learning, machine learning, and natural language processing. He has published several papers in these areas in leading journals and IEEE international conferences. He is also a reviewer for various international journals and conferences.

Jan Platos received his Master's degree in Computer Science from the VSB-Technical University of Ostrava in the Czech Republic in 2006 and his Ph.D. in Applied Mathematics from the same university in 2010. Currently, Jan is an Assistant Professor at the Department of Computer Science, Faculty of Electrical Engineering and Computer Science at the VSB-Technical University of Ostrava. Jan is interested in various areas of computer science, including data compression and bio-inspired algorithms, information retrieval, data mining, data structures and data prediction. Jan is the co-author of more than 160 scientific papers published in proceedings and journals.

Chinmoy Kar is an Assistant Professor at the Department of Computer Science and Engineering, Sikkim Manipal Institute of Technology, and is pursuing his Ph.D. in Image Recognition at the Maulana Abul Kalam Azad University of Technology. His research interests include image processing and computational intelligence, and he actively publishes in these areas.

Kalpana Sharma is a Professor and Head of the Department of Computer Science and Engineering at SMI. She completed her Ph.D. Wireless Sensor Network Security at Manipal University in 2011. Her research interests include wireless sensor networks, security and real time systems.

Content-Based Medical Visual Information Retrieval

Pranjit Das and Arambam Neelima

Abstract Medical imaging gives essential visual data of interior body portions for the purpose of clinical investigation and restorative intercession which empowers doctors for analyzing and treating an assortment of maladies. Content-based medical visual information retrieval (CBMVIR) or content-based medical image retrieval (CBMIR) is an active computer vision and medical imaging research area. This is due to the steady increase in visual data production and the expanding assortment of restorative imaging information. CBMIR helps medical specialists in recovering related medical pictures plus case chronicles to understand the particular patient's infection or damage status and furthermore clarifies about the therapeutic result to the patient. It likewise assists the medical specialist for setting up the report of specific determination all the more precisely. Other than the diagnostics, CBMIR likewise helps with instructing and research. This chapter briefly surveys the various techniques for retrieval of medical images and different methodologies adopted in various retrieval techniques are focused. For retrieval of medical images, a new feature descriptor based on Krawtchouk moment and histogram of oriented gradient (HOG) is also proposed in this chapter. Krawtchouk polynomial-based moments set defines the Krawtchouk moment, and it provides a fine description of the image shape. HOG features compute the gradient orientation happenings in the local region of an image. It mainly extracts local shape information from an image based on the intensity gradient distribution or edge direction. All the experimentations are conducted on high-resolution computed tomography (HRCT) lungs pictures and MRI images of different part of the body. Some other techniques are considered for the comparison with proposed method.

Keywords Medical imaging · Image retrieval · HOG · Feature descriptor

---

P. Das (✉) · A. Neelima
Department of CSE, NIT Nagaland, Dimapur, India
e-mail: 91pranjitdas@gmail.com

A. Neelima
e-mail: neelimaarambam@yahoo.in

S. Bhattacharyya et al. (eds.), *Hybrid Machine Intelligence for Medical Image Analysis*, Studies in Computational Intelligence 841,
https://doi.org/10.1007/978-981-13-8930-6_1

1 Introduction

Medical or restorative imaging has historically targeted on the acquirement of anatomical and physiological data about patients. Medical imaging first began with the discovery of X-ray in 1895 by Wilhelm Conrad Roentgen who got the first Nobel Prize in 1901. Diagnosis of many diseases is now performed by computed tomography (CT) which was previously done by fluoroscopy. Computed tomography is also called computed axial tomography (CAT scanning). In 1972, two scientists invented CT scan working independently. Godfrey Hounsfield from EMI labs invented in England and Allan Cormack from Tufts University, Massachusetts, invented in the USA. In the early 1970s at Atkinson Morley's Hospital, first patient head scan was performed by Hounsfield and Ambrose [1, 2] (Fig. 1).

X-ray beam is likewise the premise of mammography, which captures the breast images. Mammography captures high-resolution breast images and is mainly used for detecting breast cancer. For detection of brain tumors, an ultrasound beam was passed through the human skull by Neurologist Karl Dussik in 1942 and became the first person to utilize sonography in medical diagnoses. Nuclear medicine imaging is another medical imaging modality in which radiopharmaceuticals are provided to the patient's body intravenously or orally. Radiopharmaceuticals emit radiation, which is detected with the help of gamma cameras and generates images. Unlike X-ray, this process does not give any radiation to the patient's body for capturing the images. Techniques of diagnostic nuclear medicine include scintigraphy (2D) and SPECT (3D). Scintigraphy uses the internal radionuclides for generating 2D images, and SPECT is three-dimensional tomographic process in which gamma camera data is used for different projections [3, 4].

In the 1970s, magnetic resonance imaging (MRI) was included in radiology and performed a vital role in medicine, and it is mainly used to generate images for

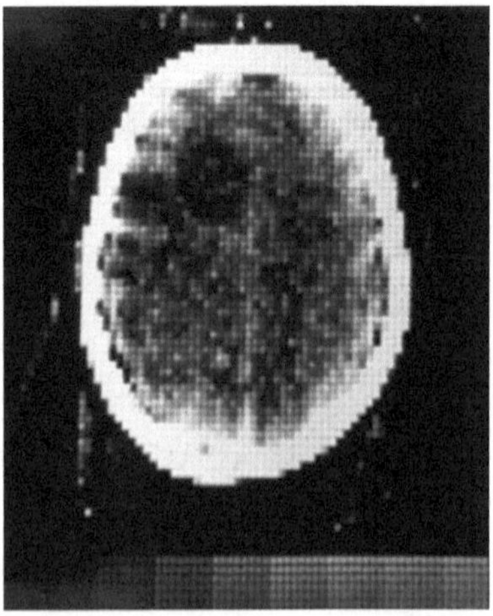

Fig. 1 First clinical CT scan (head portion) (*Courtesy* ImPACT groups)

Table 1 Human organs with their respective problems which can be detected by MRI

Organs	Respective problems
Spinal cord, brain	Blood vessel injury, stroke, brain injury, cancer, hardening or thickening of any tissue, spinal cord injuries, etc.
Heart, blood vessels	Blockage of blood vessels, injury happen by a heart attack, heart disease, heart structure troubles, etc.
Bones, joints	Bone contagions, cancer, joint problems, trouble in spine disc, etc.
Breasts, liver, kidneys, ovaries, pancreas, prostate	To check the organ wellness, etc.

detecting various diseases throughout the body. It is mostly used when other testing technique fails to generate adequate information about the patient. Radio waves, magnetic fields with high strength and magnetic field gradients are used in MRI to generate biomedical images of different internal body structures. Unlike CT scans and positron emission tomography (PET) scan, MRI does not use X-rays or ionizing radiation. MRI is a nuclear magnetic resonance (NMR) application. In NMR, very high frequency radio waves are absorbed in atomic nuclei which is subject to a stationary magnetic field of high strength. In 1946, Felix Bloch and Edward M. Purcell were the first who observed this marvel independently of each other. The dis-functioning of an organ can be detected through MRI, and Table 1 highlights different organs along with their specific problems. A specified MRI was also introduced to measure brain activity based on blood flow, which is known as functional MRI (fMRI). It can find brain problems like effects of a brain stoke and also helps in brain surgery for epilepsy or tumors [5, 6].

To check the stiffness of soft tissue, a new biomedical imaging technique used which is known as elastography. Elastography mainly uses ultrasound or MRI for generating anatomical pictures and stiffness map for comparison. It is used for decease investigation of liver, diagnosis of breast, thyroid, prostate cancers, musculoskeletal imaging, muscle condition, tendons condition, stiffness of the brain, etc. [7–11]. Apart from this biomedical imaging, some other imaging techniques are also included in imaging modalities like photoacoustic imaging, echocardiography, functional near-infrared spectroscopy, magnetic particle imaging, etc. Later, PACS and DICOM took in biomedical imaging which brought revolution in digital imaging. PACS mainly provides storage, sharing, presentation and recovery of pictures. DICOM is the imaging standard for communicating and storing the information in biomedical imaging [12–14].

In this period, with the ever-growing wide variety of hospitals and medical establishments, the variety of medical images associated with affected person diagnosis is substantially growing. For inquiry and recovery of medical images, productive usage of this medical image database is essential for precise patient finding and for any other research-related development in medicine. Image retrieval becomes a very important and effective tool in radiology for setting up the report of specific determination all the more precisely. Other than the diagnostics, CBMIR likewise helps in instructing and research.

Image retrieval is of two categories: based on text (concept-based retrieval) and based on the image content which is called content-based image retrieval (CBIR). In the 1970s, concept-based was introduced in which images are retrieved using text information that is associated with the images. Each image is index with some text like caption, keywords, subject heading, natural language text, etc., and this information is utilized for image retrieval purpose. The description-based techniques are reliable and quick when images are substantially annotated. Text-based image retrieval cannot be applied for unannotated image dataset, and annotation process of these techniques is very time consuming and immanent which are the major disadvantages of text-based methods. Text-based methods mostly retrieve many irrelevant images. To overcome the drawbacks of text-based methods and to provide an efficient image search and retrieval, CBIR techniques were introduced. In CBIR techniques, pictures are retrieved using features like shape, color, texture, etc. Later, many CBIR systems were developed for the commercial purpose like: Netra [15], QBIC [16], VisualSEEK [17], Virage [18], Photobook [19], etc. [20].

2 Content-Based Medical Image Retrieval

Initially, retrieval method for biomedical images was included in normal CBIR methods for trails [21, 22] but the retrieval accuracy was not good. This is because of the complex structures present in the medical images, region of interest (ROI) is present in very minute area, mostly the medical images are of gray scale, etc. Later, separate CBIR for medical purpose was introduced which is known as CBMIR, and it performs a vital role in the field of medicine. Many CBMIR systems were then developed viz. FIRE [23], BRICS [24], IRMA [25], etc. In CBMIR systems, the selection of the features and methodology is very crucial part. The efficiency of the CBMIR systems depends on the feature selection and methodology used. Various concepts and theories are adopted to develop efficient CBMIR techniques.

The local feature descriptors are very essential in the field of biomedical imaging for extracting robust visual features. In 1994, a visual descriptor namely local binary pattern (LBP) was first described by Ojala et al. [26]. The texture spectrum model includes the LBP as a specific case of the model which was introduced in 1990 [27, 28]. LBP mainly extracts the particular target pixel and their nearby pixels' relationships. Then, a modified version of LBP is proposed in [29] which is known as local ternary pattern (LTP). Unlike LBP, it uses three values (-1, 0, 1) to threshold the image pixel. LBP and LTP utilize hand-determined coding that is constrained to little spatial backings and common gray level examinations. Another descriptor namely local quantized pattern (LQP) [30] was proposed for large or deep patterns. LQP uses vector quantization with a lookup table, and it acquires a portion of the adaptability and intensity of visual word portrayals with good run time complexity and simple local pattern. For CBIR system, a new descriptor was presented by Murala et al. [31] which compute direction-based edge features in the direction of 0°, 45°, 90° and 135° in an image. With the concept of LQP and DLEP, Rao and Rao [32] introduced another local descriptor which is known as local quantized extrema patterns (LQEPs). Initially, quantized information of an input image is extracted, and from this quantized information, directional extrema are computed. The computed LQEP is then integrated with the RGB histogram (for collecting the color information) to generate the feature vector. For retrieval and indexing, they again developed another feature descriptor [33] which gathers the spatial relationship as large or profound texture pattern depending on gray value variation. For a specific center pixel, the directional local extrema are explored with the direction 0°, 45°, 90° and 135°. From quantized extrema, a mesh structure is generated to compute the important texture information. RGB histogram finally combined with the generated LMeQEP to enhance retrieval performance.

Deep et al. [34] presented DLTerQEP based on LBP, LTP and LQEP. DLTerQEP is a local pattern descriptor for extracting the spatial relationship among all the nearby pixels of a given center pixel. For more performance, additional spatial structure information is extracted, and it utilizes the ternary pattern from local extrema values that are computed in $HVDA_7$ structure from the given input image. In the same year, they introduced another pattern namely local mesh ternary pattern [35]. This pattern computes the relationships between the neighborhoods for a reference pixel, and it utilizes the ternary pattern of the extracted mesh pattern from a given image. Later, they developed another feature descriptor for medical image retrieval named local quantized extrema quinary pattern (LQEQryP) [36]. LQEQryP likewise gives a critical increment in discriminative power by permitting more neighbor pixels which result in bigger local pattern. With the concept of LBP and local neighborhood difference pattern (LNDP), a new retrieval technique was introduced by Verma et al. [37]. In LNDP, the relationships among each neighbor pixels are transformed into a binary pattern, and it is combined with LBP to capture more information from the images. Recently, Bhunia et al. [38] proposed a new feature descriptor which combines the texture and color features. For color component, it explores the relationship among Hue and Saturation channels in HSV color model, and for texture component, it considers the diagonally symmetric co-occurrence texture pattern with a 3×3 window.

Another local pattern [39] based on binary wavelet transform (BWT) was introduced in 2012. To generate the multiresolution binary images, BWT is executed on every bit plane, and from the generated BWT sub-bands, local binary patterns are computed which represents the final feature vector of the input image. Later, Murala et al. [40] introduced another new local pattern descriptor named local tetra patterns (LTrP) for image retrieval. In 2013, a new local pattern LTCoP was presented in [41]. LTCoP extracts ternary edges co-occurrence which is computed using gray values and later it is combined with Gabor transform to enhance the performance. An image retrieval technique with local mesh pattern (LMeP) [42] was proposed in 2014. Unlike LBP, it computes relationship between the neighbor pixels. It is also combined with Gabor transform to enhance the performance and results in Gabor local mesh pattern (GLMeP).

In retrieval of medical image, another descriptor concept namely "barcode" was added. Initially, Tizhoosh [43] introduced radon barcodes (RBC) which were used to annotate the images and their target regions like affected area, organs, tumors, etc. LBP and local radon binary pattern (LRBP) were also executed as barcode. They conducted the experimentation on IMRA X-ray dataset. The radon barcode possesses few demerits like many unique information may get lost in barcode and it also does not capture the general curvature of the projection. MinMax radon barcodes [44] were introduced to overcome the demerits of the radon barcodes. To remove low peaks (minimum)/valleys (maximum), a moving average is employed by the smoothing function of MaxMin barcode. Values present between peaks and valleys are detected after computing all the peaks and valleys. Based on the corresponding values of zeros or ones, the projections were computed using 0s' and 1s' associated values. Tizhoosh et al. [45] presented a new radon barcode which is called "autoencoded radon barcodes (ARBC)." It is a compressed radon projection which possesses minimum data loss when contrasted with basic thresholding of neighborhood. ARBC additionally decreases the radon projection redundancies for accomplishing better accuracy in image retrieval. With the concept of radon barcode and Gabor transform, two new algorithms were proposed by Nouredanesh et al. [46]. In the first one, radon transform was initially applied on input image and later it was combined with Gabor transform and lastly, binarization was employed over generated feature vector. In guided-radon-of-Gabor barcodes, Gabor transform was applied first and then radon transform utilized to an unlike orientation. An improvement of radon barcode was introduced by Babauie et al. [47] which was compressed and fast for image retrieval in huge dataset, but because of thresholding part, it may lose some important information.

In image analysis and computer vision, convolution neural network (CNN) and deep convolution network of CNN became very popular and effective technique. Initially, Ivakhnenko et al. [48] began the concept of deep learning and later many researchers performed various work [49–51] based on deep learning. Deep learning concept was also employed on CBIR techniques [52–54] which accomplished an impressive retrieval performance. Medical imaging adopted the CNNs concept in the 1990s for achieving better performance [55]. Anthimopoulos et al. [56] proposed a five convolution layered deep network to perform the classification of interstitial lung diseases pattern. The proposed network composed of five-convolution layer, 2×2 kernels and LeakyReLU activation, average pooling and three-dense layers. By combining a discriminative and a generative learning objective, Tulder et al. [57] presented a restricted Boltzmann machine for airway detection and respiratory organ texture classification in computed tomography pictures. The concept of deep learning was also implemented in biomedical image retrieval to achieve a better retrieval rate. For biomedical image retrieval, a multimodal using statistical graphical model with deep learning was proposed in [58]. For coverage of semantic gap, they combine visual and textual data. Then, for deriving losing modality, the joint density model of multimodal information was learned by proposing a new multimodal using deep Boltzmann machine. Sun et al. [59] developed an algorithm based on deep learning and also combined with RISC/PACS to enhance the retrieval speed and accuracy. Qayyum et al. [60] designed a deep learning framework for CBMIR using deep convolution neural network which specially trained for biomedical image classification. Recently, some other techniques [61–63] for retrieval of biomedical images using deep learning were designed.

Apart from above concepts and methods, many other methodologies are also adopted in CBMIR system viz. support vector machine, unified medical language system, fuzzy and other neural network techniques, Gaussian mixture model, etc. Many literature surveys [20, 64–66] were conducted on CBMIR systems describes various techniques and analyzed in details. In the proposed method, the concept of Krawtchouk moment is used to extract the global shape features. Yap et al. proposed the Krawtchouk moment [67, 68]. They presented new orthogonal moments dependent on the discrete traditional Krawtchouk polynomials. In 2007, Venkataramana et al. [69] used radial Krawtchouk moment for pattern recognition. They utilize the polar portrayal of a picture with a 1D symmetrical discrete weighted Krawtchouk polynomials and a roundabout function. Radial Krawtchouk moments were utilized by Zhu et al. [70] for retrieval of medical images. Tziola et al. [71] presented a new retrieval system based on Krawtchouk chromaticity distribution moments (KCDMs). Through a KCDMs set employed over the related chromaticity dissemination work in a shading space, the chromaticity depicted in the proposed method. For developing a new CBIR system, Raj et al. [72] used the concept of KCDM. Krawtchouk moment in 3D form was defined by Xiang et al. [73] for the analysis of 3D shape and their retrieval. Apart from the above-mentioned works, many other works [74–78] based on Krawtchouk moment has been done.

3 An Efficient Feature Descriptor for Biomedical Image Retrieval

For retrieval and indexing of images, a new and efficient feature descriptor is proposed which used the concept of Krawtchouk moments and HOG. Unlike other orthogonal moments, Krawtchouk moments extract local features from images with minimum information reduction. Krawtchouk moment features with various order is shown in Table 2.

Input: Image pixel values (intensity value) and order (value for m and n)
Output: moment features for different order Q_{mn} computed using Eq. 1.

Table 2 Krawtchouk moment features for different orders

	$n = 0$	$n = 1$	$n = 2$
$m = 0$	Q_{00}	Q_{01}	Q_{02}
$m = 1$	Q_{10}	Q_{11}	Q_{12}
$m = 2$	Q_{20}	Q_{21}	Q_{22}

Krawtchouk moment Q_{mn} is defined by:

$$Q_{mn} = \sum_{i=0}^{M-1} \sum_{j=0}^{N-1} \overline{K}_n(i; q_1, M-1)\overline{K}_n(j; q_2, N-1) f(x, y) \tag{1}$$

where

m, n	order of the moment
$f(i, j)$	image pixel values
$M \times N$	Image size
$\overline{K}_m(i; q_1, M-1)$	weighted Krawtchouk polynomial with mth order defined as.

$$\overline{K}_{m+1}(i; q, M-1) = \left(\frac{\sqrt{\frac{q(M-1-m)}{(1-q)(m+1)}}((M-1)q - 2mq + m - i)}{q(M-1-m)} \overline{K}_m(i; q, M-1) \right) - \left(\frac{\sqrt{\frac{q^2(M-1-m)(M-m)}{(1-q)^2(m+1)m}}(m(1-q))}{q(M-1-m)} \overline{K}_{m-1}(i; q, M-1) \right) \tag{2}$$

for $n = 1, 2, \ldots, M-2$

With

$$\overline{K}_0(i; q, M-1) = \sqrt{w(i; q, M-1)} \tag{3}$$

and

$$\overline{K}_1(i; q, m-1) = \left(1 - \frac{i}{q(M-1)}\right) \sqrt{\frac{q(m-1)}{(1-q)} \sqrt{w(i; q, M-1)}} \tag{4}$$

$w(i; q, M-1)$ is recursively computed as follows:

$$w(i+1; q, M-1) = \left(\frac{M-1-i}{i+1}\right) \frac{q}{1-q} w(i; q, M-1) \tag{5}$$

with $w(0; q, M-1) = (1-q)^{M-1} q \in (0, 1)$

Reconstructed form of Krawtchouk moment with order till $M-1$ is computed as follows:

$$f(i, j) = \sum_{i=0}^{M-1} \sum_{j=0}^{N-1} Q_{mn} \overline{K}_m(i; q_1, M-1)\overline{K}_n(j; q_2, N-1) \tag{6}$$

After that, computation of HOG features is done from the input medical image which mainly computes the happenings of gradient orientation of local areas in input picture. Input picture is first divided into many small areas which are known as "cells/patches" and cells are of two categories: rectangular HOG and circular HOG. Then, based on the edge orientation, a histogram is constructed within the cells. Likewise, histogram is computed for every cell, and all histograms are integrated to compute the HOG features for given picture. At last, the computed HOG feature vector of the input image is integrated with Krawtchouk moment features for generating the final feature vector which represents the proposed feature descriptor for an image.

3.1 Histogram of Oriented Gradients (HOG)

HOG is the image feature mainly utilized on computer vision with the end goal of object detection and many other similar tasks. The fundamental idea behind the HOG descriptor is that neighborhood object's visual aspects and shape inside the picture can be portrayed by the dispersion of force inclinations or edge headings. The image gradient is a vector quantity which has magnitude as well as direction, and image gradient at any pixel of an image describes the intensity change in that area. When the orientation of gradients is considered, it is known as oriented gradients. Gradient orientations are computed for a small area (cell/patch), and their orientations are plotted in the histogram which describes the likelihood of the presence of an inclination with an explicit introduction in that cell. The HOG descriptor has a few key point of interest over various descriptors. Since it takes a shot at neighborhood patches, it is geometric invariant and invariant to luminous flux changes, alongside thing demonstration. Such switches would simply appear in greater spatial regions. The algorithm for extracting the HOG features is as follows:

1. Split the image into several patches and histogram of gradient directions are computed for each patch
2. Based on the gradient orientation, each patch is discretized into angular bins
3. Angular bins filled with weighted gradients
4. Grouping of patches to form several blocks
5. Formation of block histogram by normalizing histogram groups.

4 Proposed Biomedical Image Retrieval System

The proposed feature descriptor is used in this retrieval system for extracting the features. The proposed descriptor is based on Krawtchouk moment and HOG. Krawtchouk polynomial-based moments set defines the Krawtchouk moment, and it provides a fine description of the image shape information. HOG features compute

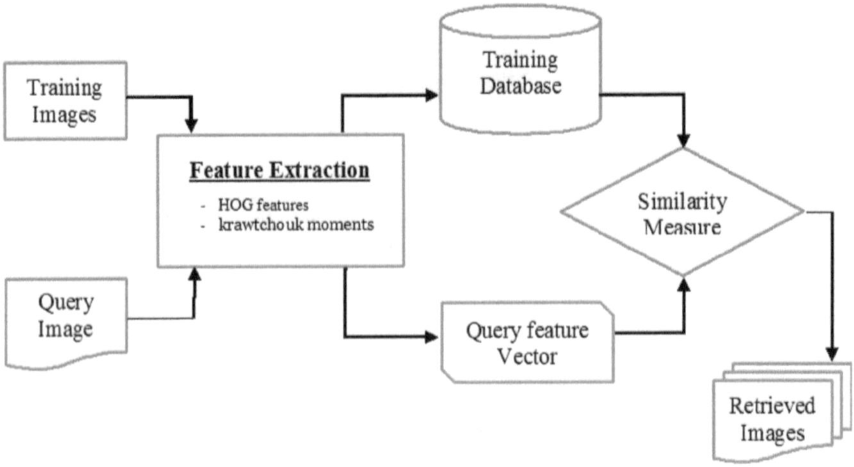

Fig. 2 Framework of the proposed system

the gradient orientation happenings in the local region of an image. It mainly extracts local shape information from an image based on the intensity gradient distribution or edge direction. The proposed retrieval system model is depicted in Fig. 2 and the algorithm is described below:

Algorithm:

1. Covert input image into gray scale using Eq. 7 if the input image is not gray scale.
2. Compute Krawtchouk moment features of the image.
3. Compute HOG features of the image.
4. Krawtchouk moment and HOG feature vectors are combined to represent the feature descriptor.
5. The distance of query image from all database images is computed using Eq. (8).
6. Top n best match images are retrieved.

In order to covert image into gray scale, grayscale values are computed for every pixel (i,j) by computing weighted sum for each color component of RGB is computed using Eq. 7.

$$\text{Grayscale\_value}(i, j) = 0.2989 \times R(i, j) + 0.5870 \times G(i, j) + 0.1140 \times B(i, j) \tag{7}$$

where R is red component, G is green component and B is blue component.

4.1 Similarity Measure

After extraction of features, the query biomedical image is represented by V_q. Training database images are represented as $V_{dB} = (V_{DB1}, V_{DB2}, V_{DB3}, \ldots, V_{DBn})$; training database DB contains "n" images. For m most related images retrieval, the distance of query image from all database images computed using Eq. (8). The similarity measure used for the comparison is represented as follows:

$$D(f_q, f_{\mathrm{db}_j}) = \sum_{i=1}^{L_g} \left| \frac{V_{\mathrm{db}_{ij}} - V_{qj}}{1 + V_{\mathrm{db}_{ij}} + V_{qj}} \right| \quad (8)$$

where

$V_{\mathrm{db}_{ij}}$ jth feature of ith image of training dataset db,
V_{qj} jth feature of the query image,
L_g size of feature vector.

5 Experiment Results and Discussion

All the experiments of new retrieval system are conducted on HRCT dataset [79] and NEMA MRI dataset [80]. The dataset HRCT contains images from several cases with interstitial lung diseases (ILDs). The dataset contains 128 patients affected by one among thirteen histologic diagnoses of ILD, 108 image sets with annotated respiratory organ, tissue patterns of 41 L. And the NEMA MRI dataset contains MRI images of various body parts. Figure 3 presents some random sample images of HRCT dataset and NEMA MRI dataset. Average retrieval precision (ARP), average retrieval rate (ARR), F_score and mean average precision (mAP) are considered for evaluating the retrieval accuracy of the system.

$$\text{precision } (p) = \frac{\text{no.of relavent images retrieved}}{\text{Total no.of images retrieved}} = \frac{r(\mathrm{db}_i)}{n} \quad (9)$$

$$\text{recall } (r) = \frac{\text{no.of relavent images retrieved}}{\text{no.of relevent images in database}} = \frac{r(\mathrm{db}_i)}{g(\mathrm{db}_i)} \quad (10)$$

$$\mathrm{ARP} = \frac{100}{w} \sum_{i=1}^{w} p_i \quad (11)$$

$$\mathrm{ARR} = \frac{100}{w} \sum_{i=1}^{w} r_i \quad (12)$$

$$\text{F\_score} = \frac{2 \times \mathrm{ARP} \times \mathrm{ARR}}{\mathrm{ARP} + \mathrm{ARR}} \quad (13)$$

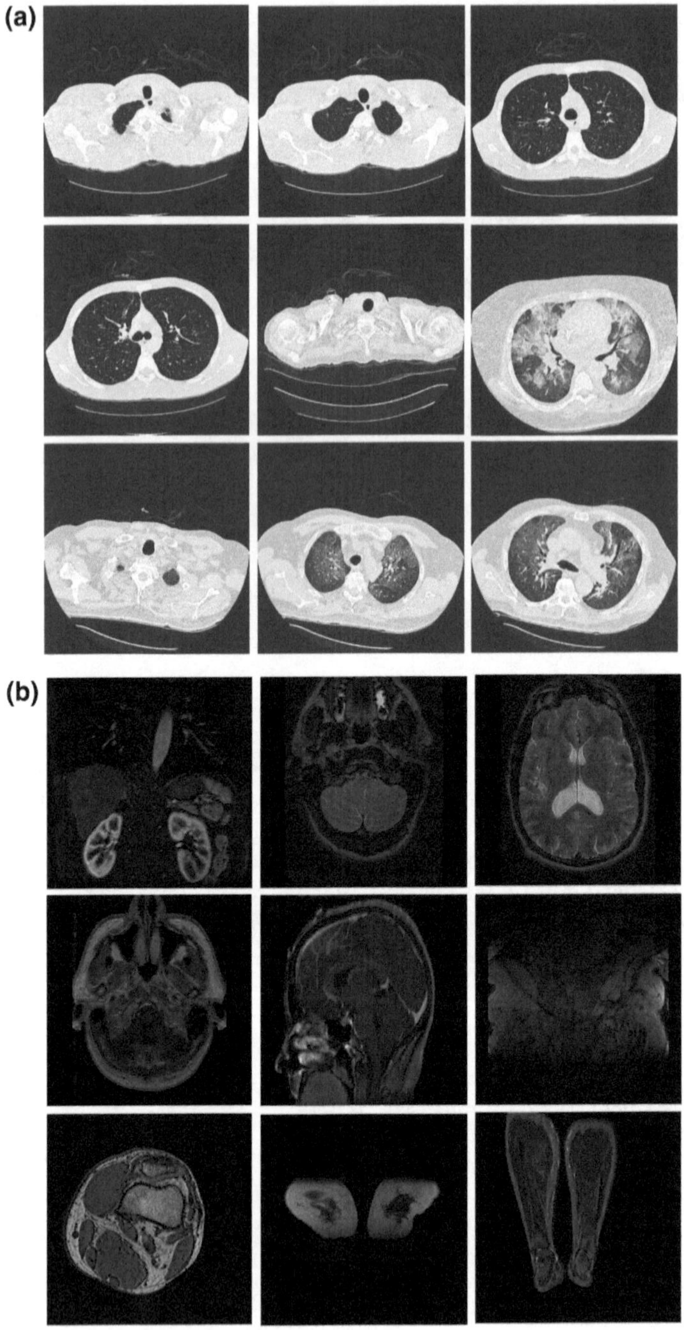

Fig. 3 Sample biomedical images from **a** HRCT dataset and **b** NEMA MRI dataset

$$\text{mAP} = \frac{100}{w} \sum_{i=1}^{w} \sum_{n=1}^{g(\text{db}_i)} \frac{r(\text{db}_i)}{n} \tag{14}$$

where

i	query image "i" in the database db,
p_i	precision for "i,"
r_i	precision for "i,"
w	size of database "db,"
n	Total no. of images retrieved for "i,"
$r(\text{db}_i)$	No. of relevant images retrieved for "i,"
$g(\text{db}_i)$	No. of relevant images in the database for "i."

To obtain the result, an image from the database is given as an input query to the retrieval system and processed through each step, and finally, distance of query image from all database images are computed in order to retrieve the relevant images.

From the Figs. 4 and 5, it is ascertained that the proposed CBMIR system accomplishes better performance on HRCT dataset as compared to other existing methods viz. DLTerQEP, LBP, LTrP, LTCoP and LMeP. Performance comparison with other techniques on NEMA MRI dataset is shown in Figs. 6 and 7. Table 3 summarizes the ARP and ARR, and Table 4 presents the F_score and mAP of various methods on HRCT dataset. Tables 5 and 6 show accuracy comparison of various methods including proposed technique on NEMA MRI dataset using different metrics: ARP, ARR, F_score and mAP.

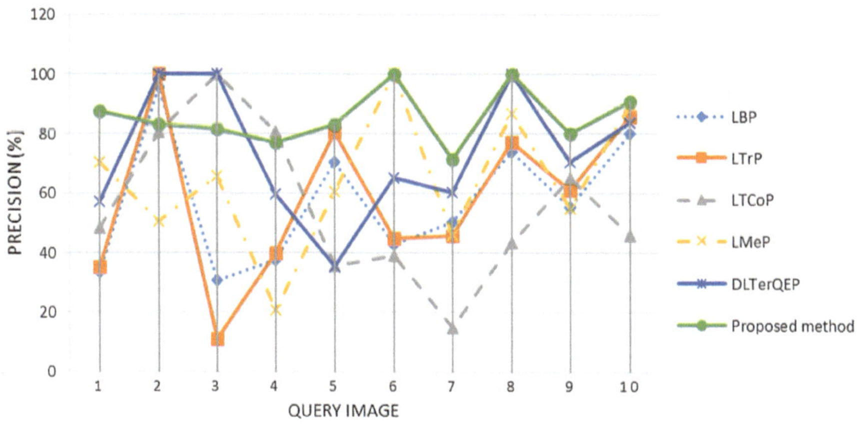

Fig. 4 Comparison of precision (%) with other techniques on HRCT dataset

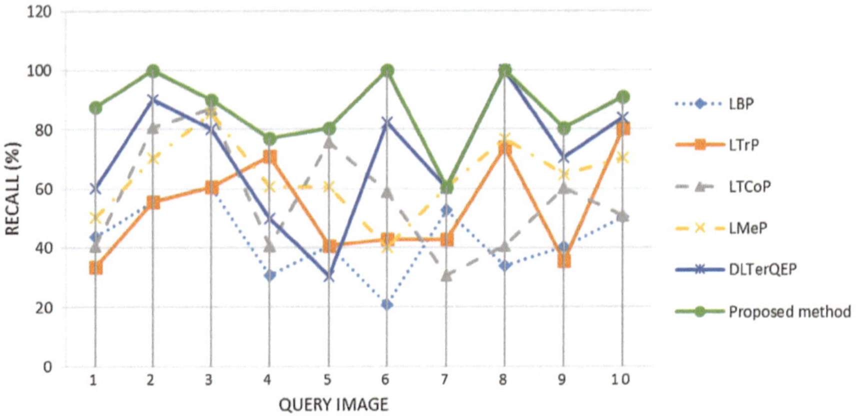

Fig. 5 Comparison of recall (%) with other techniques on HRCT dataset

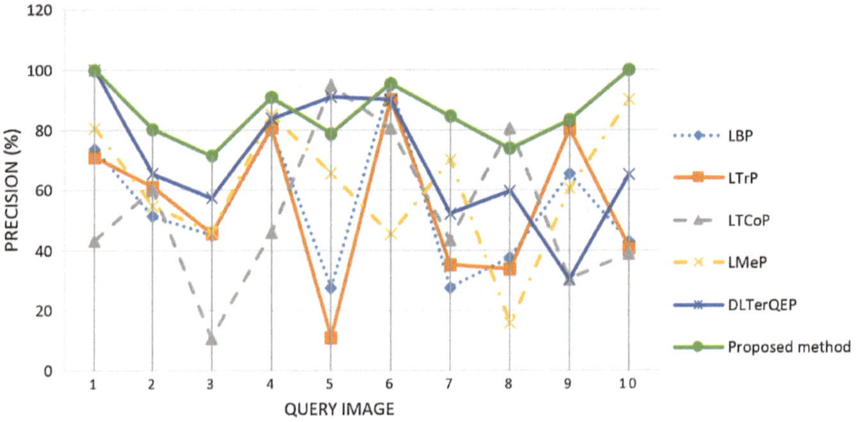

Fig. 6 Comparison of precision (%) with other techniques on NEMA MRI dataset

Table 3 Comparison of ARP (%) and ARR (%) with various techniques on HRCT dataset

Methods	LBP	LTrP	LTCoP	LMeP	DLTerQEP	Proposed method
ARP (%)	46.65	52.96	52.71	58.56	68.70	79.84
ARR (%)	55.35	57.27	55.00	61.33	79.33	87.20

Table 4 Comparison of F_score (%) and mAP (%) with various techniques on HRCT dataset

Methods	LBP	LTrP	LTCoP	LMeP	DLTerQEP	Proposed method
F_score (%)	50.62	55.03	53.83	59.91	73.63	83.35
mAP (%)	54.39	55.16	54.67	60.41	77.45	85.90

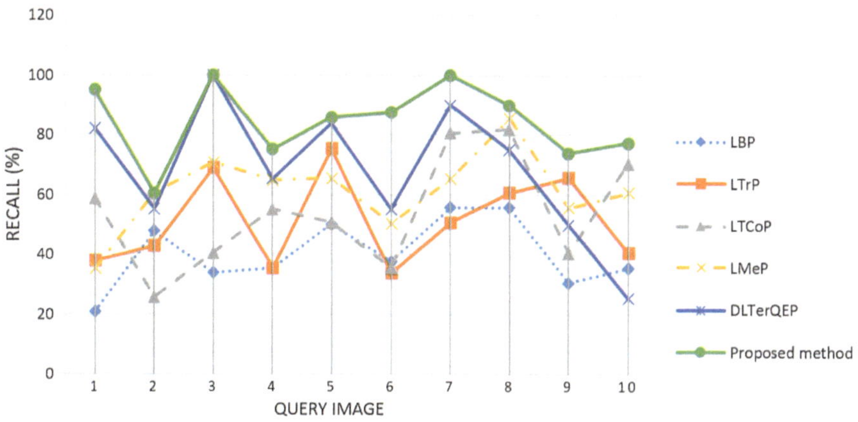

Fig. 7 Comparison of recall (%) with other techniques on NEMA MRI dataset

Table 5 Comparison of ARP (%) and ARR (%) with various techniques on NEMA MRI dataset

Methods	LBP	LTrP	LTCoP	LMeP	DLTerQEP	Proposed method
ARP (%)	37.25	42.77	42.36	49.69	60.70	71.82
ARR (%)	46.35	47.80	44.88	53.41	71.33	80.75

Table 6 Comparison of F_score (%) and mAP (%) with various techniques on NEMA MRI dataset

Methods	LBP	LTrP	LTCoP	LMeP	DLTerQEP	Proposed method
F_score (%)	41.30	45.15	43.58	51.48	65.58	76.02
mAP (%)	48.69	50.16	44.10	55.14	72.76	82.83

6 Conclusion

In this chapter, the recent advancements in medical image retrieval and different methodologies adopted in various medical image retrieval techniques are summarized in details. For retrieval of medical images, a Krawtchouk moment and HOG feature-based descriptor are also proposed in this chapter. The Krawtchouk moment is a set of moments based on Krawtchouk polynomials, and it gives fine description of the shape of an image. HOG features compute the gradient orientation happenings in the local region of an image. It mainly extracts local shape information from an image based on the intensity gradient distribution or edge direction. For performance evaluation, all the experiments are conducted on HRCT dataset and NEMA MRI dataset. The performance is compared with some of the existing descriptors and achieved a good retrieval rate with 79.84% of ARP and 87.20% of ARR on HRCT dataset. And 71.82% of ARP and 80.75% of ARR are achieved on NEMA MRI dataset.

Restorative pictures have consistently been utilized for recovery frameworks, and the therapeutic space is a great part of the time alluded to as a major area for the substance-based access advances as far as potential effect. The biomedical image retrieval has gone far from simply hypothetical research center style advancements, where without any clinical application single pictures were characterized into few, toward devices or tools that join visual and clinical information to truly help conclusion and convey important data to the clinicians. Numerous as of now CBIR frameworks just utilize little datasets but not refresh and it essential for genuine application with a coordination of the frameworks into clinical work process, or, in other words. Several specialized methodologies are presently accessible for applications with huge scale yet additional task is expected in really incorporate the apparatuses for clinical effect, and this incorporates utilization of deep learning and discussing the overall performance with the doctors.

References

1. Igbaseimokumo, U.: Brain CT Scans in Clinical Practice. Springer Science & Business Media, Berlin (2009)
2. Philbin, T.: The 100 Greatest Inventions of All Time: A Ranking Past and Present. Citadel Press, New York (2005)
3. Dorland, W.: Dorland's Medical Dictionary for Health Consumers. Saunders, an imprint of Elsevier, Philadelphia (2007)
4. Van de Wiele, C., et al.: Nuclear medicine imaging to predict response to radiotherapy: a review. Int. J. Radiat. Oncol. Biol. Phys. **55**(1), 5–15 (2003)
5. McRobbie, D.W., et al.: MRI from Picture to Proton. Cambridge University Press, Cambridge (2017)
6. Mansfield, P., Grannell, P.K.: "Diffraction" and microscopy in solids and liquids by NMR. Phys. Rev. B **12**(9), 3618 (1975)
7. Filler, A.G.: The history, development and impact of computed imaging in neurological diagnosis and neurosurgery: CT, MRI, and DTI. Internet J. Neurosurg. **7**(1), 5–35 (2010)
8. Wells, P.N., Liang, H.D.: Medical ultrasound: imaging of soft tissue strain and elasticity. J. R. Soc. Interface **8**(64), 1521–1549 (2011)
9. Sarvazyan, A., et al.: An overview of elastography—an emerging branch of medical imaging. Curr. Med. Imaging Rev. **7**(4), 255–282 (2011)
10. Ophir, J., et al.: Elastography: a quantitative method for imaging the elasticity of biological tissues. Ultrason. Imaging **13**(2), 111–134 (1991). https://doi.org/10.1016/0161-7346(91)90079-W.PMID1858217
11. Garra, B.S., et al.: Elastography of breast lesions: initial clinical results. Radiology **202**(1), 79–86 (1997)
12. Parker, K.J., et al.: Imaging the elastic properties of tissue: the 20-year perspective. Phys. Med. Biol. **56**(1), R1 (2010). https://doi.org/10.1088/0031-9155/57/16/5359
13. Huang, H.K., et al.: Picture archiving and communications systems (PACS). In: Radiological Society of North America 73rd Scientific Assembly and Annual Meeting (Abstracts) (1987)
14. Innovative Flemish In Vivo Imaging Technology. A History of Medical Imaging. Ghent University. http://www.infinityugent.be/research-development/a-history-of-medical-imaging
15. Ma, W.Y., Manjunath, B.S.: Netra: a toolbox for navigating large image databases. Multimedia Syst. **7**(3), 184–198 (1999)
16. Faloutsos, C., et al.: Efficient and effective querying by image content. J. Intell. Inf. Syst. **3**(3–4), 231–262 (1994)

17. Smith, J.R., Chang, S.F.: VisualSEEk: a fully automated content-based image query system. In: Proceedings of the Fourth ACM International Conference on Multimedia, ACM, Boston, MA, USA, 18–22 November 1996, pp. 87–98 (1997)
18. Chang, S.F., et al.: Visual information retrieval from large distributed online repositories. Communications of the ACM, New York, NY, USA **40**(12), 63–71 (1997)
19. Pentland, A.P., et al.: Photobook: tools for content-based manipulation of image databases. In: Storage and Retrieval for Image and Video Databases II, vol. 2185, pp. 34–48. International Society for Optics and Photonics (1994)
20. Hwang, K.H., et al.: Medical image retrieval: past and present. Healthc. Inform. Res. **18**(1), 3–9 (2012)
21. Kelly, P.M., Cannon, T.M., Hush, D.R.: Query by image example: the comparison algorithm for navigating digital image databases (CANDID) approach. In: Storage and Retrieval for Image and Video Databases III, vol. 2420, pp. 238–249. International Society for Optics and Photonics (1995)
22. Orphanoudakis, S.C., et al.: I2C: a system for the indexing, storage, and retrieval of medical images by content. Med. Inform. **19**(2), 109–122 (1994)
23. Deselaers, T., Keysers, D., et al.: FIRE–flexible image retrieval engine: ImageCLEF 2004 evaluation. In: Workshop of the Cross-Language Evaluation Forum for European Languages, pp. 688–698. Springer, Berlin (2004)
24. Lam, M.O., et al.: BRISC—an open source pulmonary nodule image retrieval framework. J. Digit. Imaging **20**(1), 63–71 (2007)
25. Keysers, D., et al.: Statistical framework for model-based image retrieval in medical applications. J. Electron. Imaging **12**(1), 59–68 (2003)
26. Ojala, T., et al.: Performance evaluation of texture measures with classification based on Kullback discrimination of distributions. In: Conference A: Computer Vision & Image Processing, Proceedings of the 12th IAPR International Conference on Pattern Recognition, vol. 1, pp. 582–585, IEEE (1994)
27. He, D.C., Wang, L.: Texture unit, texture spectrum, and texture analysis. IEEE Trans. Geosci. Remote Sens. **28**(4), 509–512 (1990)
28. Wang, L., He, D.C.: Texture classification using texture spectrum. Pattern Recogn. **23**(8), 905–910 (1990)
29. Tan, X., Triggs, B.: Enhanced local texture feature sets for face recognition under difficult lighting conditions. IEEE Trans. Image Process. **19**(6), 1635–1650 (2010)
30. ul Hussain, S., Triggs, B.: Visual recognition using local quantized patterns. In: Computer Vision—ECCV 2012, pp. 716–729. Springer, Berlin (2012)
31. Murala, S., et al.: Directional local extrema patterns: a new descriptor for content based image retrieval. Int. J. Multimedia Inf. Retrieval **1**(3), 191–203 (2012)
32. Rao, L.K., Rao, D.V.: Local quantized extrema patterns for content-based natural and texture image retrieval. Human-centric Comput. Inf. Sci. **5**(1), 26 (2015)
33. Rao, L.K., et al.: Local mesh quantized extrema patterns for image retrieval. SpringerPlus **5**(1), 976 (2016)
34. Deep, G., et al.: Directional local ternary quantized extrema pattern: a new descriptor for biomedical image indexing and retrieval. Eng. Sci. Technol. Int. J. **19**(4), 1895–1909 (2016)
35. Deep, G., et al.: Local mesh ternary patterns: a new descriptor for MRI and CT biomedical image indexing and retrieval. Comput. Meth. Biomech. Biomed. Eng. Imaging Vis. **6**(2), 155–169 (2018)
36. Deep, G., et al.: Local quantized extrema quinary pattern: a new descriptor for biomedical image indexing and retrieval. Comput. Meth. Biomech. Biomed. Eng. Imaging Vis. 1–17 (2017)
37. Verma, M., Raman, B.: Local neighborhood difference pattern: a new feature descriptor for natural and texture image retrieval. Multimedia Tools Appl. **77**(10), 11843–11866 (2018)
38. Bhunia, A. K., et al.: A novel feature descriptor for image retrieval by combining modified color histogram and diagonally symmetric co-occurrence texture pattern. arXiv preprint arXiv:1801.00879 (2018)

39. Murala, S., et al.: Directional binary wavelet patterns for biomedical image indexing and retrieval. J. Med. Syst. **36**(5), 2865–2879 (2012)
40. Murala, S., et al.: Local tetra patterns: a new feature descriptor for content-based image retrieval. IEEE Trans. Image Process. **21**(5), 2874–2886 (2012)
41. Murala, S., Wu, Q.J.: Local ternary co-occurrence patterns: a new feature descriptor for MRI and CT image retrieval. Neurocomputing **119**, 399–412 (2013)
42. Murala, S., Wu, Q.J.: Local mesh patterns versus local binary patterns: biomedical image indexing and retrieval. IEEE J. Biomed. Health Informatics **18**(3), 929–938 (2014)
43. Tizhoosh, H.R.: Barcode annotations for medical image retrieval: a preliminary investigation. In: 2015 IEEE International Conference on Image Processing (ICIP), pp. 818–822. IEEE (2015)
44. Tizhoosh, H.R., et al.: Minmax radon barcodes for medical image retrieval. In: International Symposium on Visual Computing, pp. 617–627. Springer, Cham (2016)
45. Tizhoosh, H.R., et al.: Barcodes for medical image retrieval using autoencoded radon transform. arXiv preprint arXiv:1609.05112 (2016)
46. Nouredanesh, M., et al.: Radon-Gabor barcodes for medical image retrieval. In: 2016 23rd International Conference on Pattern Recognition (ICPR), pp. 1309–1314. IEEE (2016)
47. Babaie, M., et al.: Retrieving similar x-ray images from big image data using radon barcodes with single projections. arXiv preprint arXiv:1701.00449 (2017)
48. Lo, S.C., et al.: Artificial convolution neural network techniques and applications for lung nodule detection. IEEE Trans. Med. Imaging **14**(4), 711–718 (1995)
49. Yamaguchi, Y., et al.: U.S. Patent No. 3,701,095. U.S. Patent and Trademark Office, Washington, DC (1972)
50. Bagheri, E., Cheung, J.C.: Advances in artificial intelligence. In: Bagheri, Ebrahim, Cheung, Jackie C.K. (eds.) Springer International Publishing (2018). https://www.springer.com/gp/book/9783319896557
51. Cireşan, D.C., et al.: Deep, big, simple neural nets for handwritten digit recognition. Neural Comput. **22**(12), 3207–3220 (2010)
52. Krizhevsky, A., et al.: Imagenet classification with deep convolutional neural networks. In: Advances in Neural Information Processing Systems, pp. 1097–1105 (2012)
53. Wu, P., Hoi, S.C.H., et al.: Online multimodal deep similarity learning with application to image retrieval. In: Proceedings of the 21st ACM International Conference on Multimedia, pp. 153–162. ACM (2013)
54. Wan, J., et al.: Deep learning for content-based image retrieval: a comprehensive study. In: Proceedings of the 22nd ACM International Conference on Multimedia, pp. 157–166. ACM (2014)
55. Anthimopoulos, M., et al.: Lung pattern classification for interstitial lung diseases using a deep convolutional neural network. IEEE Trans. Med. Imaging **35**(5), 1207–1216 (2016)
56. Ivakhnenko, A.G., Lapa, V.G.: Cybernetic Predicting Devices. CCM Information Corporation (1965)
57. van Tulder, G., de Bruijne, M.: Combining generative and discriminative representation learning for lung CT analysis with convolutional restricted Boltzmann machines. IEEE Trans. Med. Imaging **35**(5), 1262–1272 (2016)
58. Cao, Y., et al.: Medical image retrieval: a multimodal approach. Cancer Inform. **13**, CIN-S14053 (2014)
59. Sun, Q., et al.: Using deep learning for content-based medical image retrieval. In: Medical Imaging 2017: Imaging Informatics for Healthcare, Research, and Applications, vol. 10138, p. 1013812. International Society for Optics and Photonics (2017)
60. Qayyum, A., et al.: Medical image retrieval using deep convolutional neural network. Neurocomputing **266**, 8–20 (2017)
61. Khatami, A., et al.: A sequential search-space shrinking using CNN transfer learning and a Radon projection pool for medical image retrieval. Expert Syst. Appl. **100**, 224–233 (2018)
62. Khatami, A., et al.: Parallel deep solutions for image retrieval from imbalanced medical imaging archives. Appl. Soft Comput. **63**, 197–205 (2018)

63. Chung, Y.A., Weng, W.H.: Learning deep representations of medical images using siamese CNNs with application to content-based image retrieval. arXiv preprint arXiv:1711.08490 (2017)
64. Akgül, C.B., et al.: Content-based image retrieval in radiology: current status and future directions. J. Digit. Imaging **24**(2), 208–222 (2011)
65. Deep, G., et al.: Biomedical image indexing and retrieval descriptors: a comparative study. Procedia Comput. Sci. **85**, 954–961 (2016)
66. Das, P., Neelima, A.: An overview of approaches for content-based medical image retrieval. Int. J. Multimedia Inf. Retrieval **6**, 271–280 (2017)
67. Yap, P.T., et al.: Image analysis by Krawtchouk moments. IEEE Trans. Image Process. **12**(11), 1367–1377 (2003)
68. WU, K., Zhong, S.H.: Image texture segmentation based on Krawtchouk moment and SVM. J. Appl. Sci. **26**(5), 521–525 (2008)
69. Venkataramana, A.: Radial Krawtchouk moments for rotational invariant pattern recognition. In: 2007 6th International Conference on Information, Communications & Signal Processing, pp. 1–5. IEEE (2007)
70. Zhu, N., et al.: Medical image retrieval by radial Krawtchouk moments. J. Biomed. Eng. Res. **27**, 40–44 (2008)
71. Tziola, E., et al.: Image retrieval using Krawtchouk chromaticity distribution moments. In: VISAPP, no. 2, pp. 248–251 (2008)
72. Raj, P.A., Venkataramana, A.: Krawtchouk chromaticity distribution moments for content based image retrieval. In: NCC 2009, 16–18 January, IIT Guwahati
73. Xiang, P., et al.: Content-based 3d retrieval by krawtchouk moments. In: International Conference Image Analysis and Recognition, pp. 217–224. Springer, Berlin (2006)
74. Wu, Y., Wu, Y.: Image retrieval based on NSCT and Krawtchouk moments. Geomatics Inf. Sci. Wuhan Univ. **36**(6), 691–694 (2011)
75. Priyal, S.P., Bora, P.K.: A robust static hand gesture recognition system using geometry based normalizations and Krawtchouk moments. Pattern Recogn. **46**(8), 2202–2219 (2013)
76. Giakoumis, D., et al.: Automatic recognition of boredom in video games using novel bio signal moment-based features. IEEE Trans. Affect. Comput. **2**(3), 119–133 (2011)
77. Batioua, I., et al.: 3D image analysis by separable discrete orthogonal moments based on Krawtchouk and Tchebichef polynomials. Pattern Recogn. **71**, 264–277 (2017)
78. Clemente, C., et al.: Automatic target recognition of military vehicles with Krawtchouk moments. IEEE Trans. Aerosp. Electron. Syst. **53**(1), 493–500 (2017)
79. Depeursinge, A., et al.: Building a reference multimedia database for interstitial lung diseases. Comput. Med. Imaging Graph. **36**(3), 227–238 (2012)
80. NEMA MR Image Database. ftp://medical.nema.org/medical/Dicom/Multiframe/MR

Pixel N-Grams Representation for Medical Image Classification

Pradnya Kulkarni and Andrew Stranieri

Abstract Image classification has wide applications in many fields including medical imaging. A major aspect of classification is to extract features that can correctly represent important variations in an image. Global image features commonly used for classification include Intensity Histograms, Haralick's features based on Gray-level co-occurrence matrix, Local Binary Patterns and Gabor filters. A novel feature extraction and image representation technique 'Pixel N-grams' inspired from 'Character N-grams' concept in text categorization is described in this chapter. The classification performance of Pixel N-grams is tested on the various datasets including UIUC texture dataset, binary shapes dataset, miniMIAS dataset of mammography, and real-world high-resolution mammography dataset provided by an Australian radiology practice. The results are compared with other feature extraction techniques such as co-occurrence matrix features, intensity histogram, and bag of visual words. The results demonstrate promising classification accuracy in addition to reduced computational costs, enabling a new way for mammographic classification on low resource computers.

Keywords Pixel N-grams · Image classification · Mammograms · Texture · Shape · Bag of visual words

1 Introduction

Breast cancer is the most common cause of death among women (140 out of 184 countries) [40]. Early detection of breast cancer is essential for cancer treatment and reduction in mortality rate [8]. However, interpretation of mammographic images is a tedious and difficult task for radiologists due to complexities like variability

P. Kulkarni (✉)
MIT World Peace University, Pune, India
e-mail: pradnya.kulkarni@mitwpu.edu.in; p.kulkarni@federation.edu.au

P. Kulkarni · A. Stranieri
Federation University, Ballarat, Australia

S. Bhattacharyya et al. (eds.), *Hybrid Machine Intelligence for Medical Image Analysis*, Studies in Computational Intelligence 841,
https://doi.org/10.1007/978-981-13-8930-6_2

in appearance of lesions and density of the surrounding tissue [25]. In addition to enormous number of mammograms generated every day, radiologists are under tremendous workload pressure. Therefore, they are prone to make errors in diagnosis simply due to tiredness, fatigue or insufficient experience levels [9, 34].

In recent years, computer-aided systems are deployed to help radiologists perform accurate and efficient detection and diagnosis for breast cancer. This is plausible as computers can perform image analysis [6] consistently and repetitively. Computer-aided detection/diagnosis (CAD) systems are mainly grouped into two categories, namely CADe (computer-aided detection) and CADx (computer-aided diagnosis) [14]. However, the development of CADx systems has been challenging for the following reasons. Firstly, very high accuracy is needed as misclassifications can have serious effects on patient care. Secondly, the abnormalities can be occluded by dense tissue. Finally, the appearance of the abnormalities may differ drastically making it difficult to analyse [15]. More information about computer-aided detection/diagnosis of breast cancer can be found in the review article [14].

Despite many efforts in designing CAD systems, high sensitivity is only achieved at the cost of lower specificity resulting in many false positives [14]. Considering the emotional stress for patients and higher healthcare costs due to unnecessary biopsies, there is a great need to improve CAD systems in terms of achieving high accuracy, specificity and sensitivity.

There are various types of abnormalities that indicate progress of breast cancer such as calcifications and mass/tumours. Basically, mass lesions have three important characteristics, namely texture, shape and margins [38]. Therefore, CADx systems are designed to extract the features which can model these important characteristics. These features determine whether the lesion is benign or malignant [36]. Thus, the breast cancer detection can be thought of as an image classification problem based on the shape, texture and margin/boundary features. Work done by various researchers in the field of mammographic lesion classification is summarized below.

Early work in classification of mammograms includes texture, shape and bag of visual words (BoVW) features. Texture can be modelled using various approaches, namely statistical, structural and spectral. An intensity histogram is one of the commonly used, simple statistical approaches for modelling texture. Statistical features based on histogram were used by Islam et al. [13]. The first-order statistical features provide information related to the Gray-level distribution of the image. However, they do not provide any information about the relative positions of the various Gray levels. Thus, in order to model the spatial information between the neighbouring pixels second-order statistical features were developed. Second-order histogram is also known as GLCM (Gray-level co-occurrence matrix) [10]. It is clear that better classification performance using selective GLCM features can be obtained than with first-order statistical features. However, GLCM features are computationally expensive. Another statistical feature for texture representation is the local binary pattern (LBP) [22]. Results indicate that false positive reduction can be achieved using LBP in a much better way for lesions of large size than the smaller sized lesions. For uniform image regions, LBP is sensitive to noise. Local ternary patterns (LTP) overcome this limitation [29].

Another approach for texture feature extraction is a spectral approach. The Gabor filter is a commonly used spectral approach for mammogram classification [12]. Fourier transform has been used for benign/malignant classification [30]. It has been observed that Fourier transform lacks the spatial localization and hence performs poorly [24]. To overcome the limitations of Fourier transform method, wavelet transform was proposed [26]. Wavelet transform is found to be superior to Gabor filters as it allows the representation of textures at suitable scale by varying the spatial resolution. In structural approaches, texture is defined by primitives and spatial distributions [23].

Shape features are also known as geometric features or morphological features. Shape feature computation can be classified broadly into two main categories: (1) contour-based methods and (2) region-based methods [41]. A boundary or region can also be described using moments. Zernike moments [31] are robust to noise at the cost of computational complexity. Features calculated directly from the boundary include margin speculation, margin sharpness, area, circularity measure, convexity, rectangularity, perimeter, perimeter to area ratio and acutance measure [11]. Normalized radial length shape features have been tried for mammographic classification [32]. Normalized chord length has been used by El-Faramawy et al. [7] who demonstrated an accuracy of 92.3% for circumscribed/speculation and 95% accuracy for benign/malignant classification. Wei et al. [37, 38] proposed a mammogram retrieval system based on similar mass lesions. A combination of LBP, Haar wavelet and Haralick features has been used for classification of mammograms into benign and malignant categories by Joseph and Balakrishnan [16]. Shape features were found to be superior to gradient and texture features (Haralick and wavelet) for mass classification [27] as well as microcalcification classification. However, shape and margin features of masses/calcifications are highly dependent upon segmentation accuracy, and hence, any errors in segmentation can diminish the classification accuracy, sensitivity and specificity values. Further, shape and boundary feature extraction is computationally complex.

The aforementioned low-level features, namely texture and shape features, represent global image features and represent the image as a whole but cannot represent local pattern variations very well. The idea of capturing local pattern variations in an image gave rise to the use of bag of visual words models (BoVW) for image representation [33]. BoVW approach also called the textons approach has been used for classification of mammograms into benign or malignant classes [21]. BoVW approach was used for mammographic image classification using a histogram intersection method [4].

Although, the BoVW model has proven to be better than models using low-level global features such as histogram, Haralick's features and LBP [35], it has major drawbacks. The BoVW model does not consider spatial relationships among visual words. Another, BoVW drawback involves the high computational cost to generate vocabularies from low-level features. Further, the vocabulary construction process often results in noisy words that diminish classification. Convolutional neural network (CNN) is a neural network which shares connections between hidden units leading to low computational time and translational invariance properties. Use of convolutional neural networks for learning the features for mammographic mass lesions is found in Arevalo et al. [2]. A cascade of two-level deep convolutional neural networks and random forest classifiers was used for mammographic mass detection [5]. Deep belief networks have recently been used for breast cancer detection [1]. The limitation of the deep learning neural networks is that it requires huge amount of data for training.

The proposed Pixel N-grams [19] method on the other hand does not require quantization step for vocabulary construction process. Hence, it has some advantages over the BoVW model. One is that it is computationally cost effective as the number of occurrences of sequences of Gray-level pixels has to be counted. The other advantage is that noisy words are not created as the quantization is not necessary. Also, the spatial relationships among the pixels are taken care of by computing Pixel N-grams in various directions. Further, the vocabulary size for Pixel N-grams is dependent upon the Gray levels present in an image (usually 256) and hence is constant. The Pixel N-grams approach has been tested on four different datasets (UIUC texture dataset, binary shapes dataset, miniMIAS dataset of mammography, lake imaging dataset of mammography).

The chapter is organized as follows. Pixel N-grams model for image representation is described in Sect. 2 and Sect. 3 describes materials and methods, experimental details are given in Sect. 4. The results are then analysed in Sect. 5, and the Sect. 6 finally concludes the chapter providing ideas of future scope for Pixel N-grams.

2 Pixel N-Grams Model for Image Representation

Pixel N-grams model is inspired by the character N-gram model in the text categorization domain. Essentially, character N-grams are formed by a sequence of N consecutive characters. Similarly, an image can be represented using sequences of adjacent visual characters/pixels. This is called *Pixel N-gram Model* for representation of image. In this model, every Gray-level value of a pixel is considered a visual character. The images are then represented with the histogram of sequence of intensity levels of pixels. The generation of 3-gram features in horizontal and vertical direction is shown in the Fig. 1.

3 Materials and Methods

3.1 Materials

All the algorithms for Gray scale reduction, Pixel N-gram feature extraction, intensity histogram feature extraction, co-occurrence matrix-based feature extraction as well as computational time comparison and shape database creation were imple-

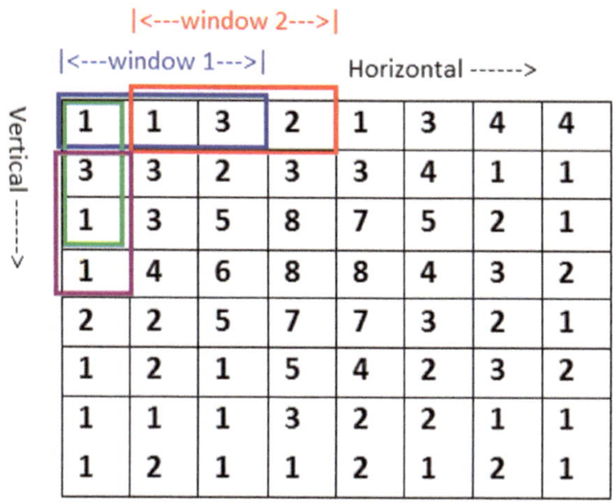

Fig. 1 Sliding window for 3-gram computation

Table 1 Characteristics of datasets used for this study

Dataset name	Number of examples	Number of predictive attributes	Number of classes	% examples in majority class (%)
UIUC texture	1000	318	25	33.3
Basic shapes	240	07	3	33.3
miniMIAS	270	147	3	76.7
LakeImaging	80	150	3	50.0

mented using MATLAB 7.9 software. Weka 3.6 [39] software was used for all the classification experiments.

3.2 *Datasets*

Performance of Pixel *N*-gram features was evaluated using four different datasets. These datasets include UIUC texture dataset, basic shapes dataset, miniMIAS mammographic dataset and LakeImaging mammographic dataset. Table 1 shows the characteristics of the datasets used for the study.

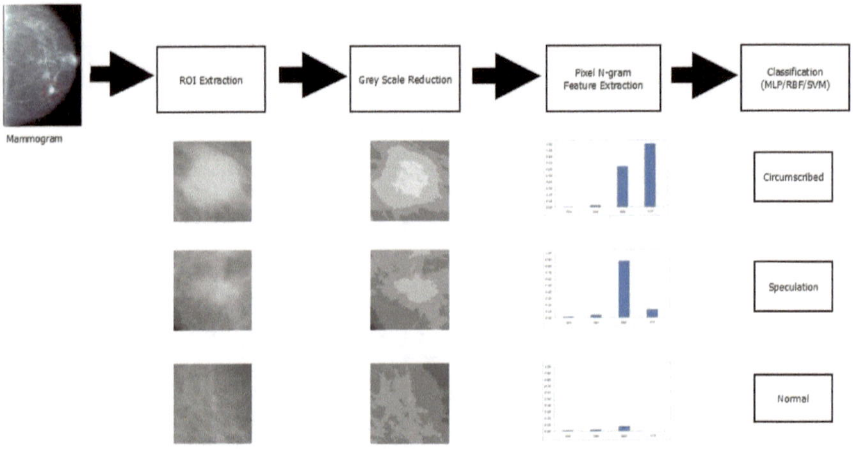

Fig. 2 Schematic overview of mammographic classification

3.3 *Method*

An empirical approach has been adopted, and a series of classification experiments have been designed to test the efficacy of Pixel *N*-grams for mammograms classification.

Figure 2 shows the schematic overview of the classification process. The first stage in applying the Pixel *N*-gram approach is to assemble a training set of mammograms with a region of interest (ROI) identified by an experienced radiologist. The images are then reduced in Gray scale for reduction of computational complexity and reduction in noise level (Fig. 3).

After Gray scale reduction of the ROI images, the next step is to calculate the pixel *N*-gram features of an image. The pixel *N*-gram features are then normalized using min-max, tf-idf methods and fed to the classifier. Various classifiers are used in order to figure out the optimum solution. Lesions are classified into three categories, namely circumscribed, speculation and normal.

4 Experimental Work

The main aim of this research is to classify mammographic lesions using the Pixel *N*-grams technique. Fine-grained classification (circumscribed lesion, speculated lesion and normal) is useful for automated diagnosis of lesions. Two mammographic datasets were used for evaluating classification performance. These two databases are a benchmark dataset miniMIAS and a real-world mammography dataset provided by Victoria-based radiology firm LakeImaging Pvt. Ltd. Inc. for this project.

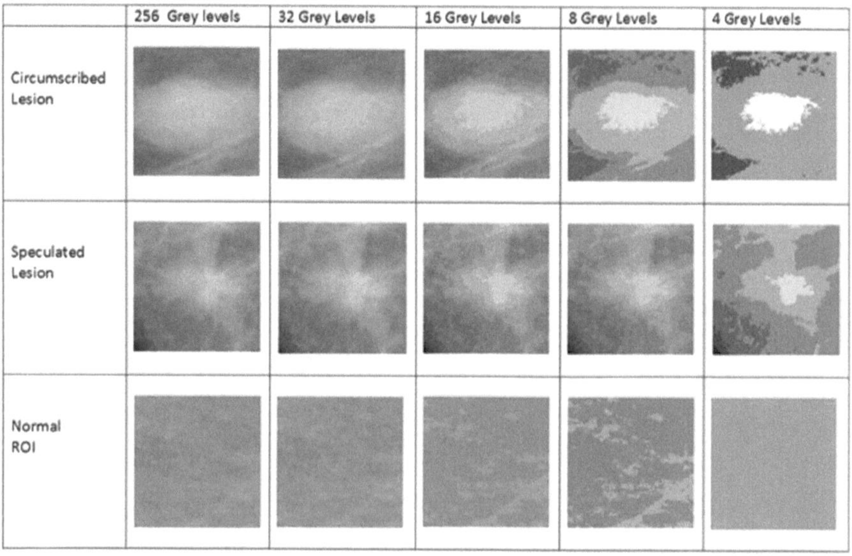

Fig. 3 Gray scale reduced ROIs

The LakeImaging dataset consisted of high-resolution truly digital mammograms[1] generated via patient scans during 2013–2014. The miniMIAS dataset consists of secondary digital mammograms.[2]

Texture and shape are the two important features of mammographic lesions. The effectiveness of pixel *N*-grams for classification of texture and basic shapes is tested using UIUC texture dataset and binary shapes dataset. Various experiments are conducted in order to find out the optimum value of *N* and optimum classifier. Finally, the experiments were conducted in order to compare the pixel *N*-grams classification performance with the existing feature extraction and image representation techniques such as intensity histograms, Haralick's features based on co-occurrence matrix and bag of visual words.

4.1 Experiments on the Texture Dataset

Experiments were conducted on the benchmark texture image database (UIUC) (http://www-cvr.ai.uiuc.edu/ponce_grp) to explore the use of Pixel *N*-gram fea-

[1]Truly digital mammograms/primary digital mammograms are digital mammograms directly generated with the help of advanced imaging equipment.

[2]Secondary digital mammograms are conventional film-based mammograms digitised with the help of a scanner.

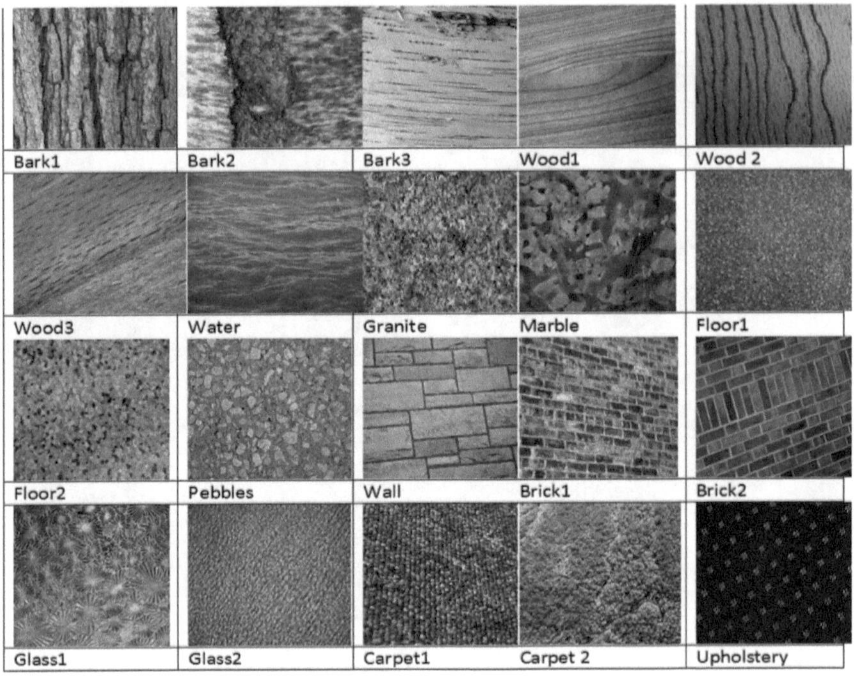

Fig. 4 UIUC texture dataset

tures for texture classification. Figure 4 shows the sample images from UIUC dataset.

It is very difficult to analyse texture. It is defined as quantitative measure of arrangement of intensities in an image. There are three main approaches to model texture, namely statistical, spectral and structural [42].

4.1.1 Finding Optimum Value of *N*

The objective of this experiment is to use Pixel *N*-grams for the classification of texture images and analyse the effect of varying *N* on classification performance. The value of $N = 1, 2, 3, 4$ and 5 was considered. *N*-gram features were computed and normalized using min-max normalization [28]. Classification was performed using SVM classifier, and 10-fold cross-validation was used to validate the results. The detailed results of this experiment have been published in the Kulkami et al. [17]. The overall Fscore is mentioned in Table 2.

Table 2 Effect of varying *N* on texture image classification

	1-gram	2-gram	3-gram	4-gram	5-gram
Overall Fscore (%)	53.7	66.1	80.3	**89.5**	85.4

Table 3 Texture image classification: Fscore for various techniques

	Histogram	Haralick	BoVW	4-gram
Overall Fscore (%)	53.7	21.9	84.4	**89.5**

4.1.2 Comparison with Existing Techniques (Classification Performance)

In this experiment, the *N*-gram classification performance for texture image classification is compared with other techniques such as bag of visual words, histogram and Haralick's features.

To compute histogram features, the number of occurrences of Gray-level pixels in an image is counted. Normalization is achieved using min-max method, and SVM classifier is used for classification.

Haralick's features [10] are very successful for texture classification. Haralick's features (contrast, correlation, energy and homogeneity) were computed using MATLAB code.[3] After normalizing these features were used for classification using SVM. The experiments on UIUC dataset for texture classification using BoVW approach have been conducted by Lazebnik et al. [20]. The classification results for BoVW are summarized from this work. Table 3 provides the overall Fscore using various techniques. For detailed results, please refer to Kulkarni and Stranieri [18].

4.2 Experiments on the Shapes Dataset

Lesions in mammographic images are of different shapes and sizes. They could be located at different sites in an image. Above all, mammograms produced at different imaging stations could be of different resolutions. Motivated by these properties of mammographic images, it becomes important to evaluate the performance of Pixel *N*-gram features for classification of lesions considering various shapes, shape sizes, shape locations and different resolutions of images. Experiments were conducted on a database of binary shape images to examine the extent to which classification using Pixel *N*-grams may be independent of size and location of a shape in an image. Further, verification of the resolution independency of these features is also important. Sample images from basic shapes dataset can be seen in Fig. 5.

[3] MATLAB 7.9.0.

Fig. 5 Sample images from basic shapes dataset

Table 4 Optimum value of N for shapes dataset

	Triangle			Circle			Square		
Grams	Precision	Recall	Fscore	Precision	Recall	Fscore	Precision	Recall	Fscore
1	79.8	98.8	88.3	93.3	87.5	90.3	93.9	77.5	84.9
2	97.5	96.3	96.9	100	95	97.4	94.1	100	97
3	100	100	100	100	100	100	100	100	100

4.2.1 Finding Optimum Value of N for Shapes

The classification of basic shapes using different values of N was performed in order to discover the optimum value of N. The database consisting of 240 binary images of three basic shapes (80 images each) was used for this experiment. The classification performance was measured by considering precision, recall and Fscore results which are illustrated in Table 4. This led to the use of 3-grams for the remaining experiments.

4.2.2 Size Invariance

For this experiment, the same 60 images from the above-mentioned shapes database were used. Three shapes (circle, triangle and square) of 20 different sizes each at centre location were considered. These images were of size 512×512 pixels. Precision, recall and Fscore results are illustrated in Table 5.

Table 5 Classification results: shapes of different sizes

Features	Triangle			Circle			Square		
	Fscore	Precision	Recall	Fscore	Precision	Recall	Fscore	Precision	Recall
Histogram	78	67	95	67	100	50	62	59	65
Haralick	50	50	50	100	100	100	50	50	50
N-gram	100	100	100	100	100	100	100	100	100

4.2.3 Location Invariance

The objective of this experiment is to see how accurately *N*-gram features can classify shapes located at different sites in an image. For this experiment, various images were created by changing the shape locations. This was achieved by varying the shape's centre coordinates along the *X* axis, along the *Y*-axis and then along the diagonals. The dataset thus consisted of 80 circles, 80 triangles and 80 squares at different locations. All images were of size 512 × 512 pixels. The classification results are shown in Table 6.

4.2.4 Resolution Invariance

Images of three basic shapes (triangle, circle and square) of 10 different resolutions (512 × 512, 1024 × 1024, 1536 × 1536, 2048 × 2048, 2560 × 2560, 3072 × 3072, 3584 × 3584, 4096 × 4096, 4608 × 4608, 5120 × 5120) were generated for this experiment. Table 7 provides the results.

Table 6 Classification results (shapes at various locations)

Features	Triangle			Circle			Square		
	Fscore	Precision	Recall	Fscore	Precision	Recall	Fscore	Precision	Recall
Histogram	89	80	100	89	93	85	83	91	76
Haralick	67	50	100	100	100	100	0	0	0
N-gram	100	100	100	100	100	100	100	100	100

Table 7 Classification results (shapes of various resolutions)

Features	Triangle			Circle			Square		
	Fscore	Precision	Recall	Fscore	Precision	Recall	Fscore	Precision	Recall
Histogram	84	89	80	76	73	80	74	78	70
Haralick	0	0	0	100	100	100	9.5	9.1	10
N-gram	90	90	90	100	91	100	90	90	90

4.3 Experiments on Mammographic Datasets (MiniMIAS and LakeImaging)

Classification of mammographic lesions is an important step for automated breast cancer diagnosis. miniMIAS dataset is the benchmark dataset which consists of digitized mammograms. A real-time truly digitized mammographic dataset was provided by LakeImaging Pvt. Ltd. The Pixel *N*-grams technique was tested on both of the datasets. The results are provided below.

4.3.1 Finding Optimum Value of *N*

For finding the optimum value of *N*, classification performance is analysed with change in value of *N*. Values of *N* considered for the experiment were 1, 2, 3, 4 and 5. With further increase in *N*, the vocabulary size and feature vector size are increased, thus increasing the computational cost. Further, with increase in *N* the sequences are hard to find therefore resulting in features more specific to a particular image. This will make it harder for the classifier to generalize. Therefore, we stop at $N = 5$. The images were Gray scale reduced using 8 Gray levels as this was found to be the optimum Gray scale reduction value. The experimental results to determine the optimum value of *N* for miniMIAS as well as LakeImaging dataset using the circumscribed/speculated/normal classification are given in Tables 8 and 9.

4.3.2 Comparison of Different Classifiers

Various classifiers can be used for classification of images. Three most commonly used classifiers (MLP, SVM, KNN) were used for classification of mammographic lesions into circumscribed, speculation and normal categories in order to compare

Table 8 Effect of varying *N* (miniMIAS dataset)

Unit: %	1-gram	2-gram	3-gram	4-gram	5-gram
Fscore	72.5	80.2	86.4	85.2	79.6
Sensitivity	73	79.9	84.6	82.2	80.2
Specificity	88.2	88	88.9	88.3	88.5
ROC area	80.3	83.4	85.8	84.1	81.6

Table 9 Effect of varying *N* (LakeImaging dataset)

Unit: %	1-gram	2-gram	3-gram	4-gram	5-gram
Fscore	75.3	82.7	83.9	79.0	78.8
Sensitivity	75.0	83.2	84.0	79.0	78.3
Specificity	91.4	89.3	91.1	89.8	90.3
ROC area	84.5	86.8	90.6	88.0	86.2

Table 10 Optimum parameters for classifiers

MLP		SVM		KNN	
Parameter	Value	Parameter	Value	Parameter	Value
Learning rate (η)	0.3	Epsilon	1.0 e^{-12}	K	6
Momentum (α)	0.2	Complexity constant (C)	1.0		
Number of iterations	500	Number of iterations	1000		
Number of hidden units	10	Tolerance	0.001		
Cross-validation	10-fold		10-fold		10-fold

Table 11 Classification accuracy (miniMIAS)

	Classification accuracy (%)—miniMIAS			
Classifier	Circumscribed	Speculation	Normal	Overall
KNN	60.0	56.0	70.0	70.0
SVM	42.0	61.0	82.0	71.0
MLP	72.7	68.3	90.4	82.0

Table 12 Classification accuracy (LakeImaging)

	Classification accuracy (%) LakeImaging			
Classifier	Circumscribed	Speculation	Normal	Overall
KNN	66.7	70.0	91.6	79.8
SVM	66.7	71.4	96.0	82.0
MLP	80.0	82.0	94.0	84.0

Table 13 Sensitivity and specificity (miniMIAS)

	Sensitivity (%)				Specificity (%)			
Classifier	Circumscribed	Speculated	Normal	Overall	Circumscribed	Speculated	Normal	Overall
KNN	67	58	76	70	88	89	72	79
SVM	33	63	89	69	92	90	65	76
MLP	67	74	91	82	95	90	86	89

the performances of different classifiers. Optimum parameters used for the MLP, SVM and KNN classifiers are detailed in Table 10.

The results of circumscribed/speculation/normal classification using different classifiers for miniMIAS dataset are shown in Tables 11, 13 and for LakeImaging dataset are shown in Tables 12 and 14.

Table 14 Sensitivity and specificity (LakeImaging)

	Sensitivity (%)				Specificity (%)			
Classifier	Circumscribed	Speculated	Normal	Overall	Circumscribed	Speculated	Normal	Overall
KNN	65	67	95	80	90	92	88	89
SVM	65	71	98	83	90	90	90	91
MLP	70	71	98	84	93	92	95	95

5 Discussion

In an attempt to reduce the computational cost and effect of noise, Gray scale reduction of images as a pre-processing step was proposed. There is a trade-off between wanting to decrease Gray levels so that the computational complexity is reduced and wanting to increase Gray levels so that the features are not too coarse to diminish the classification accuracy. The optimum value of Gray scale reduction was found to be 8.

Figure 6 suggest that as N is increased the classification performance for irregular texture images is improved whereas the N has negligible effect on the classification performance of the regular textures. Pixel N-grams with smaller value of N are unable to model these irregular textures whereas longer sequences can model the irregular patterns more appropriately providing high classification performance. The 4-gram features were observed to give the best performance for texture classification from UIUC texture database. Thus, the optimum value of N for UIUC texture dataset was noted as $N = 4$. Further, the classification performance of Pixel N-grams for texture classification was found to be comparable with that of the BoVW approach.

Experiments on the binary shapes dataset demonstrate that the Pixel N-gram features were able to discern among various basic shapes accurately (100% accuracy, 100% precision and 100% recall). For the shapes dataset, $N = 3$ provided the best classification performance, and therefore, $N = 3$ was noted as the optimum value of N for binary shapes dataset. Moreover, the classification using Pixel N-grams was found to be independent of the location and size of the shapes in an image. Further,

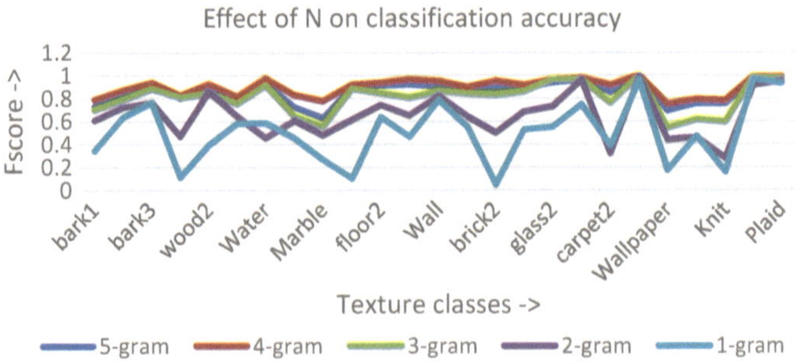

Fig. 6 Classification Fscore for 5, 4, 3, 2, 1 N-grams for UIUC texture dataset

the shape classification using Pixel *N*-grams was noted to be resolution independent (Fig. 7).

As the value of *N* is increased, the Pixel *N*-gram representation of image becomes more and more complete. The complete representation of an image is obtained if the spatial relationship of every pixel with every other pixel is modelled while generating features. However, with increase in *N* the feature vector dimensionality is increased producing the risk of overfitting which is normally referred to as 'Curse of dimensionality problem' [3]. Due to this, the classification performance could be degraded. Additionally, as the longer sequences are hardly observed, with increase in *N* the feature vector becomes too specific to a particular image making it hard for the classifier to generalize well. Moreover, the computational cost is increased with the increase in *N*. Thus, a balance has to be achieved between increasing *N* for achieving the complete image representation (and therefore better classification performance) and decreasing *N* for reducing the feature vector dimensionality (improve classifier generalization, avoid increase in computational cost). It is therefore necessary to find an optimum value of *N*.

It is evident that the classification accuracy increases when *N* is increased up to 3, however starts to fall as '*N*' is further increased to 4 and 5 for both the datasets (miniMIAS and LakeImaging). Similar observation could be found in case of sensitivity. The sensitivity has higher value for $N = 3$ as compared to that for $N = 1$, 2, 4 and 5. Also, receiver operating characteristic (ROC) curve area increases with increase in *N* up to $N = 3$ and then starts decreasing. Similarly, specificity values are found to be higher at $N = 3$ for LakeImaging dataset. However, the specificity values seem to have small variation with change in *N* for miniMIAS dataset. Thus, the optimum value of *N* for miniMIAS as well as LakeImaging dataset is 3 (Fig. 8).

For unbalanced datasets, classification accuracy is not the perfect measure for classifier performance. Hence, sensitivity and specificity parameters along with the Fscore were considered for comparing the performances of the classifiers. Figure 9 shows the graphical representation of various parameters (Fscore, sensitivity and specificity) using three different classifiers (MLP, SVM and KNN) for miniMIAS as

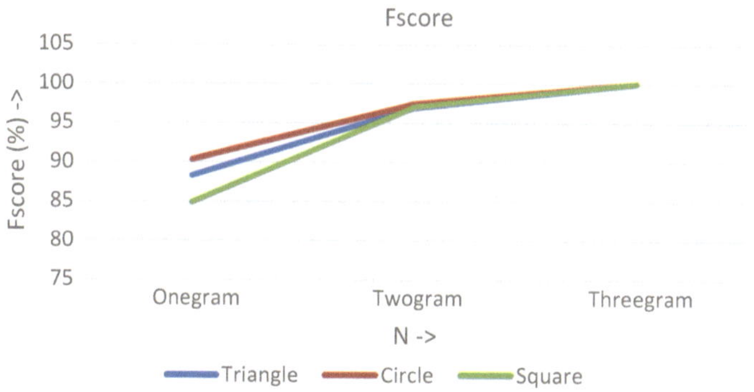

Fig. 7 Effect of *N* (shapes dataset)

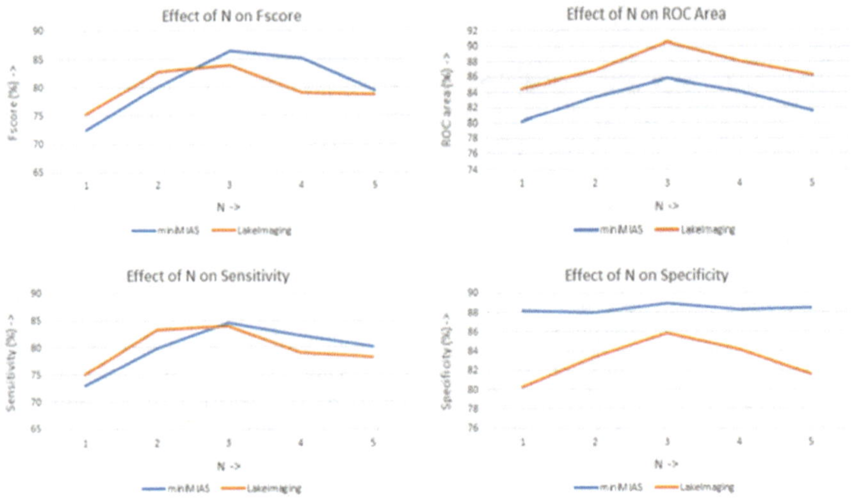

Fig. 8 Effect of *N* (miniMIAS and LakeImaging)

well as LakeImaging datasets. It is observed that the MLP classifier has performed better than the SVM and KNN classifiers considering Fscore, specificity and sensitivity for both the datasets.

Observation is that the novel Pixel *N*-gram features are quite effective in distinguishing various textures and shapes. Moreover, the performance of pixel *N*-grams was significantly better than intensity histogram and Haralick's features for texture, shape and mammograms classification. Further, they performed equally well as compared to the BoVW features with an added benefit of ease and low computational complexity. Another advantage of using Pixel *N*-grams features is that the segmentation is not required and thus technique is not dependent upon the accuracy of the segmentation algorithm. Thus, the results of mammographic lesion classification using Pixel *N*-grams are found to be quite promising for breast cancer detection as well as diagnosis applications.

6 Conclusion and Future Work

Classification of mammographic lesions (circumscribed/speculation/normal) is useful for breast cancer diagnosis. Texture and shape are the two important characteristics of mammographic lesions. The chapter analyses the efficacy of Pixel *N*-gram features for classification of texture, shape and mammographic images. Various parameters, namely Fscore, classification accuracy, sensitivity and specificity, were used for measuring the classification performance. In case of cancer detection or diagnosis, false

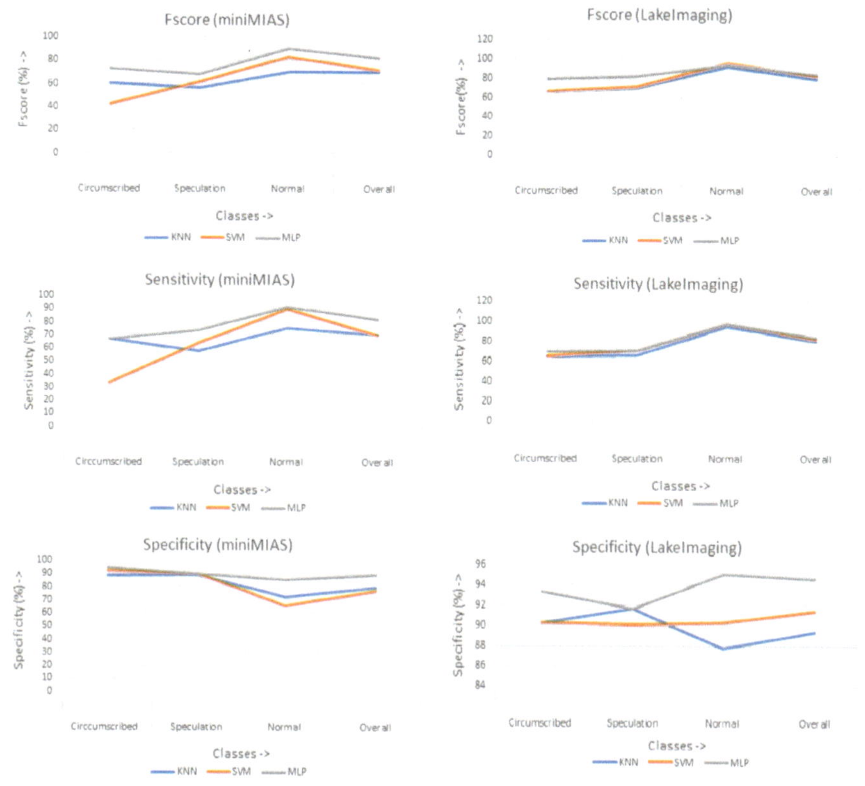

Fig. 9 Classifier comparison (miniMIAS and LakeImaging)

positives increase the patient's anxiety and healthcare costs whereas false negatives can be fatal.

Pixel *N*-grams provide excellent classification results for texture dataset. Shape is another important characteristic of breast lesions. Results have demonstrated that the Pixel *N*-grams can classify various shapes irrespective of the size of the shape, location of the shape in an image as well as resolution of the image.

Experiments revealed that Pixel *N*-grams classification performance increased with *N* up to certain value and then started dropping. The optimum value of *N* is different for different datasets.

The classification results were compared with various other methods such as histogram and co-occurrence matrix features and BoVW. It is evident that the Pixel *N*-grams performance improvement over histogram, Haralick's features was significant with respect to Fscore, sensitivity (true positive rate) and specificity for texture, shape as well as mammographic lesion classification. Also, texture classification performance using Pixel *N*-gram features was comparable with the texture classification using state-of-the-art BoVW technique (SIFT features).

Although, the experiments were conducted on the mammographic images, the pixel N-grams technique could easily be deployed for other medical as well as non-medical image classification and retrieval applications. Thus, future work will involve exploring the extent to which Pixel N-grams are effective for other medical image classification tasks outside mammogram classification such as diabetic retinopathy, histopathological images, wound images, dentistry X-ray images, detection of lung tumour and brain tumour.

Apart from the biomedical image classification for disease detection and diagnosis, Pixel N-grams can be used for several other applications such as satellite image segmentation and analysis of documents.

References

1. Abdel-Zaher, A.M., Eldeib, A.M.: Breast cancer classification using deep belief networks. Expert Syst. Appl. **46**, 139–144 (2016)
2. Arevalo, J., González, F.A., Ramos-Pollán, R., Oliveira, J.L., Lopez, M.A.G.: Convolutional neural networks for mammography mass lesion classification. In: Engineering in Medicine and Biology Society (EMBC), 2015 37th Annual International Conference of the IEEE, pp. 797–800. IEEE (2015)
3. Bankman, I.: Handbook of Medical Image Processing and Analysis. Academic Press (2008)
4. Cheng, E., Xie, N., Ling, H., Bakic, P.R., Maidment, A.D., Megalooikonomou, V.: Mammographic image classification using histogram intersection. In: 2010 IEEE International Symposium on Biomedical Imaging: From Nano to Macro, pp. 197–200. IEEE (2010)
5. Dhungel, N., Carneiro, G., Bradley, A.P.: Automated mass detection in mammograms using cascaded deep learning and random forests. In: 2015 International Conference on Digital Image Computing: Techniques and Applications (DICTA), pp. 1–8. IEEE (2015)
6. Eberl, M.M., Fox, C.H., Edge, S.B., Carter, C.A., Mahoney, M.C.: BI-RADS classification for management of abnormal mammograms. J. Am. Board Family Med. **19**, 161–164 (2006)
7. El-Faramawy, N., Rangayyan, R., Desautels, J., Alim, O.: Shape factors for analysis of breast tumors in mammograms. In: 1996 Canadian Conference on Electrical and Computer Engineering, pp. 355–358. IEEE (1996)
8. Engeland, S.V.: Detection of Mass Lesions in Mammograms by Using Multiple Views, [Sl: sn] (2006)
9. Evans, K.K., Birdwell, R.L., Wolfe, J.M.: If you don't find it often, you often don't find it: why some cancers are missed in breast cancer screening. PLoS ONE **8**, e64366 (2013)
10. Haralick, R.M., Shanmugam, K., Dinstein, I.H.: Textural features for image classification. In: IEEE Transactions on Systems, Man and Cybernetics, pp. 610–621 (1973)
11. Huo, Z., Giger, M.L., Vyborny, C.J., Wolverton, D.E., Schmidt, R.A., Doi, K.: Automated computerized classification of malignant and benign masses on digitized mammograms. Acad. Radiol. **5**, 155–168 (1998)
12. Hussain, M., Khan, S., Muhammad, G., Berbar, M., Bebis, G.: Mass detection in digital mammograms using gabor filter bank. In: IET Conference on Image Processing (IPR 2012), pp. 1–5. IET (2012)
13. Islam, M.J., Ahmadi, M., Sid-Ahmed, M.A.: An efficient automatic mass classification method in digitized mammograms using artificial neural network. arXiv preprint arXiv:1007.5129 (2010)
14. Jalalian, A., Mashohor, S.B., Mahmud, H.R., Saripan, M.I.B., Ramli, A.R.B., Karasfi, B.: Computer-aided detection/diagnosis of breast cancer in mammography and ultrasound: a review. Clin. Imaging **37**, 420–426 (2013)

15. Jirari, M.: Computer Aided System For Detecting Masses in Mammograms. Kent State University (2008)
16. Joseph, S., Balakrishnan, K.: Local binary patterns, haar wavelet features and haralick texture features for mammogram image classification using artificial neural networks. In: Advances in Computing and Information Technology. Springer, Berlin (2011)
17. Kulkami, P., Stranieri, A., Ugon, J.: Texture image classification using pixel N-grams. In: IEEE International Conference on Signal and Image Processing (ICSIP), pp. 137–141. IEEE (2016)
18. Kulkarni, P., Stranieri, A.: Comparison of pixel N-grams with histogram, haralick's features and bag-of-visual-words for texture image classification. In: 2018 3rd International Conference for Convergence in Technology (I2CT), pp. 1–5. IEEE (2018)
19. Kulkarni, P., Stranieri, A., Kulkarni, S., Ugon, J., Mittal, M.: Visual character n-grams for classification and retrieval of radiological images. Int. J. Multimed. Appl. **6**, 35 (2014)
20. Lazebnik, S., Schmid, C., Ponce, J.: A sparse texture representation using local affine regions. IEEE Trans. Pattern Anal. Mach. Intell. **27**, 1265–1278 (2005)
21. Li, Y., Chen, H., Rohde, G.K., Yao, C., Cheng, L.: Texton analysis for mass classification in mammograms. Pattern Recogn. Lett. **52**, 87–93 (2015)
22. Lladó, X., Oliver, A., Freixenet, J., Martí, R., Martí, J.: A textural approach for mass false positive reduction in mammography. Comput. Med. Imaging Graph. **33**, 415–422 (2009)
23. Lu, S., Bottema, M.J.: Structural image texture and early detection of breast cancer. In: Proceedings of the 2003 APRS Workshop on Digital Image Computing, pp. 15–20 (2003)
24. Materka, A., Strzelecki, M.: Texture analysis methods—a review. Technical University of Lodz, Institute of Electronics, COST B11 Report, Brussels, pp. 9–11 (1998)
25. Mckenzie, E.: Breast Cancer Screening (2014)
26. Mousa, R., Munib, Q., Moussa, A.: Breast cancer diagnosis system based on wavelet analysis and fuzzy-neural. Expert Syst. Appl. **28**, 713–723 (2005)
27. Mu, T., Nandi, A.K., Rangayyan, R.M.: Classification of breast masses using selected shape, edge-sharpness, and texture features with linear and kernel-based classifiers. J. Digit. Imaging **21**, 153–169 (2008)
28. Myatt, G.J.: Making Sense of Data: A Practical Guide to Exploratory Data Analysis and Data Mining. Wiley (2007)
29. Nanni, L., Brahnam, S., Lumini, A.: A very high performing system to discriminate tissues in mammograms as benign and malignant. Expert Syst. Appl. **39**, 1968–1971 (2012)
30. Ojansivu, V., Heikkilä, J.: Blur insensitive texture classification using local phase quantization. In: International Conference on Image and Signal Processing, pp. 236–243. Springer, Berlin (2008)
31. Papakostas, G.A., Boutalis, Y.S., Karras, D.A., Mertzios, B.G.: A new class of Zernike moments for computer vision applications. Inf. Sci. **177**, 2802–2819 (2007)
32. Petrick, N., Chan, H.P., Sahiner, B., Helvie, M.A.: Combined adaptive enhancement and region-growing segmentation of breast masses on digitized mammograms. Med. Phys. **26**, 1642–1654 (1999)
33. Sivic, J., Zisserman, A.: Video Google: A Text Retrieval Approach to Object Matching in Videos. USA (2003)
34. te Brake, G.M., Karssemeijer, N., Hendriks, J.: Automated detection of breast carcinomas not detected in a screening program. Radiology **207**, 465–471 (1998)
35. Tsai, C.-F.: Bag-of-words representation in image annotation: a review. ISRN Artificial Intell. **2012**, 1–19 (2012)
36. Varela, C., Timp, S., Karssemeijer, N.: Use of border information in the classification of mammographic masses. Phys. Med. Biol. **51**, 425 (2006)
37. Wei, C.-H., Chen, S.Y., Liu, X.: Mammogram retrieval on similar mass lesions. Comput. Methods Progr. Biomed. **106**, 234–248 (2012)
38. Wei, C.-H., Li, Y., Huang, P.J.: Mammogram retrieval through machine learning within BI-RADS standards. J. Biomed. Inform. **44**, 607–614 (2011)
39. Weka, W.: 3: Data Mining Software in Java. University of Waikato, Hamilton, New Zealand (www.cs.waikato.ac.nz/ml/weka) (2011)

40. WHO.: Latest World Cancer Statistics, International Agency for Research on Cancer (IARC) [Online]. World Health Organisation. [Accessed] (2016)
41. Yang, M., Kpalma, K., Ronsin, J.: A survey of shape feature extraction techniques. Pattern Recogn. 43–90 (2008)
42. Zhang, J., Tan, T.: Brief review of invariant texture analysis methods. Pattern Recogn. **35**, 735–747 (2002)

Penalized Fuzzy C-Means Enabled Hybrid Region Growing in Segmenting Medical Images

Shouvik Chakraborty, Sankhadeep Chatterjee, Ajanta Das and Kalyani Mali

Abstract Segmentation is considered as one of the challenging and important processes in the field of digital image processing, and there are numerous applications like medical image analysis and satellite data processing where the digital image processing can beneficial. Various algorithms have been developed by many researchers to analyze different medical images like MRI and X-rays. Nuclear image analysis and interpretation also a promising research topic in medical image analysis. For example, positron emission tomography (PET) image can be used to accurately localize disease to help doctors in providing the right treatment and saving valuable time. In recent years, there is a significant advancement in the biomedical imaging domain with the increasing accessibility of computational power as well as automated systems; medical image analysis has become one of the most interesting research areas. Microscopic image analysis is also valuable in the domain of medical imaging as well as medicine. For example, different cell determination, identification, and counting are an important and almost unavoidable step that assists to diagnose some precise diseases. Many computer vision and digital image analysis instances require a basic segmentation phase to find different objects or separate the test image into distinct segments, which can be treated homogeneous depending upon a given property, such as color and texture. Region growing and fuzzy C-Means are two efficient and popular segmentation techniques. In this chapter, we have proposed a hybrid scheme for image segmentation using fuzzy C-Means clustering, region growing method, and thresholding. The fuzzy C-Means (FCM) clustering is used as a preprocessing

S. Chakraborty · K. Mali
Department of Computer Science and Engineering, University of Kalyani, Kolkata, India
e-mail: shouvikchakraborty51@gmail.com

K. Mali
e-mail: kalyanimali1992@gmail.com

S. Chatterjee · A. Das (✉)
Department of Computer Science and Engineering, University of Engineering & Management, Kolkata, India
e-mail: ajanta.desarkar@gmail.com

S. Chatterjee
e-mail: chatterjeesankhadeep.cu@gmail.com

S. Bhattacharyya et al. (eds.), *Hybrid Machine Intelligence for Medical Image Analysis*, Studies in Computational Intelligence 841,
https://doi.org/10.1007/978-981-13-8930-6_3

step. It helps to process the image more accurately in further stages. To eliminate the noise sensitivity property of fuzzy *C*-Means (FCM), we have used the PFCM, i.e., penalized FCM clustering algorithm. In this work, to find the appropriate region, we have used the region growing segmentation technique coupled with the thresholding. In the proposed work, a similarity feature depending on pixel intensity is used. The threshold value can be calculated using different techniques, such as iterative approach, Otsu's technique, local thresholding, and manual selection, to determine the optimal threshold. The results are obtained and derived by applying the proposed method on different images obtained from publicly available benchmark datasets of human brain MRI images. Experimental results indicated that the PFCM supported hybrid model is far more superior in segmenting medical images than traditional image segmentation methods.

Keywords Medical images · MRI · PFCM · Hybrid region growing · Otsu's method · Local thresholding

1 Introduction

The image segmentation technique is focused on separating an image which is independent of domain. The prime goal of such an algorithm is to segment the image into some regions which are visually different and homogeneous in nature. The individual regions must be meaningful which is to be measured based on some attributes like texture or color. Segmentation helps in easy image analysis. Segmentation process segregates an image into different regions. The extent of the segregation level depends on the problem under consideration. Image segmentation is very useful in different domains and applied as a tool in various sectors like industry, health care, and various other domains. Segmentation is conceptually very simple and easy to implement. By observing an image, anyone can determine the different regions contained in an image. Any human being can easily determine whether an object is a building, or an animal, a person, a tree, or any background scene? Human vision is efficient enough to detect the area of interest. If we want to achieve the same thing by a computer algorithm, then it will be something tricky. It is little difficult to say or define a region. To perform the segmentation using a computer algorithm, one must specify some feature or properties that are used to distinguish one region from another. Another challenge is to find the number of regions in a given digital image. Many image segmentation algorithms depend on the discontinuity and similarity or homogeneity of the pixels to their local surroundings. There are two broad categories of segmentation methods. One is edge based. It is dependent on the discontinuity attribute. Another one is known as region-based method which is based on similarity or homogeneity. Thresholding-based image segmentation methods divide an image into separate segments depending upon a threshold value which can be computed locally or globally [1].

Clustering process divides an image in such a way that object with some similar characteristics falls in the same category. Similarity measure can be determined depending upon the problem. There are various clustering methods found in the literature. Clustering approaches can be broadly classified into two classes such as the hard clustering approach and the fuzzy clustering approach. Both of them have their own efficiency and drawbacks. The hard clustering technique compels each point to fall under one specific class resulting in a very crisp segmentation; i.e., every point falls under exactly one region. This method has a severe drawback and not applicable for some images with bad contrast, noise, poor resolution, etc. This method fails on those images and detects false segments.

Now fuzzy set theory [2] comes into picture. It basically works on the concept of partial membership. Fuzzy set-based clustering methods are frequently used as a part of soft segmentation technique and successfully applied on various segmentation processes on different datasets [3–10]. In this work, a penalized fuzzy C-Means clustering scheme is used to preprocess the image. After that, an integrated region growing and thresholding approach has been proposed that takes the preprocessed image as input and detects segments from this.

Seeded region growing (SRG) is a well-known segmentation technique proposed by Adams and Bischof [11]. It is hybrid in nature. This scheme begins by taking **n** initial seed points A_1, A_2, …, A_n, and, at each iteration, it increases the size of the region by incorporating a pixel x that satisfies the homogeneity criteria into its surrounding. The SRG scheme is robust, speedy, and not dependent on any controlling parameters [12]. Seed points are grown to construct a region, and at last, similar or small regions are merged to form an exact region for which we are interested about. In [12], seed points are generated by employing a tree method. The first scheme divides the image into some fixed rectangular regions first and then considers their centers as seeds. Next approach detects the edges of the subject and then finds the basic seed points along with their centroid. At last, the third method is an extension of the second method. In this scheme, a smoothing filter is employed to manage the noise. Tremeau and Borel [13] proposed a new scheme with both growing and merging technique, and this can be applied to the color image also. This algorithm first applies the region growing algorithm. It considers the color similarity and spatial proximity as the homogeneity measure. After that, the resulting regions are merged. In this step, only color similarity has taken into consideration.

Thresholding is an important and widely used method that basically works on the distribution of the gray level. In this method, the first goal is to identify and extract an object of interest from its background. To perform this job, the procedure depends on the arrangement of gray-level values or texture in different objects in the subject image. Two major factors on which most of the thresholding algorithms depend are: (i) frequency distribution of the one-dimensional (1D) gray levels (ii) the bi-dimensional (2D) co-occurrence matrix of the image under test. Various one-dimensional thresholding methods have been studied [14–22]. Parametric or nonparametric approaches can be taken to compute the threshold value [14, 17, 23]. In parametric approaches, the details of the gray levels in different regions are the key point for computing the threshold value. For example, in Wang and Haralick's scheme [18], the pixel intensities of an image are segregated in two categories, e.g., edge pixels and non-edge pixels. Edge pixels are then divided into darker or brighter classes depending on their local surroundings and following this step histogram of these two classes are obtained. Then the threshold is computed from these two histograms, and the highest peaks of the histograms are considered as the threshold. On the other hand, moment preserving thresholding is an example of the parametric method. In this process, at first, segment the image into separate regions. Segmentation is performed depending on the criteria of equal moment which is to be satisfied by the actual and segmented image [16].

2 Penalized Fuzzy *C*-Means-Based Preprocessing

The fuzzy *C*-Means clustering approach forces every pixel to fall in some class by using fuzzy membership functions. Let $X = (x_1, x_2, \ldots, x_N)$ represents an image where a count of pixels is denoted by N. Assume this image is to be segmented in c clusters. Here x_i represents the multispectral data. This algorithm optimizes the cost function by iterative process. The cost function is given in Eq. (1).

$$J = \sum_{j=1}^{N} \sum_{i=1}^{c} u_{ij}^{m} \left\| x_j - v_i \right\|^2 \tag{1}$$

where u_{ij} denotes the associativity of a pixel x_j in the ith cluster and ith cluster center is given by v_i, ||.|| denotes a norm metric, and m is a constant. The fuzziness of the resulting segments is controlled by the parameter m. Here $m = 2$ is considered for this work.

In the FCM scheme, the distance of every pixel and the corresponding cluster center is the key factor for determining the probability. The membership functions along with the cluster centers are modified using the following equations:

$$u_{ij} = \frac{1}{\sum_{k=1}^{c} \left(\frac{\left\| x_j - v_i \right\|}{\left\| x_j - v_k \right\|} \right)^{2/(m-1)}} \tag{2}$$

And,

$$v_i = \frac{\sum_{j=1}^{N} u_{ij}^{m} x_j}{\sum_{j=1}^{N} u_{ij}^{m}} \tag{3}$$

FCM process starts with some initial points for each cluster centers. The convergence of the process can be checked by either observing the change in the cluster centers or by the change occurred in the membership function. The original FCM algorithm can be applied to any images, but some major issues are associated with it. For example, the image should be noise-free. It does not work well on noisy images. Moreover, any information about spatial context is not taken into consideration. Hence, this algorithm is not useful for noisy images and sensitive to some image artifacts.

One simplest and natural solution to the above-mentioned problem is that image can be smoothened before application of the FCM algorithm. For that purpose, we have to employ some filters. But conventional filters may lose some important image information. It can lose some vital information like edges. The balance between smoothing and clustering cannot be handled appropriately. To overcome this issue, many different techniques have been developed [4–9]. Tolias et al. [4] developed a method which was based on fuzzy rule and named the rule-based neighborhood enhancement system. This algorithm performs some post-processing on the obtained results to add spatial continuity in general FCM algorithm. In another method of the same authors [5], multi-resolution information was attached. Because of this operation, the spatial constraint is imposed on the image. Noordam et al. [6] suggested a new FCM method which is geometrically guided (GG-FCM). The main concept behind this algorithm was a semi-supervised FCM technique. In their proposed method, the membership values of neighboring play a major role to determine the condition of each pixel. Then the pixel is either included to or eliminated from the obtained cluster.

In the proposed work, a FCM algorithm with some modification to the conventional FCM clustering algorithm is applied for preprocessing. This process is called penalized FCM (PFCM) [24] algorithm. The penalty term is based on Neighborhood Expectation-Maximization (NEM) concept [25]. The penalty parameter considers the spatial dependence of the objects. The fitness parameter of the PFCM process is given in Eq. (4);

$$J_{\mathrm{PFCM}} = \sum_{j=1}^{N}\sum_{i=1}^{c} u_{ij}^{m}\left\|x_j - v_i\right\|^2 + \gamma\sum_{j=1}^{N}\sum_{k=1}^{N}\sum_{i=1}^{c} u_{ij}^{m}(1 - u_{ik})^{m_{w_{jk}}} \tag{4}$$

where w_{kj} is defined as Eq. (5). The parameter γ ($\geq$0) is the controlling parameter to handle the consequence of the penalty term efficiently.

$$w_{jk} = \begin{cases} 1 \text{ if } x_j \text{ and } x_k \text{ are neighbours and } j \neq k \\ 0 \text{ otherwise} \end{cases} \tag{5}$$

$$u_{ij} = \frac{1}{\sum_{l=1}^{c}\left(\frac{d^2(x_j,v_i)+\gamma\sum_{k=1}^{n}(1-u_{ik})^q w_{jk}}{d^2(x_j,v_l)+\gamma\sum_{k=1}^{n}(1-u_{lk})^q w_{jk}}\right)^{\frac{1}{q-1}}} \tag{6}$$

$$v_i = \frac{\sum_{k=1}^{n}(u_{ik})^q x_k}{\sum_{k=1}^{n}(u_{ik})^q} \tag{7}$$

In this work, images of interest first preprocessed using the above-mentioned penalized fuzzy *C*-Means algorithm. The resultant image will be given as an input to the hybrid region growing algorithm.

3 Thresholding

The main objective of thresholding method in digital image processing is to segregate image pixels into two or more values. Here, each pixel value is checked for the similarity with the pre-specified or predetermined threshold value *T*. If the value of the threshold is no correctly decided, then there can be following three scenarios may occur [1]

- The separated regions may not match with the actual one. It can be smaller or larger than the actual one.
- There can be some discontinuity in the detected image; i.e., they may not be connected.
- The resultant image may be over segmented, or it can be under segmented.

The value of the threshold can be calculated by various techniques. For example, we can determine the threshold value manually or we can use some of the well-known methods like iterative approach, Otsu's Method or local thresholding techniques.

4 Region Growing Process

The prime goal of the segmentation algorithms is to segregate an image into different regions. Thresholding achieves this goal by detecting edges. Edges are basically depended on the discontinuities in gray levels or in the color attributes. In contrast, region growing approach detects the regions directly without depending on the edges.

The initial step of the region growing process is to choose a set of seeds. Selection of seeds is dependent on the need of the user. In this approach, the initial seed can be chosen interactively. So a particular area of interest can be chosen separately. The initial region begins at the selected seed or the collection of seeds. The regions are increased by incorporating the surrounding seed points. Neighboring points can be incorporated depending on homogeneity criteria. There are various criteria depending on which regions can be grown like pixel intensity. Due to this fact, the information of the image is considered to be important. For example, the histogram of the image is very important where the intensity threshold value is considered to be the criteria because a suitable threshold can be useful for segmentation process [1].

The seed point can be chosen by many ways. For example, one can choose a seed point by taking a random position. In this proposed scheme, the seed point has been chosen interactively. The seed point can be chosen in such a way so that one can get a specific segment of interest. It will be very effective in different kind of images; for example, medical image where a doctor may wish to separately analyze a portion of the image under test. In this work, we have combined the fuzzy C-Means, region growing method, and thresholding method. We have computed the threshold by using various techniques and used those values to determine the region [26].

5 Proposed System

5.1 Preprocessing

In this work, the subject image is first passed through the penalized fuzzy C-Means algorithm. For all the images under test, the value of the parameter γ is considered as 400 (unless otherwise stated). To select the value of this parameter, the experimental results are considered as the basis. Experiments proved that 400 is the most suitable value of this parameter that gives the most efficient and appropriate results. But if the image is noisy, then we may need to increase the value of this parameter. In this work, the spatial effect on a particular pixel is determined by the eight neighbors, and for that reasons, a 3×3 window of pixels is taken under consideration.

Algorithm: Penalized fuzzy C-Means algorithm [24]
Step 1: Set the cluster parameters
Step 2: Calculate the degree of the membership using Eq. (6)
Step 3: Determine the centroid of the clusters using Eq. (7)
Step 4: Go to step 2 and repeat the total procedure while it not converges

Now, after convergence of the process, defuzzification should be performed. This process is required for conversion on the fuzzy partition into crisp form.

5.2 Stack-Based Seeded Region Growing

At the very beginning, manual intervention is necessary for an initial selection of the seed. So it begins with a semiautomatic process. So the user needs to specify some feature of interest. After that, it is the responsibility of the region growing process to locate other points those are related with the same properties and incorporate them to form a larger region. This method proved to be very useful in several biomedical applications like detecting high-quality blood vessels in MRA and CTA data. To generate precise models for examining the vascular structures from X-rays or other radiological image sources, a pipeline of processing stages has been developed. These models are very useful to study the different branching patterns and various measurements.

x-1, y-1	x, y-1	x+1, y-1
x-1, y	x, y	x+1, y
x-1, y+1	x, y+1	x+1, y+1

Fig. 1 Four neighbors of the center pixel considered to grow the region

In this proposed work, two-dimensional SRG is implemented with the help of stacks. The stack is a very simple and elegant data structure. Moreover, data access is very simple, and therefore, surrounding pixel traversal around the seed location is simple. In this work, four neighbors are considered during the region growing process and the corresponding pseudocode is as follows (Fig. 1):

Algorithm: Seeded region growing (stack-based implementation)
Initialize the stack with default value
for every seed points considered
Push the location of the seed into the stack
While (some elements present in the stack)
pop location (of a pixel)
select the location of the pixel as a *Region*
Select the pixel as an already visited node
if homogeneity criteria/region growing attribute is similar for the immediate left neighbor of the pixel
if the pixel at the left neighbor is not visited
push the pixel (of the left neighbor) into the stack
if homogeneity property/region growing attribute matches for the pixel at location's right neighbor pixel
if the pixel at the right neighbor is not visited
push the pixel (of the right neighbor) into the stack
if homogeneity property/region growing attribute matches for the pixel at location's top neighbor pixel
if the pixel at the top neighbor is not visited
push the pixel (of the top neighbor) into the stack
if homogeneity property/region growing attribute matches for the pixel at location's bottom neighbor pixel
if the pixel at the bottom neighbor is not visited
push the pixel (of the bottom neighbor) into the stack
end

To include a neighbor inside the region, we have checked whether new region mean including that point is less than or equal to a predetermined threshold value. The threshold value is very difficult to choose. Various approaches have been studied to compute the threshold value and perform segmentation.

Here, the homogeneity criterion is based on the region mean. The new pixel becomes the part of the new region if the newly obtained region mean of the region is satisfying a particular threshold value. This threshold value can be selected in various ways like manual selection, Otsu's method, iterative method, and local thresholding.

The initial seed selection and the base condition on which the region growing process terminates are the two key factors that influence the performance of the region growing algorithm. Therefore, these things should be carefully chosen to achieve an efficient result and obtain the desired region. Generally, region growing process is based on some key factors like homogeneity of the region, contrast of the desired object from its background, strength of the region edges, size, texture, shape, and color. The proposed method has used a method based on region mean. The value of the homogeneity attribute is computed as given below:

Set the value of the region_mean with the seed value (i.e., pixel value of the seed point)

$$\begin{aligned} \text{new\_region\_mean} = {} & (\text{old\_region\_mean} * \text{no\_of\_points} \\ & + \text{pixel value of new entry})/(\text{no\_of\_points} + 1) \end{aligned} \tag{8}$$

5.3 *Thresholding*

There are two basic categories are possible to generate from the image thresholding process:

1. Those pixels, to which, some attribute computed falls below a threshold and those at which the attribute computed from the image equals or exceeds a threshold.
2. Thresholding binarizes the image and can perform cell counts in digitized histological images.

As we have previously mentioned that, the threshold can be selected in various ways. In this proposed work, we have used local thresholding on experimental basis as it provides a better result. We have also tested our system by providing some manual inputs as threshold. After experiment, we found that threshold value 0.2 is working well for most of the images. The structuring element disk is used to test the integrated thresholding method. Other structuring elements are also tested, but the threshold value was not satisfactory for most of the images.

5.3.1 Manual Method

This is a simple approach in which the selection of the threshold value is arbitrary, i.e., any random can be given to test the system. Here the problem is, it completely dependent on the human and better result can be achieved after some attempts.

5.3.2 Iterative Method

In this case, the threshold value is calculated from the following algorithm:

Algorithm: Iterative thresholding method	
1	**Initial** estimate of T
2	**Segmentation** using T
3	G_1, pixels having higher intensity than T
4	G_2, pixels having lower (or equal to) intensity than T
5	**Computation** of the average intensities m_1 and m_2 of G_1 and G_2
6	**New** threshold value: $T_{\text{new}} = (m_1 + m_2)/2$
7	**If** $\lvert T - T_{\text{new}} \rvert > \Delta T$, back to step 2, otherwise stop

This is a simple approach in which the selection of the threshold value is arbitrary, i.e., any random can be given to test the system. Here the problem is, it completely dependent on the human and better result can be achieved after some attempts.

5.3.3 Otsu's Method

In this method, the threshold that optimizes the weights in class variance is the key factor.

The main assumptions of this method are:

(1) Images and the histogram of the images are considered to be bimodal.
(2) Spatial coherence should not be used. Moreover, no other information about the object structure can be used.
(3) Mainly statistics are stationary in nature, but addictiveness can be achieved locally.
(4) Illumination should be uniform throughout the image. It is important to recognize the differences that arise from the bimodal brightness behavior.

Detailed description of this method can be found in [15].

5.3.4 Local Thresholding

Normally $T = T[i, j, \text{feat}(i, j), f(i, j)]$.

If T is a function which is only based on $f(i, j)$, then this type of threshold is considered as the global threshold.

If T is a function of both $f(x, y)$ and local features feat(x, y), then this type of threshold is considered as the local threshold.

We have created a morphological structuring element. Different structuring elements are available like:

Arbitrary It is one of the flat-structuring elements. Here NHOOD is a parameter is used to determine the neighborhood. NHOOD is nothing but a matrix. It is a two-dimensional array consisting of only binary values. The location of the 1's is used to identify the neighborhood. This information is useful for morphological operations. The origin of NHOOD is the center element of the matrix which can be expressed as: floor((size(NHOOD) + 1)/2) (Fig. 2).

Diamond It is also a flat-structuring element. It has diamond shape. Here R is a scalar parameter that denotes the Euclidean distance between different points enclosed within the diamond shape and origin of the SE. R cannot be negative, and it should be an integer scalar (Fig. 3).

Disk: This structuring element is also flat in nature. It has a disk-like shape. Here R is a scalar parameter that is used to specify the radius. R must be nonnegative integer because it used to specify the distance. N is a scalar parameter. It can take a value among 0, 4, 6 and 8 ($N > 0$). This structuring element can be approximately represented as a sequence of N lines which are periodic in nature. No approximation is required when $N = 0$. The members of the structuring element contain all pixels with centers within the distance R from the center. If no value is specified for the parameter N, then the default value is considered as 4 (Fig. 4).

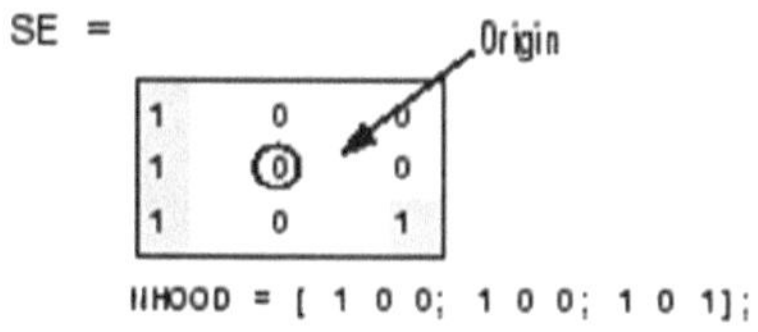

Fig. 2 Arbitrary structuring element

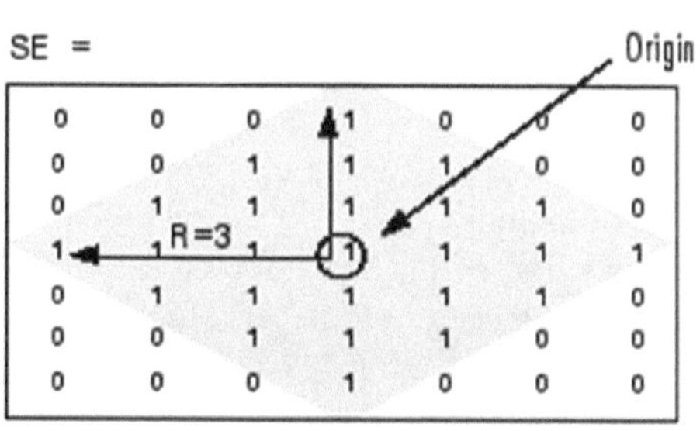

Fig. 3 Diamond structuring element

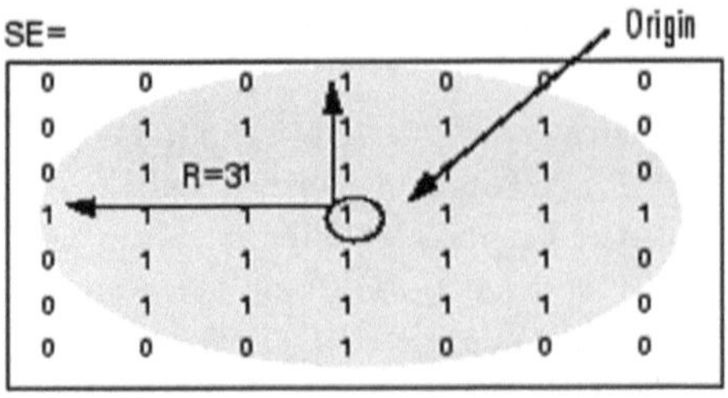

Fig. 4 Disk structuring element

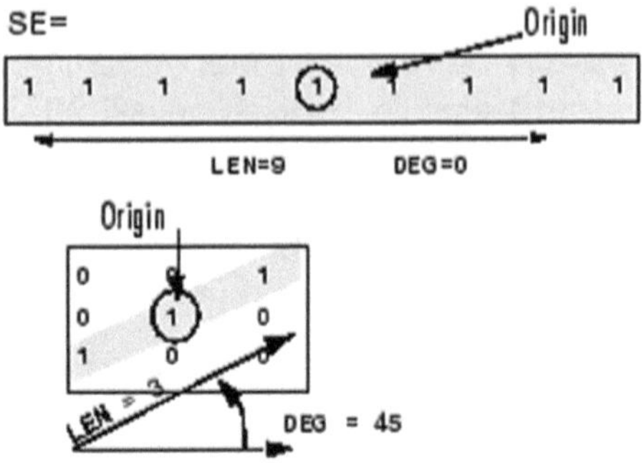

Fig. 5 Line structuring element

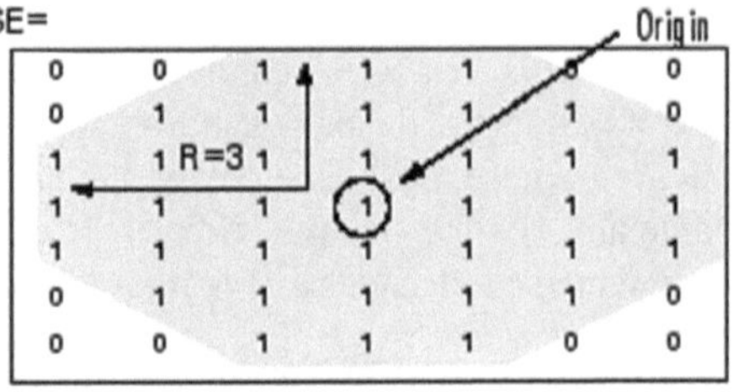

Fig. 6 Octagon structuring element

Line: This structuring element is also flat in nature. It possesses symmetric behavior measured from the origin. DEG is a scalar parameter that is used to represent the angle formed by the lines. The angle is computed in a counterclockwise where the baseline is the horizontal axis. LEN is also a scalar parameter. The separation length of the members of the SEs from the origin is given in Fig. 5.

Octagon: This structuring element is also flat in nature. It is octagonal in shape. Here R is a parameter which is used to specify the Euclidian distance between sides and the origin. Horizontal and vertical axes are used as the baseline of this computation. R must be an integer greater than or equal to 0. Moreover, R must be divisible by 3 (Fig. 6).

Pair: This structuring element is also flat in nature. It consists of two elements. One of them is placed at the center. OFFSET is a parameter that is used to represent the position of the second vector. OFFSET must consist of two elements, and it should be a vector of integers (Fig. 7).

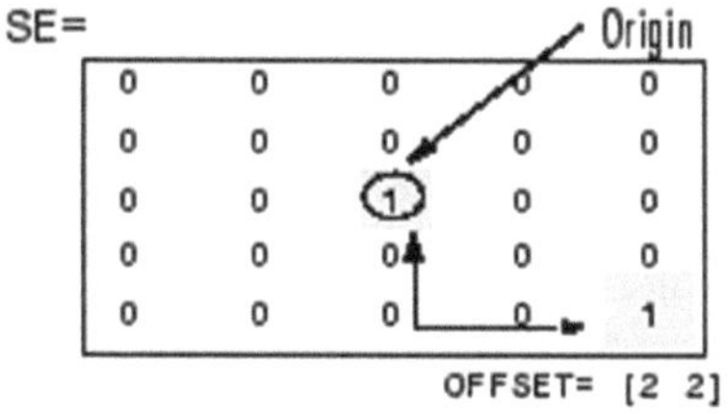

Fig. 7 Pair structuring element

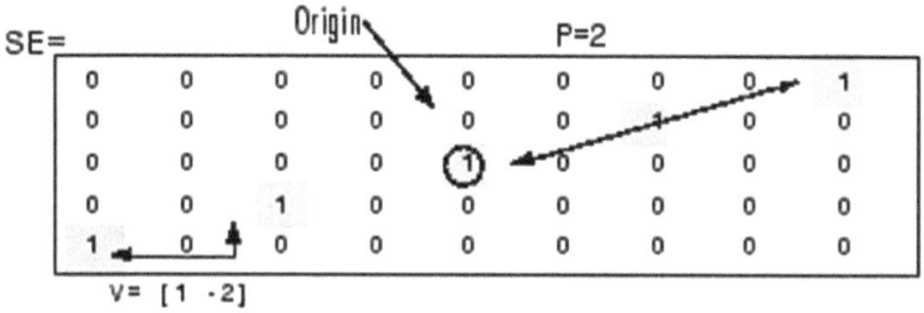

Fig. 8 Periodic line structuring element

Fig. 9 Rectangle structuring element

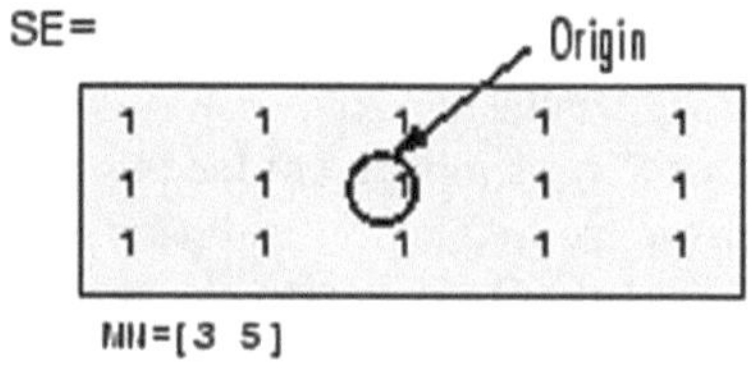

Fig. 10 Square structuring element

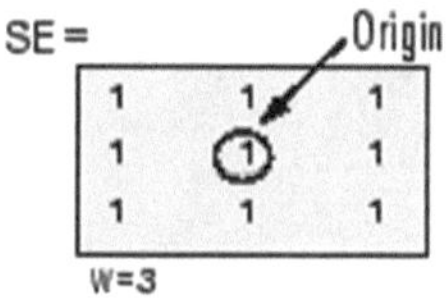

Periodic line: This structuring element is also flat in nature. It is built with of $2 * P + 1$ elements. V is a vector consists of two members. V is built with integral values offsets that represent. One of the SE members is located at the center. The other members are placed at $1 * V$, $-1 * V$, $2 * V$, $-2 * V$, $3 * V$, $-3 * V$, …, $P * V$, $-P * V$ (Fig. 8).

Rectangle: This structuring element is also flat in nature. It is rectangular in shape. Here MN is a vector that is used to specify the size. MN must be a vector consists of two elements. It must consist of integers greater than or equal to 0. The initial element of MN row count in the structuring element surrounding; the second element is the column count (Fig. 9).

Square: It is a square-shaped structuring element. Its width is denoted by W pixels. Here W must be an integer greater than or equal to 0 of type scalar (Fig. 10).

After creating a structuring element, top-hat filtering has been applied on the grayscale or binary input image. This operation is a morphological operation. Morphological opening of the image is performed and computed by the top-hat filtering and then subtracts the obtained result from the actual image. To perform the operation, a structuring element is needed. It should be a single-structuring element in nature. It cannot be an array that contains multiple structuring elements. After that, threshold value is computed (Fig. 11).

6 Results and Discussions

At the very beginning, the segmentation process is semiautomatic because of the seed selection. Then the image is preprocessed by PFCM algorithm followed by region growing and thresholding integrated method.

The MedPix Medical (radiological) image database [27] is used for the experimental results. MedPix is an online database of medical images and clinical topics, integrating images, and textual information including over 19,000 case scenarios and nearly 54,000 images. Different types of regions have been considered to the efficiency of the proposed approach.

The precision–recall is well-known and efficient way to measure the performance. Due to the simplicity of the implementation and efficiency, precision, recall, and *F*-Score are frequently used parameters. Here the main target is to define the optimal boundary. The precision can be demonstrated as the probability of correct detection of boundary pixel (for a detected boundary pixel) [28]. The probability that an exact boundary pixel is found is known as recall rate [29]. The harmonic mean of the precision and the recall values is represented as the *F*-Score [29]. Precision is given by Eq. (9), recall is given by Eq. (10), and *F*-score is given by Eq. (11).

$$\text{Precision} = \frac{\text{detected true boundary pixels}}{\text{detected all boundary pixels}} \tag{9}$$

$$\text{Recall} = \frac{\text{detected true boundary pixels}}{\text{all true boundary pixels}} \tag{10}$$

$$F\text{-Score} = \frac{2 \times \text{Precision} \times \text{Recall}}{\text{Precision} + \text{Recall}} \tag{11}$$

The results in Table 1 as well as Figs. 12, 13, 14, 15 and 16, illustrate a comparative study of different metrics for the proposed method.

The proposed approach has better recall value than the values obtained using the direct application of the algorithm without preprocessing. The same phenomena are observed in the case of the *F*-score and precision results.

- In respect of the precision, the developed method is better and obtained higher values than that obtained using the direct application of the algorithm without preprocessing.

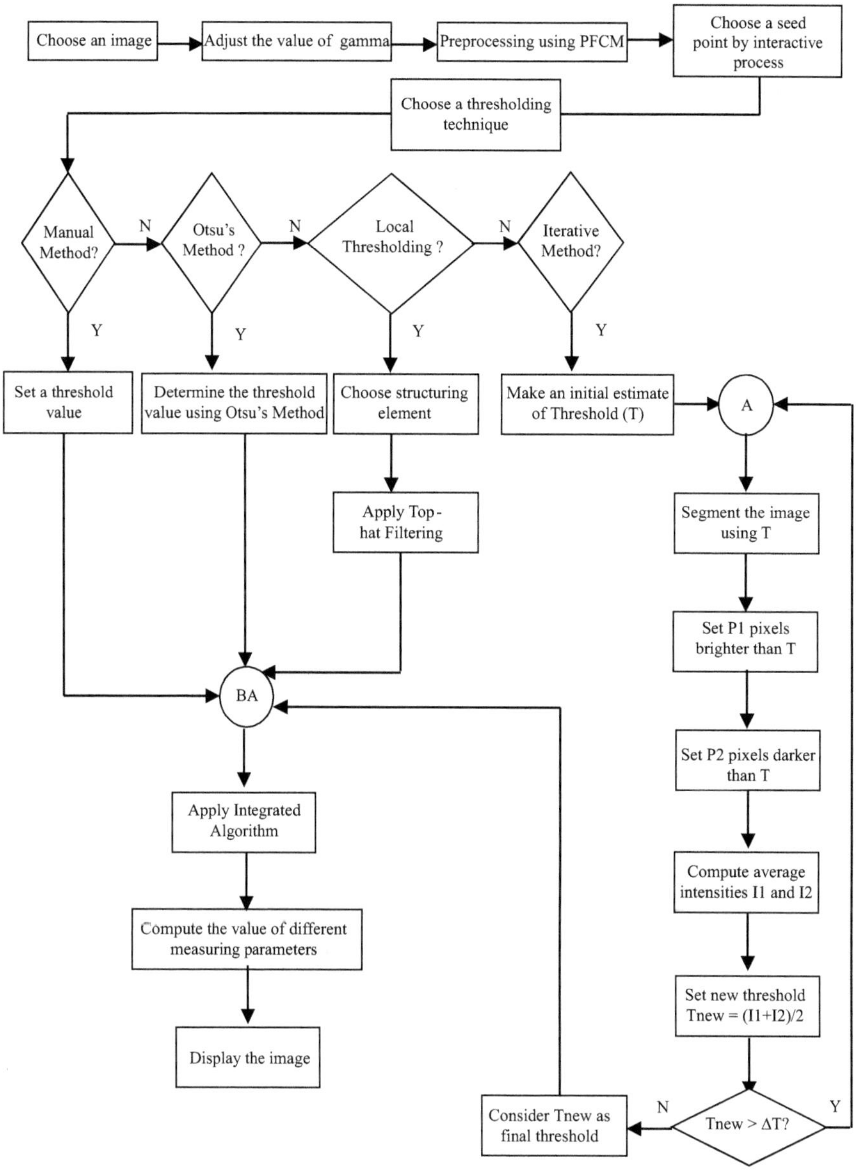

Fig. 11 Schematic diagram of the proposed approach

Table 1 Comparative analysis of the proposed work

S. No.	Proposed			Without PFCM		
	Precision	Recall	*F*-score	Precision	Recall	*F*-score
1	0.6489	0.62149	0.634899	0.663521	0.62145	0.641797
2	0.57145	0.86549	0.688385	0.56247	0.87456	0.684626
3	0.59126	0.888654	0.710076	0.65984	0.83214	0.736041
4	0.64254	0.88475	0.744439	0.63247	0.86231	0.72972
5	0.799658	0.89598	0.845083	0.79854	0.83745	0.817532
6	0.8945	0.8123	0.851421	0.7894	0.76589	0.777467
7	0.68745	0.78452	0.732784	0.64587	0.69875	0.67127
8	0.8456	0.9014	0.872609	0.8315	0.85697	0.844043
9	0.86548	0.76485	0.812059	0.8462	0.76842	0.805437
10	0.8456	0.963	0.90049	0.7984	0.912	0.851428
11	0.900145	0.874512	0.887143	0.8965	0.8759	0.88608
12	0.78946	0.886197	0.835036	0.86571	0.87413	0.8699
Average	0.75684	0.84526	0.79287	0.7492	0.815	0.77628

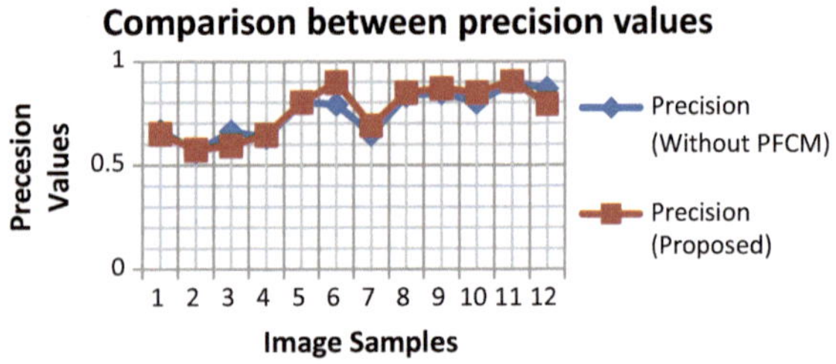

Fig. 12 Comparison between precision values

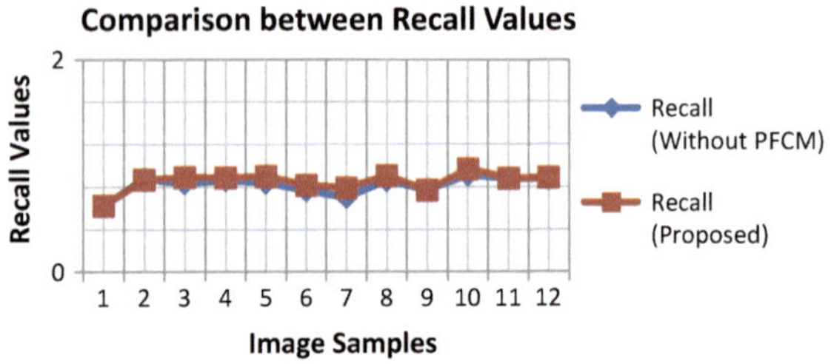

Fig. 13 Comparison between recall values

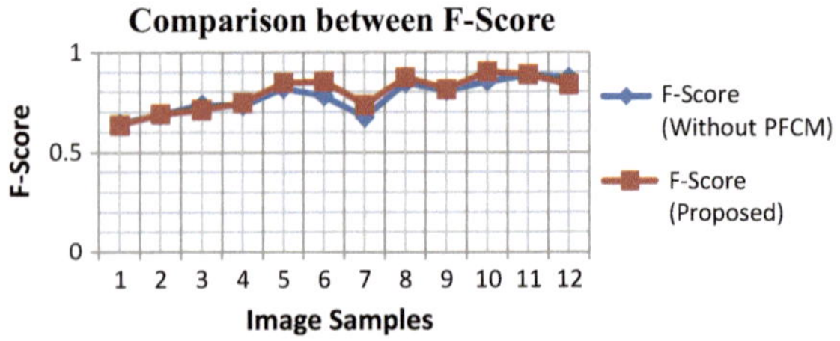

Fig. 14 Comparison between *F*-score values

Fig. 15 Comparison among *F*-score, recall, and precision without preprocessing

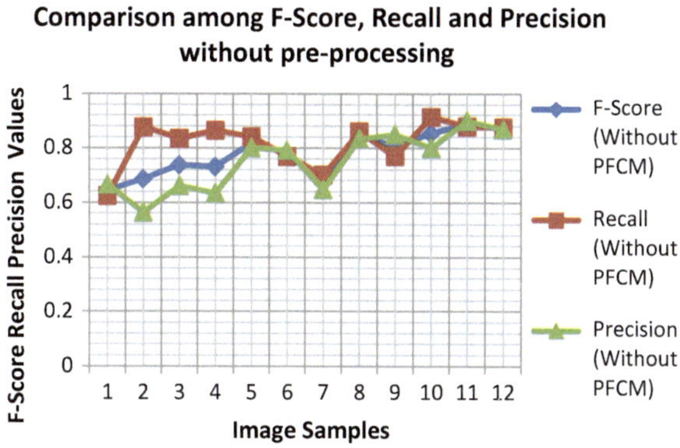

Fig. 16 Comparison among *F*-score, recall, and precision in the proposed approach

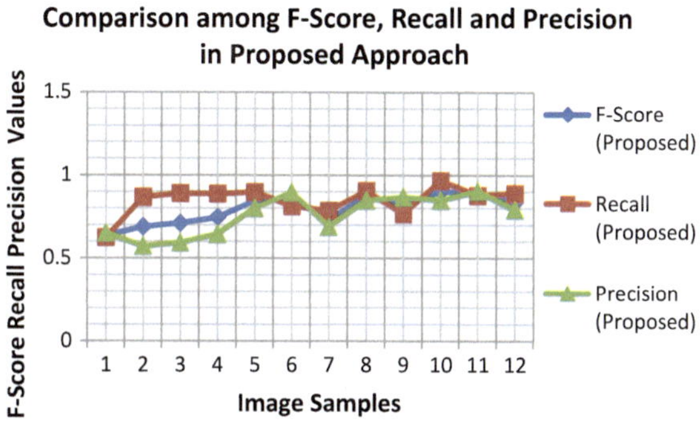

- The computed average values of the performance metrics establish the same and demonstrate that even the proposed method has superior average recall, precision, and *F*-Score values.
- It can be observed from the medical images under consideration that the proposed approach performs well in case of different images in terms of shape, parts of the body, etc.

Experimental results have been shown in Figs. 17 and 18. Both figures contain 12 test images where in Fig. 17 column a indicates the original image, column b indicates contour detected image, and column c indicates the image after segmentation where as in Fig. 18 column a indicates the preprocessed image, column b indicates contour detected image after preprocessing, and column c indicates the image after segmentation after preprocessing.

7 Conclusion

The proposed approach is employed to the biomedical images to exploit its capability to segment the images. It is a hybrid approach that uses the advantages of PFCM,

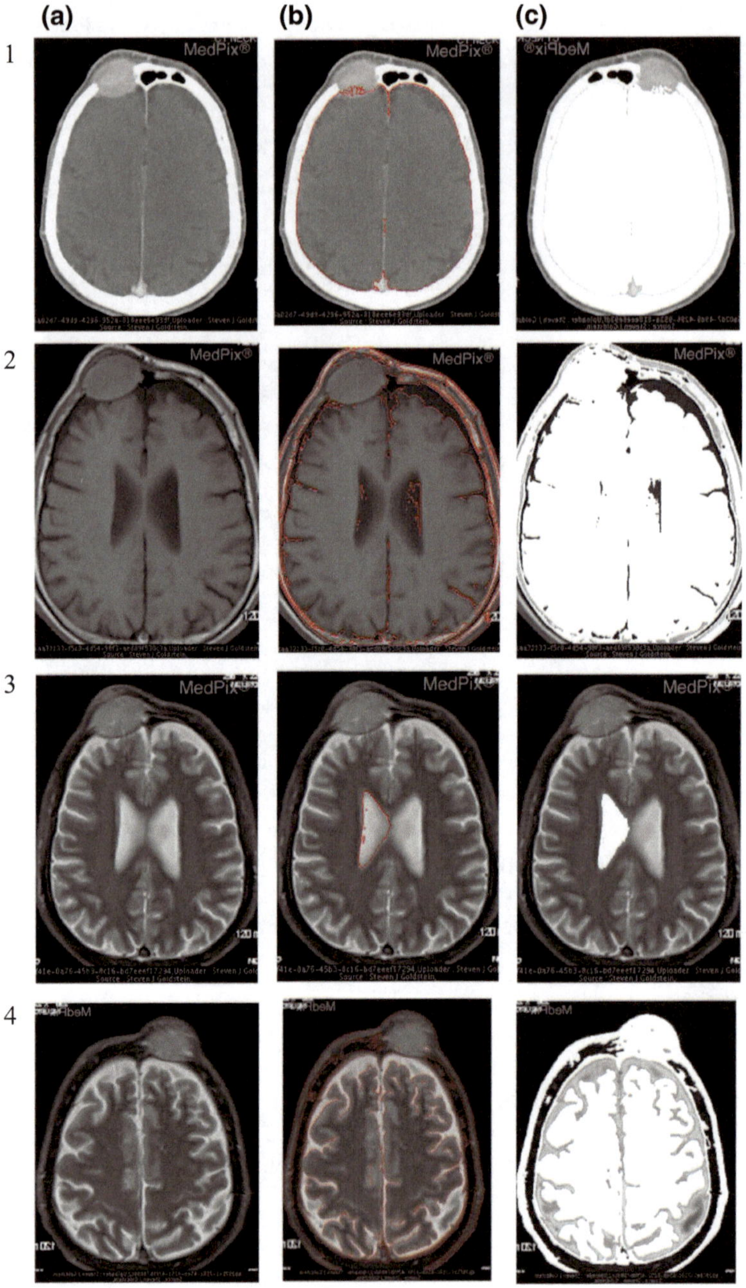

Fig. 17 **a** Original image, **b** contour detected image, **c** segmented image

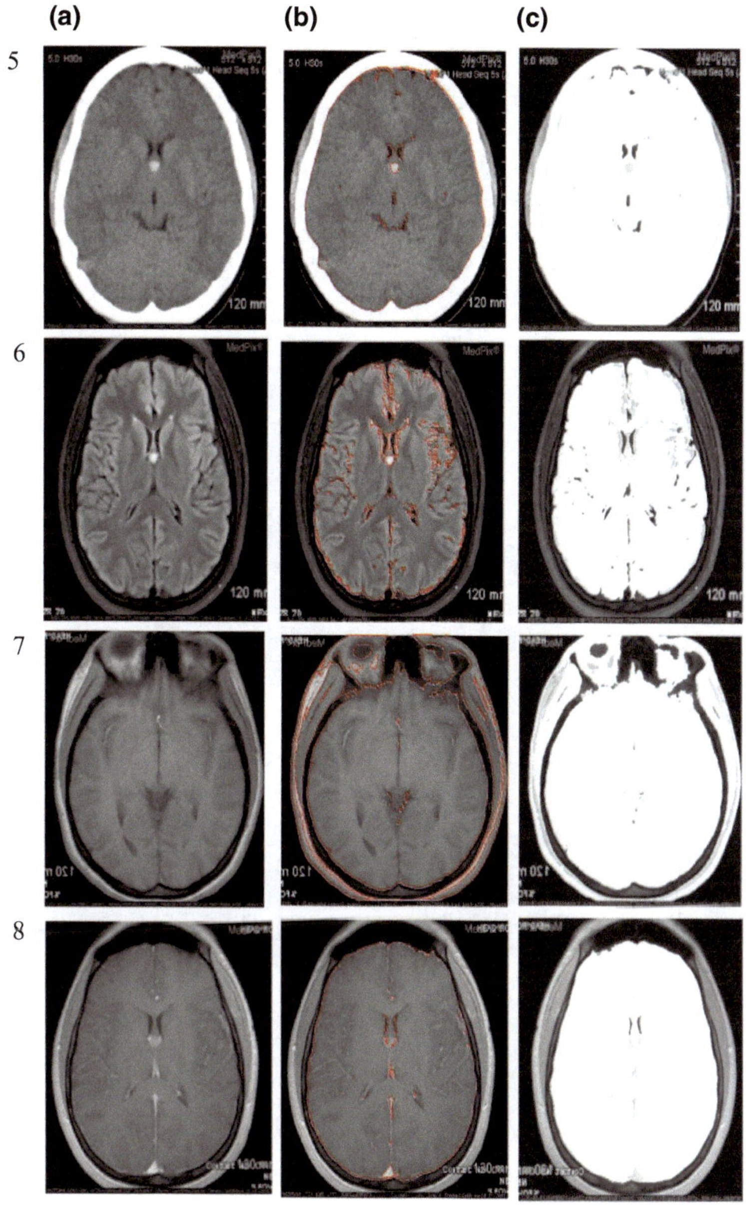

Fig. 17 (continued)

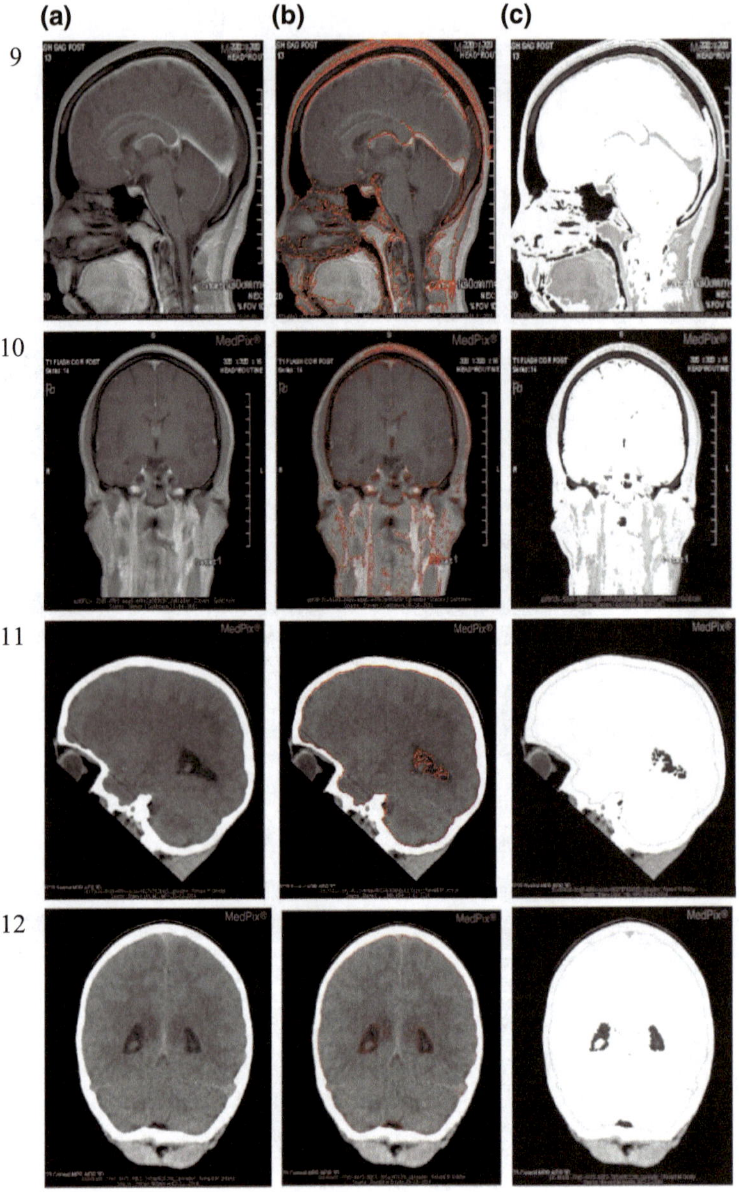

Fig. 17 (continued)

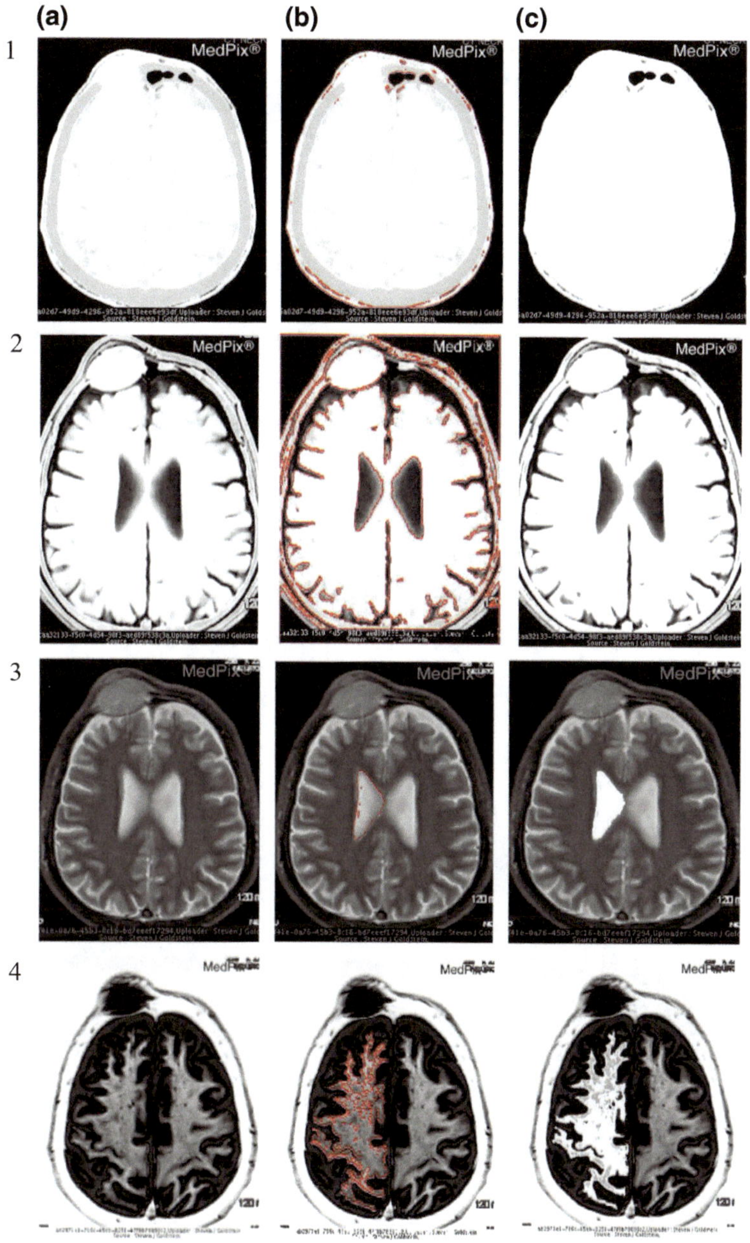

Fig. 18 **a** Preprocessed image, **b** contour detected image after preprocessing, **c** segmented image after preprocessing

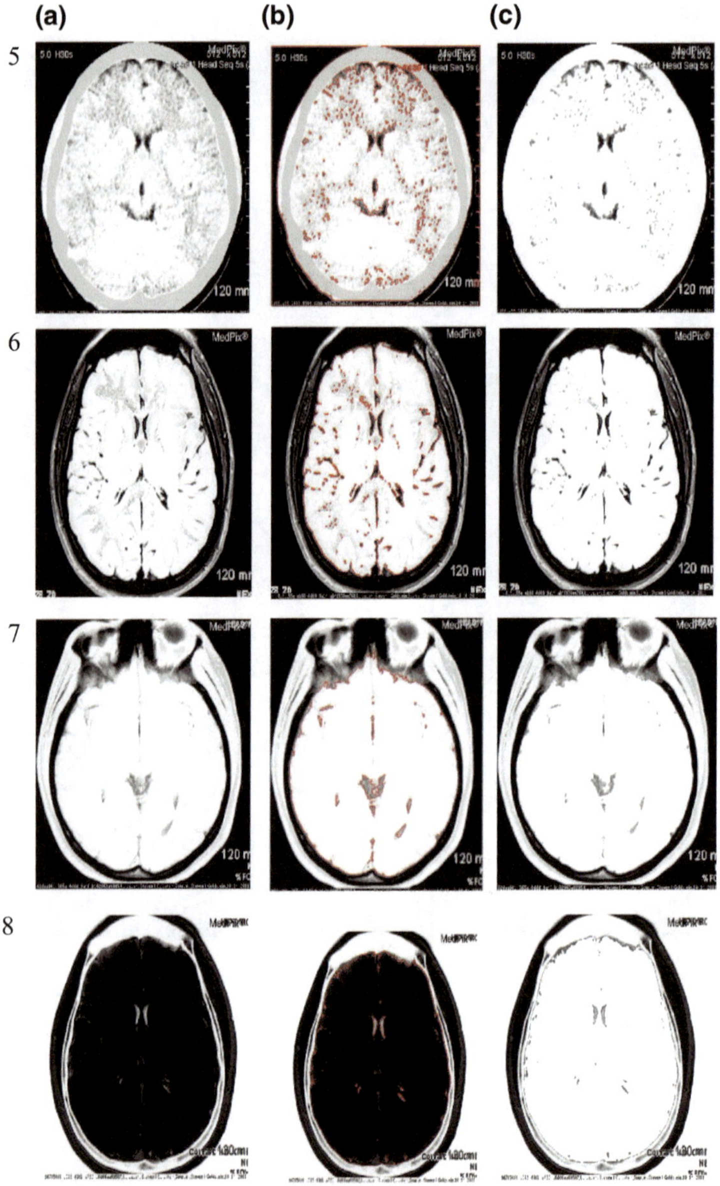

Fig. 18 (continued)

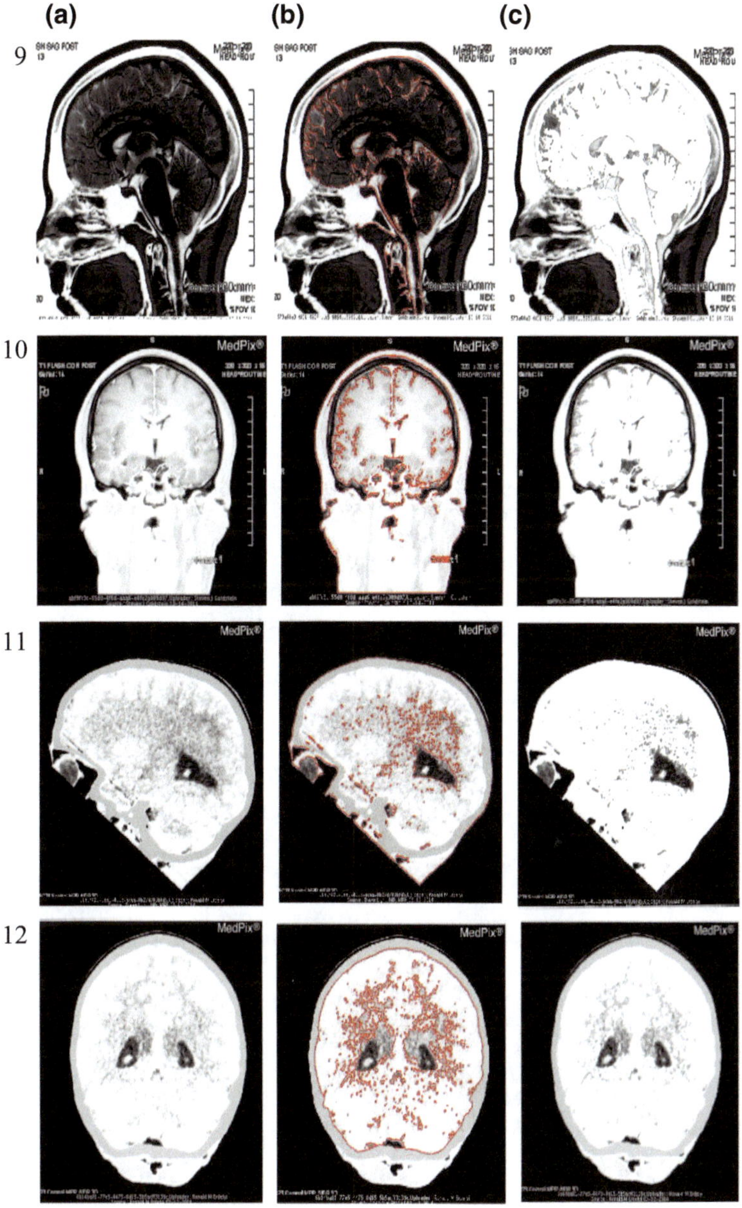

Fig. 18 (continued)

thresholding, and region growing methods to construct a better technique compared to the individual methods for image segmentation.

Moreover, this method provides a method by which medical practitioners can interact with the system manually. It can help users to choose a seed point according to their interest. The efficiency can be measured with three parameters (e.g., recall, precision, and F-score).

In the further work, semiautomatic methods can be implemented for different modalities like PET, fine and small blood vessels in magnetic resonance angiogram (MRA) or retinal images, and computed tomography angiography (CTA) volume data.

References

1. Kamdi, S., Krishna, R.K.: Image segmentation and region growing algorithm. Int. J. Comput. Technol. Electron. Eng. (IJCTEE) **2**(1), (2012)
2. Zadeh, L.A.: Fuzzy sets. Inf. Control **8**, 338–353 (1965)
3. Kwon, M.J., Han, Y.J., Shin, I.H., Park, H.W.: Hierarchical fuzzy segmentation of brain MR images. Int. J. Imaging Syst. Technol. **13**, 115–125 (2003)
4. Tolias, Y.A., Panas, S.M.: On applying spatial constraints in fuzzy image clustering using a fuzzy rule-based system. IEEE Signal Process. Lett. **5**, 245–247 (1998)
5. Tolias, Y.A., Panas, S.M.: Image segmentation by a fuzzy clustering algorithm using adaptive spatially constrained membership functions. IEEE Trans. Syst. Man Cybern. **28**, 359–369 (1998)
6. Noordam, J.C., van den Broek, W.H.A.M., Buydens, L.M.C.: Geometrically guided fuzzy C-means clustering for multivariate image segmentation. In: Proceedings of International Conference on Pattern Recognition, vol. 1 (2000), pp. 462–465
7. Ahmed, M.N., Yamany, S.M., Mohamed, N., Farag, A.A., Moriarty, T.: A modified fuzzy C-Means algorithm for bias field estimation and segmentation of MRI data. IEEE Trans. Med. Imaging **21**, 193–199 (2002)
8. Zhang, D.Q., Chen, S.C., Pan, Z.S., Tan, K.R.: Kernel-based fuzzy clustering incorporating spatial constraints for image segmentation. In: Proceedings of International Conference on Machine Learning and Cybernetics, vol. 4 (2003), pp. 2189–2192
9. Li, X., Li, L., Lu, H., Chen, D., Liang, Z.: Inhomogeneity correction for magnetic resonance images with fuzzy C-Mean algorithm. Proc. SPIE Int. Soc. Opt. Eng. **5032**, 995–1005 (2003)
10. Pham, D.L., Prince, J.L.: Adaptive fuzzy segmentation of magnetic resonance images. IEEE Trans. Med. Imaging **18**, 737–752 (1999)
11. Adams, R., Bischof, L.: Seeded region growing. In: IEEE Transactions on Pattern Analysis and Machine Intelligence, vol. 16. IEEE Computer Society, Los Alamitos California (1994), pp. 641–647
12. Fan, J., Zeng, G., Body, M., Hacid, M.: Seeded region growing: an extensive and comparative study. Pattern Recogn. **26**, 1139–1156 (2005) (Elsevier, Amsterdam)
13. Tremeau, A., Borel, N.: A region growing and merging algorithm to color image segmentation. Pattern Recogn. **30**, 1191–1203 (1997) (Elsevier, Amsterdam)
14. Tsai, D.M., Chen, Y.H.: A fast histogram-clustering approach for multilevel thresholding. Pattern Recogn. Lett. **13**(4), 245–252 (1992)
15. Otsu, N.: A threshold selection method from gray-level histogram. IEEE Trans. Syst. Man Cybern. **SMC-9**(1), 62–66 (1979)
16. Tsai, W.H.: Moment-preserving thresholding: a new approach. Comput. Vis. Graphics Image Process. **29**, 377–393 (1985)

17. Yen, J.C., Chang, F.J., Chang, S.: A new criterion for automatic multilevel thresholding. IEEE Trans. Image Process. **4**(3), 370–378 (1995)
18. Wang, S., Haralick, R.: Automatic multithreshold selection. Comput. Vis. Graphics Image Process. **25**, 46–67 (1984)
19. Sahoo, P.K., Soltani, S., Wong, A.K.C., Chen, Y.: A survey of thresholding techniques. Comput. Vis. Graphics Image Process. **41**, 233–260 (1988)
20. Pun, T.: A new method for gray-level picture thresholding using the entropy of the histogram. Sig. Process. **2**, 223–237 (1980)
21. Kapur, J.N., Sahoo, P.K., Wong, A.K.C.: A new method for gray-level picture thresholding using the entropy of the histogram. Comput. Vis. Graphics Image Process. **29**, 273–285 (1985)
22. Lee, S.U., Chung, S.Y.: A comparative performance study of several global thresholding techniques for segmentation. Comput. Vis. Graphics Image Process. **52**, 171–190 (1990)
23. Abutaleb, A.S.: Automatic thresholding of gray-level pictures using two-entropy. Comput. Vis. Graphics Image Process. **47**, 22–32 (1989)
24. Yang, Y., Huang, S.: Image segmentation by fuzzy C-means clustering algorithm with a novel penalty term. Comput. Inf. **26**, 17–31 (2007)
25. Ambroise, C., Govaert, G.: Convergence of an EM-Type algorithm for spatial clustering. Pattern Recogn. Lett. **19**, 919–927 (1998)
26. Hore S., et al.: An integrated interactive technique for image segmentation using stack based seeded region growing and thresholding. IJECE 6(6) (2016)
27. https://medpix.nlm.nih.gov/home. Last Accessed 15 Mar 2017
28. Buckland, M., Gey, F.: The relationship between recall and precision. J. Am. Soc. Inf. Sci. **45**, 12–19 (1994)
29. Dmw, P.: Evaluation: from precision, recall and F-factor to ROC, informedness, markedness & correlation. J. Mach. Learn. Technol. **2**, 37–63 (2007)

Classification of Diabetic Retinopathy Based on Segmentation of Medical Images

Pavan Kumar Mishra and Awanish Kumar

Abstract Diabetic retinopathy (DR) is a prime reason for escapable blindness in the world. As it progresses, the eyesight of the patient starts worsening, which may lead to blindness if not treated in an early stage. Medical image segmentation and analysis techniques are used for this type of detection of abnormality in retina that correlates and defines the harshness of DR. In this chapter, we discuss the method of DR medical image segmentation to automatically detect and classify the condition of DR. Chapter also discusses the feature extraction of blood vessels, optic disk, microaneurysm, exudates, and macula. The texture of features (gray-level co-occurrence matrix features, histogram intensity features, moment invariants, and gray-level run-length matrix features) finally classifies the DR images into four classes—normal, mild, severe, and proliferative. The method was applied by many researchers on medical images for doing accurate classification and testing. Result obtained with the method is accurate and correctly classified medical images. This chapter is not only a collection of information and facts, but it explains methods/procedure for the classification of diabetic retinopathy based on segmentation of medical images and presented information in the form of result.

Keywords Diabetic retinopathy · Medical image · Segmentation · Classification

1 Introduction

Diabetes is one of the major chronic diseases on this globe. Diabetes leads to diabetic retinopathy which leads to blindness in the person. Of the patients suffering from

P. K. Mishra
Department of Information Technology, National Institute of Technology (NIT), GE Road, Raipur 492010, Chhattisgarh, India

A. Kumar (✉)
Department of Biotechnology, National Institute of Technology (NIT), GE Road, Raipur 492010, Chhattisgarh, India
e-mail: awanik.bt@nitrr.ac.in

S. Bhattacharyya et al. (eds.), *Hybrid Machine Intelligence for Medical Image Analysis*, Studies in Computational Intelligence 841,
https://doi.org/10.1007/978-981-13-8930-6_4

diabetes, diabetic retinopathy (DR) affects 80% of them and this is the chief cause of blindness in global society [1]. However, if it can be noticed in the early stage, blindness can be prevented in those suffering from diabetic retinopathy and for this, regular screening is very necessary [2]. However, many places do not have adequate medical facilities or the ratio of doctors to people is low, making this difficult. So, the automatic analysis of digital fundus image plays a big role. An eye suffering from retinopathy shows various symptoms like bright lesions (cotton wool spots and exudates), dark lesions (microaneurysms and hemorrhages), and thickening of blood vessels [3].

Various researchers have proposed different methods to detect these features, but the accurate and reliable detection and classification of diabetic retinopathy remain a challenge [4]. Detection of these features is made difficult by the presence of other normal objects in the eye, nonuniform illumination, low contrast, and varying image capturing techniques. In this work, we are proposing a method for classification of fundus images as normal or suffering from diabetic retinopathy by detecting five morphological features—blood vessels, optic disk, microaneurysm, exudates, macula, and four sets of texture features—gray-level co-occurrence matrix features, gray-level run-length matrix features, intensity histogram features, and moment invariants [5]. The objective measurements like density of blood vessels, area of optic disk, area of microaneurysms, area of exudates, and area of macula are calculated for each fundus image. These objective measurements are then fed to classifiers for the automatic classification.

2 Process of Medical Image Segmentation of DR

The process of medical image segmentation is given in Fig. 1.

2.1 Image Acquisition

The images used in the process were taken from Messidor program partners [6]. A total of 100 images were taken from the database out of which 70 were chosen for training and 30 were used for the evaluation of the system.

2.2 Preprocessing of Image

Due to their low contrast and the nonuniform illumination, fundus images have to always be corrected before features can be extracted. This is called as the preprocessing step. We have used different preprocessing methods for different features. They are listed here.

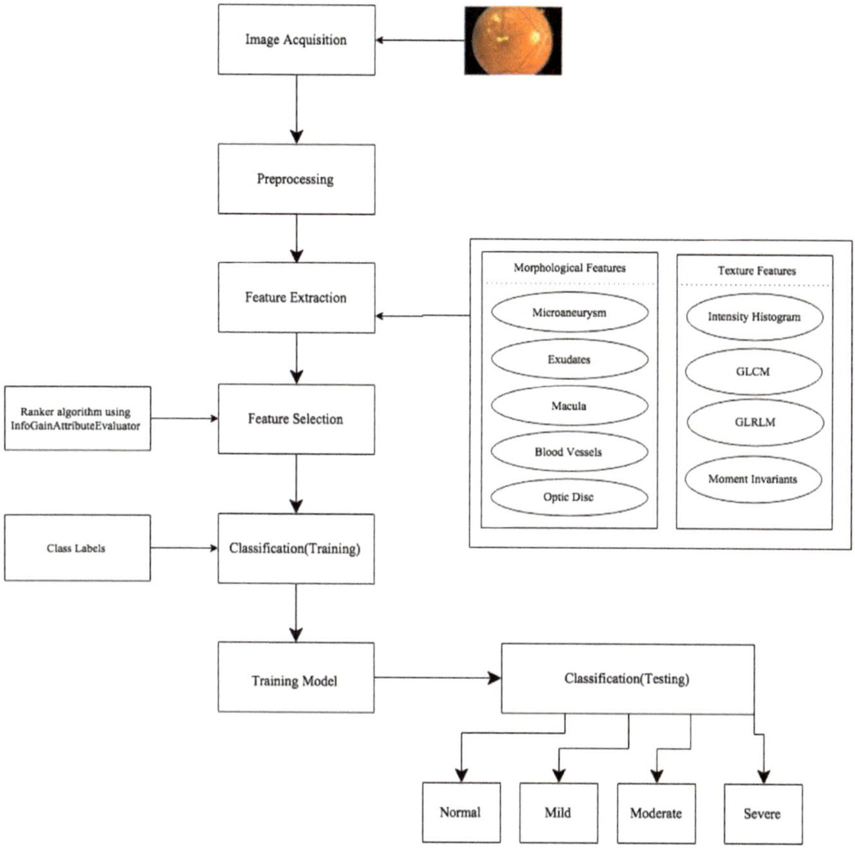

Fig. 1 Architecture of medical image processing

2.2.1 Microaneurysms and Macula

The fundus image was converted to Lab color space from the RGB color space for its preprocessing. Unlike the CMYK and RGB color space, the Lab color space was designed to the estimated vision of human and is the closest to it. It separates the intensity component from the color component and hence, the nonuniform illumination can be corrected without affecting the color. After converting the image into Lab color space, adaptive histogram equalization was administered on it [7]. Then, the image was converted back to the RGB color space for feature extraction.

2.2.2 Texture Features

Adaptive histogram equalization was applied that improve local contrast in images. First, the image was converted into grayscale and then it was applied on the fundus image before feature extraction.

2.2.3 Optic Disk, Blood Vessels, and Exudates

Median filter was applied first on the fundus image. The median filter is a nonlinear filter that is frequently used in image preprocessing to remove noise [8]. Then, adaptive histogram equalization was administered on the output of the filtered image.

3 Feature Extraction of DR Images

3.1 Morphological Features

Microaneurysm Microaneurysms are small outpouchings on the capillary vessels which give the impression as small red dots in the image of the fundus. Microaneurysms have a maximal diameter of 10–100 μm which is clinically established [9]. It is less than the diameter of the major veins. They are the earliest signs of DR and therefore, it is imperative to detect them during the mild phase of DR. Their number progressively increases with the increase in the severity of diabetic retinopathy. To detect microaneurysms, we detect the edges using canny edge detection. Then, hole filling is done to fill up the microaneurysms, however, other noise also gets filled. Since the microaneurysms have a fixed size range and are very small, the larger objects in the image are removed by detecting connected components. We have extracted the optic disk which is subtracted from the image. Now, a small bright object like exudates is left to be removed. For this, the green channel of the original preprocessed image was taken, and further histogram equalization was also done. This emphasizes the bright features which can be removed by applying AND logic. Similarly, the blood vessels are extracted and removed by applying the AND logic. Finally, we calculate the total area covered by the microaneurysms in the image (Fig. 2).

Macula Macula is oval-shaped with a diameter of around 5.5 mm and lies near the center of the retina. However, it appears at around 1.5 mm in fundus images. It has the largest concentration of cone cells in the eye and is therefore responsible for the crisp central vision. The size of macula increases with the severity of DR and leads to a loss of central vision. This does not cause complete blindness as peripheral vision may remain, however, it severely limits the functioning of the person.

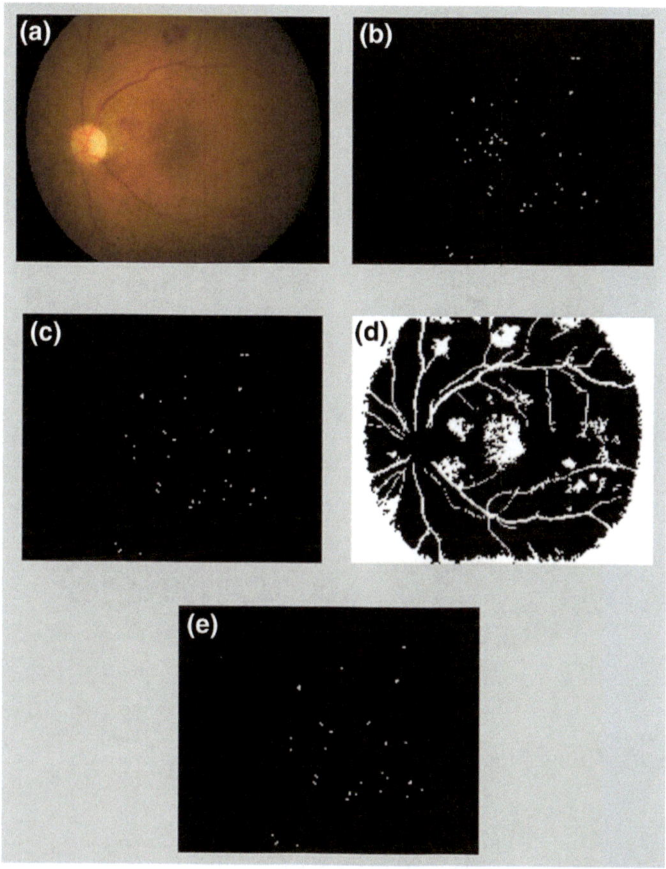

Fig. 2 Set of multimodal photographs of eye analyzed for microaneurysm detection. **a** Input image; **b** microaneurysms with exudates and noise; **c** microaneurysm after removing exudates; **d** blood vessels and noise; **e** microaneurysms

To detect macula, we take the red channel of the image as it provides the best contrast for macula and was converted to binary. For this, the threshold value which gives the best result was determined to be 0.2 after experimentation. Then, the complement of this binary image is taken. Since macula is generally the biggest object in the eye after optic disk and sometimes even bigger than optic disk, the smaller components are removed using "bwareaopen" which calculates connected components and removes all those components which are smaller than a certain P number of pixels. Here, P is taken as a total number of pixels by 4000. Then, region filling is used to fill the outer region leaving only the macula in the image and sometimes small parts of optic disk or blood vessels. To remove them again connected components are calculated and only the biggest connected component is kept and others are removed (Fig. 3).

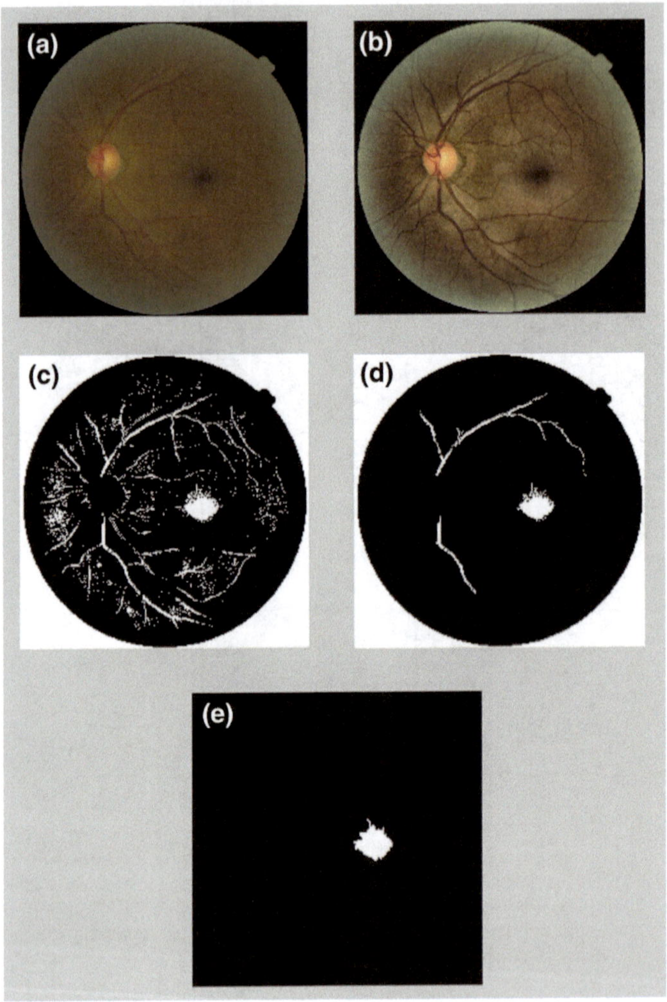

Fig. 3 Set of multimodal photographs of eye analyzed for microaneurysm detection. **a** Input image; **b** image after preprocessing; **c** complimented binary image; **d** image after removing the smaller components; **e** macula

Exudates Exudates are yellowish, fatty, and protein-based intraretinal deposits and are formed due to the leakage from damaged retinal vessels usually found in the posterior pole near the macula. While exudates may occur in other retinal vascular pathologies, diabetic retinopathy is a common cause. Since they generally accumulate near the area of macular, there is the main reason for loss in vision even in case of non-proliferative forms of DR [10]. Hence, they need to be identified at a primary stage when the loss of vision may be prevented using laser treatment.

Exudates appear as dark yellowish spots with brighter edges in the fundus image. To find them, we have first applied dilation on the green channel of the image preprocessed with a square-shaped structure element having side length 25. The dilation was performed to remove the blood vessels from the entire image. We again applied dilation using the disk as a structuring element to get an image with emphasized exudates. Then, we subtract these two images to get exudates with sharp edges and no blood vessels. Since exudates are bright objects, therefore, in the green channel, it has very high intensity. So, the area whose mean intensity in the green channel is greater than a threshold value (found through training) becomes candidate regions [11]. Now, the objects which satisfy both conditions remain in the output and rests are removed. We extract optic disk from the original image and subtract it from the output so that we are left with exudates only (Fig. 4).

Optic Disk The optic disk is the entering point for major blood vessels and the beginning of optic nerve in the eye and therefore serves as a significant landmark for the detection of other features. Its dimensions are studied for a clue of some diseases and a change in its shape size or color can indicate that the person is suffering from diabetic retinopathy. It is circular or oval-shaped and is bright yellowish or white in color.

For extraction of the optic disk, we first select the red channel of the preprocessed image as the intensity range of OD lies in red color range and apply the morphological close operation with a disk of radius 15 as structuring element. After that, segmentation was performed on the image with the help of Otsu Algorithm for automatic determination of the threshold value. Finally, we get a binary image containing the optic disk. We calculate two metrics from this image, first was the number of pixels occupied by the optic disk and second was the ratio of pixel number in the optic disk to pixel number in the eye.

Blood Vessels Through analysis of the blue, green, and red channels of the image, we have found that the green channel provides the best contrast for blood vessels. The Gabor wavelet transform is applied using the fast Fourier transform as it has directional selectiveness capability which helps in identifying oriented features. At each pixel in the image, we calculate two features. First, we calculate the difference $S = L - N$ for a square-shaped window centered at the current pixel where L denotes the maximum gray-level obtained when the line operator is convoluted up to 360° and N denotes the average gray-level. Secondly, the line orthogonal to the first feature is selected. This line is three-pixel lengths and is centered at the midpoint of the basic line operator. Its average is taken as L_0 and the strength S_0 is then given by $S_0 = L_0 - N$. The value of L_0 indicates the thin vessels as they will have a low L_0 value as opposed to the background pixels who will have a high L_0 value. We calculate two metrics for blood vessels, first is the number of pixels occupied by the blood vessels and second is the ratio of a pixel number in blood vessels to the number of pixels in the eye [12] (Fig. 5).

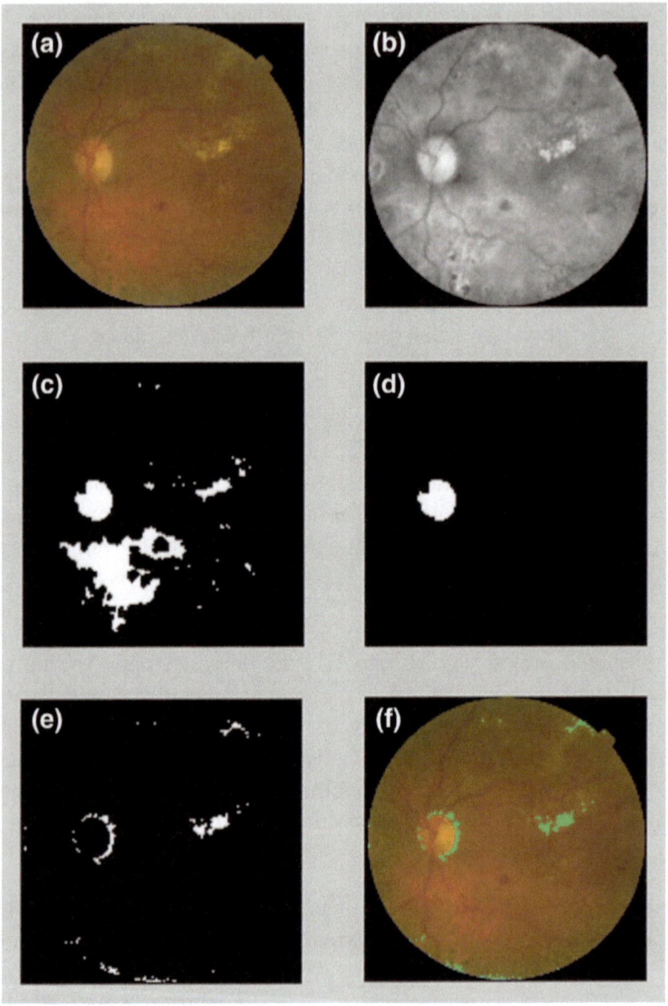

Fig. 4 Set of multimodal photographs of eye analyzed for exudate detection. **a** Input image; **b** image after preprocessing; **c** image segmentation; **d** optic disk segmentation; **e** image after subtracting optic disk; **f** exudates

3.2 Texture Features

Intensity Histogram Features The intensity histogram shows the frequency of each intensity level in the various images. It is frequently used to calculate statistical properties for an image. We have calculated six features based on the intensity histogram, which are listed in Table 1. Here "*I*" expresses the pixel intensity, "*p*(*i*)" the number of pixels with intensity "*i*". This mean gives an average level of intensity of the

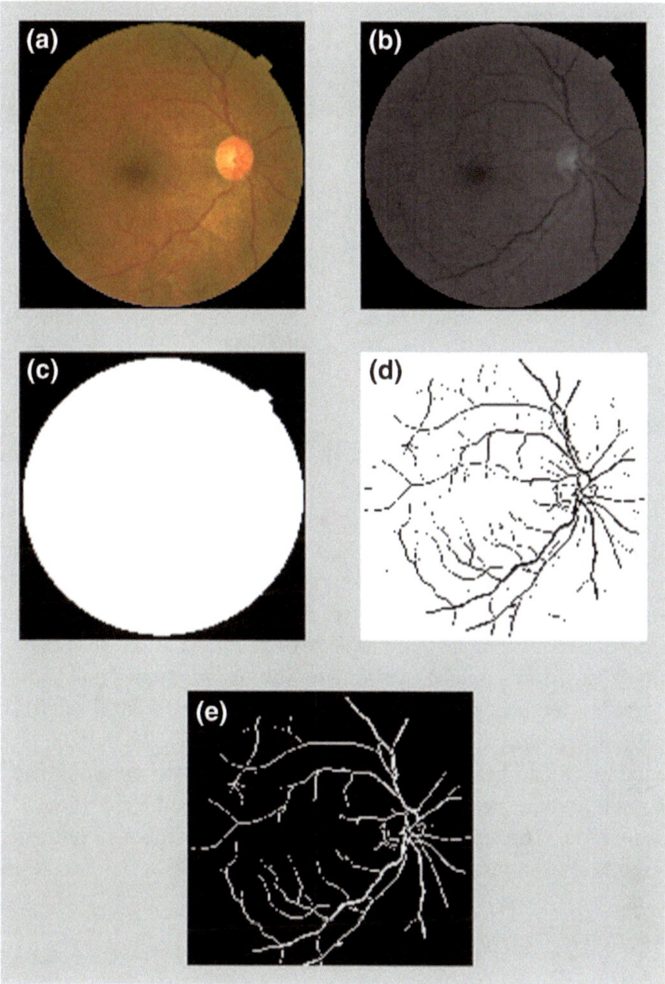

Fig. 5 Set of multimodal photographs of eye analyzed for the extraction of the blood vessel. **a** Input image; **b** processed image; **c** mask of eye; **d** gabor filter applied on the blood vessels; **e** final output

image. The kurtosis was used to measure histogram flatness [13]. The variance was directly a measure of the image roughness. Entropy tells how normal or non-normal is the gray-level distribution and energy tells how much intensity variation was in the region. Measure of the asymmetry of the distribution is skewness, i.e., it measures if there is a wider range of either darker or lighter pixels. Positive skew means that the distribution is concentrated on the left of the figure implying that there are more dark pixels than average, and a negative skew means that there are more bright pixels than average. These features are first-order texture calculations.

Table 1 Intensity histogram features

SI. No.	Name	Equation
1.	Mean	$(\mu) = \sum_{i=0}^{N-1} ip(i)$
2.	Energy	$(E) = \sum_{i=0}^{N-1} p(i)^2$
3	Variance	$(\sigma^2) = \sum_{i=0}^{N-1} (i - \mu)^2 p(i)$
4.	Entropy	$H = \sum_{i=0}^{N-1} p(i) \log_2 p(i)$
5.	Skewness	$\mu_3 = \sigma^{-3} \sum_{i=0}^{N-1} p(i - \mu)^3 p(i)$
6.	Kurtosis	$\mu_4 = \sigma^{-4} \sum_{i=0}^{N-1} p(i - \mu)^4 p(i) - 3$

Spatial Gray-Level Dependency Matrix It is also known as "gray-level co-occurrence matrix" which was one of the most widely used statistical tools in image processing for texture features extraction from the image. It estimates how frequently pair of pixels with explicit values and in a definite spatial relationship happens in an image. Three parameters need to be defined for gray-level co-occurrence matrix, first is the direction displacement, second is the orientation of the measurements, and third is the intensity-level number. The displacement parameter is important as a large displacement value would mean that proper texture information was not captured and vice versa. We tested multiple displacement values and found that a displacement value of 1 yielded the best results. There can be four different directions in which the measurements can be oriented—0°, 45°, 90°, and 135°. We calculated gray-level co-occurrence matrix and the related features in all four orientations and took their average for each feature. We have analyzed the output on different levels by scaling the image down to 8, 16, 32, 64, and 128 levels for the number of intensity levels. It is also found a loss in accuracy for these values. Hence, we calculated the matrix on 256 levels.

A total of 21 features (illustrated by Haralick et al. [14]) were calculated which was grounded on the gray-level co-occurrence matrix. These features are listed in Table 2. Here, G is the number of intensity levels, and $p(i, j)$ is the (i, j) which is the entry in the gray-level co-occurrence matrix. The distinction is an intensity quantity of gray-level differences between a pixel and their neighbor. In the co-occurrence matrix, feature of correlation shows the linear dependence of gray-level values. It shows how a reference pixel is related to its surroundings. 1 is perfectly correlated and 0 is uncorrelated. The autocorrelation function of an image can be practiced detecting repetitive patterns of textures. Cluster shade is the quantity of the matrix skewness. It is believed to gauge the perceptual concepts of uniformity. The image would be asymmetric if the cluster shade is high. Cluster distinction is also an amount of unevenness. The image would be less symmetric if the cluster prominence value is high. When cluster prominence value is low, there would be a peak in the gray-level co-occurrence matrix around the mean values. The features and gray-level co-occurrence matrix are second-order texture calculation. Homogeneity was the most commonly used measure. It also increases with less contrast in the window.

Gray-Level Run-Length Matrix A gray-level run is made by a continuous set of pixels with the same gray-level and collinear in a given direction. The number of pixels in the run is a run length. A run-length value is the number of times of a run with a particular run length that happens in an image. The gray-level run-length matrix is a 2D matrix where each element $p(i, j|\theta)$ gives the total number of occurrences of run-length "*j*" for gray-level "*i*". θ gives the direction of the measurement. Features based on Gray-level run-length matrix are texture features of higher order [15].

We have scaled down the image to eight gray-levels, constructed the gray-level run-length matrix for all four angles 0°, 45°, 90°, and 135°, and then taken their average. We have calculated 11 features which are listed in Table 3 that is founded on the gray-level run-length matrix. Short-run emphasis calculates the distribution of short runs, and its value is higher for fine textures. Similarly, long-run emphasis measures distribution of long runs. It is higher for coarse textures. Low gray-level run and high gray-level run emphasis calculate the distribution of low and high gray-level values correspondingly. The short-run high gray-level emphasis, short-run low gray-level emphasis, long-run low gray-level emphasis, and the long-run high gray-level emphasis are joint measure of run and gray-level distribution. Throughout the image, gray-level nonuniformity gives the similarity of gray-levels. Gray-levels are alike throughout the image with a low value. Run-length nonuniformity provides the similarity of length of runs throughout the image. A low value indicates that the run lengths are alike throughout the image. Run percentage measures the distribution and homogeneity of runs of an image in a specified direction. When the length of runs is 1 for all gray-level, the run percentage is highest.

Invariant Moments In 1962, Hu utilized algebraic invariants to derive a set of moment invariants which are immune to translation, scaling, or rotation. We have calculated the Hu's seven invariant moments.

4 Feature Selection

In total, we have calculated 52 features for all the images in the dataset. However, not all features are necessary and equally useful for classification. Also, the use of too many features for classification also increases the time required for training. Hence, we have used an attribute evaluator called *Info Gain Attribute Evaluator* that assesses the value of every attribute by measuring the information gain with reference to the class. Ranker search was used as a search algorithm in the attribute evaluator. Results are displayed in Table 4.

Finally, we have selected the following attributes:

1. Run-Length Nonuniformity

Table 2 Gray-level co-occurrence features of matrix

SI. No.	Name	Equation
1.	Autocorrelation	$f_1 = \sum_i \sum_j (ij) \cdot P(i, j)$
2.	Contrast	$f_2 = \sum_{n=0}^{N_9-1} n^2 \left\{ \sum_{i=1}^{N_g} \sum_{j=1}^{N_G} P(i, j) \middle\| \|i - j\| = n \right\}$
3.	Correlation	$f_3 = \frac{\left[\sum_i \sum_j (ij) P(i,j) - \mu_x \mu_y\right]}{\left[\sigma_x \sigma_y\right]}$
4.	Cluster prominence	$f_4 = \sum_i \sum_j \left(i + j - \mu_x - \mu_y\right)^4 P(i, j)$
5.	Cluster shade	$f_5 = \sum_i \sum_j \left(i + j - \mu_x - \mu_y\right)^3 P(i, j)$
6.	Dissimilarity	$f_6 = \sum_i \sum_j \|i - j\| P(i, j)$
7.	Energy	$f_7 = \sum_i \sum_j P(i, j)^2$
8.	Entropy	$f_8 = -\sum_i \sum_j P(i, j) \log(P(i, j))$
9.	Homogeneity	$f_9 = \sum_i \sum_j \frac{1}{1+(i+j)^2} P(i, j)$
10.	Extreme probability	$f_{10} = \max_{i,j} P(i, j)$
11.	Sum of squares	$f_{11} = \sum_i \sum_j P(i - \mu)^2 P(i, j)$
12.	Sum average	$f_{12} = \sum_{i=2}^{2N_g} i P_{x+y}(i)$
13.	Sum variances	$f_{13} = \sum_{i=2}^{2N_g} (i - f_{14})^2 i P_{x+y}(i)$
14.	Sum entropy	$f_{14} = \sum_{i=2}^{2N_g} P_{x+y}(i) \log\{P_{x+y}(i)\}$
15.	Difference variance	$f_{15} =$ Variance of P_{x-y}
16.	Difference entropy	$f_{16} = -\sum_{n=0}^{N_9-1} P_{x-y}(i) \log\{P_{x-y}(i)\}$
17.	Information measures of correlation-1	$f_{17} = \frac{\text{HXY}-\text{HXY1}}{\max\{\text{HX},\text{HY}\}}$ where HX and HY are entropy of P_x and P_y
18.	Information measure of correlation-2	$f_{18} = (1 - \exp[-2.0(\text{HXY2} - \text{HXY})])^{1/2}$
19.	Inverse difference	Same as homogeneity
20.	Inverse difference normalized (INN)	$C_{ij} =$ the definition of co-occurrence probability between gray-levels i and j is explained as $C_{ij} = \frac{P_{ij}}{\sum_{i,j=1}^{G} P_{ij}}$ $f_{20} = \sum_{i,j=1}^{G} P_{ij} \frac{C_{ij}}{1 + \|i - j\|^2 / G^2}$
21.	Inverse difference moment normalized	$f_{21} = \sum_{i,j=1}^{G} P_{ij} \frac{C_{ij}}{1+(i-j)^2/G^2}$

Table 3 Gray-level run-length matrix features

SI. No.	Name	Equations
1.	Short-run emphasis (SRE)	$f_1 = \frac{1}{n_r}\sum_{i=1}^{M}\sum_{j=1}^{N}\frac{p(i,j)}{j^2} = \frac{1}{n_r}\sum_{j=1}^{N}\frac{p_r(j)}{j^2}$
2.	Long-run emphasis (LRE)	$f_2 = \frac{1}{n_r}\sum_{i=1}^{M}\sum_{j=1}^{N}\left(p(i,j)\cdot j^2\right) = \frac{1}{n_r}\sum_{j=1}^{N}p_r(j)\cdot j^2$
3.	Gray-level nonuniformity (GLN)	$f_3 = \frac{1}{n_r}\sum_{i=1}^{M}\left(\sum_{j=1}^{N}(p(i,j))\right)^2 = \frac{1}{n_r}\sum_{j=1}^{N}p_g(i)^2$
4.	Run-length nonuniformity (RLN)	$f_4 = \frac{1}{n_r}\sum_{i=1}^{M}\left(\sum_{j=1}^{N}(p(i,j))\right)^2 = \frac{1}{n_r}\sum_{j=1}^{N}p_r(i)^2$
5.	Run percentage (RP)	$f_5 = \frac{n_r}{n_p}$
6.	Low gray-level run emphasis (LGRE)	$f_6 = \frac{1}{n_r}\sum_{i=1}^{M}\sum_{j=1}^{N}\frac{p(i,j)}{i^2} = \frac{1}{n_r}\sum_{j=1}^{N}\frac{p_g(j)}{i^2}$
7.	High gray-level run emphasis (HGRE)	$f_7 = \frac{1}{n_r}\sum_{i=1}^{M}\sum_{j=1}^{N}\left(p(i,j)\cdot i^2\right) = \frac{1}{n_r}\sum_{j=1}^{N}p_r(j)\cdot i^2$
8.	Short-run low gray-level emphasis (SRLGE)	$f_8 = \frac{1}{n_r}\sum_{i=1}^{M}\sum_{j=1}^{N}\frac{p(i,j)\cdot i^2}{i^2\cdot j^2}$
9.	Short-run high gray-level emphasis (SRHGE)	$f_9 = \frac{1}{n_r}\sum_{i=1}^{M}\sum_{j=1}^{N}\frac{p(i,j)\cdot i^2}{j^2}$
10.	Long-run low gray-level emphasis (LRLGE)	$f_{10} = \frac{1}{n_r}\sum_{i=1}^{M}\sum_{j=1}^{N}\frac{p(i,j)\cdot j^2}{i^2}$
11.	Long-run high gray-level emphasis (LRHGE)	$f_{11} = \frac{1}{n_r}\sum_{i=1}^{M}\sum_{j=1}^{N}p(i,j)\cdot i^2\cdot j^2$

2. Gray-Level Nonuniformity
3. Run Percentage
4. Short Run High Gray-Level Emphasis
5. Gray-Level co-occurrence matrix (Energy)
6. Information measure of correlation 2
7. High Gray-Level run Emphasis
8. Long Run High Gray-Level Emphasis
9. Inverse difference normalized
10. Difference Entropy
11. Contrast
12. Low Gray-Level Run Emphasis
13. Information measure of correlation 1
14. Gray-Level co-occurrence matrix (Entropy)

Table 4 Features ranked on the basis of their information gain

Rank	Feature name
1.8508	Run-Length Nonuniformity
1.8066	Gray-Level Nonuniformity
1.7341	Run Percentage
1.2514	Short Run High Gray-Level Emphasis
1.1697	Energy
1.1455	Information measure of correlation 2
1.1424	High Gray-Level run Emphasis
1.1118	Long Run High Gray-Level Emphasis
1.1056	Inverse difference normalized
1.0392	Difference Entropy
1.0392	Contrast
0.9991	Low Gray-Level Run Emphasis
0.9868	Information measure of correlation 1
0.9717	Entropy
0.9672	Cluster Prominence
0.9457	Correlation
0.9108	Energy
0.8783	Homogeneity
0.7272	Sum of Squares: Variance
0.7245	The inverse difference is homom
0.703	Blood Vessel 1
0.6864	Maximum Probability
0.3774	Phi1
0.3716	Sum entropy
0.3654	Macula
0.3502	Sum average
0.3502	Long Run Emphasis
0.3488	Short Run Low Gray-Level Emphasis
0.3424	Mean
0.318	Autocorrelation
0.2832	Microaneurysm
0.2673	Homogeneity: MATLAB
0.2314	Sum variance

(continued)

Table 4 (continued)

Rank	Feature name
0.1882	Optic disk 2
0.1882	Optic disk 1
0.1869	Exudates
0	Kurtosis
0	Variance
0	Skewness
0	Blood vessel 2
0	Entropy 1
0	Cluster Shade
0	Phi5
0	Phi6
0	Phi7
0	Phi4
0	Phi3
0	Phi2
0	Dissimilarity
0	Short Run Emphasis
0	Long Run Low Gray-Level Emphasis
0	Difference variance

15. Phi1
16. No of pixels occupied by Blood Vessel
17. Microaneurysms
18. Exudates
19. Macula
20. The ratio of pixels occupied by the optic disc to total pixels in the eye.

5 Classification of DR

We have classified the DR image in four classes normal, mild, severe, and proliferative DR. A total of 70 DR images were taken for training and 30 were for the purpose of testing. For classification, we tested the system on multiple classifiers and the obtained results are listed in Table 5.

Table 5 Results on different classifiers

Classifier/ Algorithm used	Correctly classified instances	Incorrectly classified instances	Kappa statistic	Mean absolute error	Root mean square error	Accuracy (%)
Naïve Bayes	25	5	0.7774	0.0879	0.2683	83.33
SVM	7	23	0	0.3833	0.6191	23.33
Multilayer perceptron	28	2	0.911	0.0634	0.1953	93.33
AdaBoost M1 method	15	15	0.3202	0.3719	0.4295	50
Random forest	28	2	0.911	0.0505	0.1618	93.33
Bagging	28	2	0.911	0.0404	0.1642	93.33

6 Conclusions

In this process, we have used bagging on REP tree classifier. Based on the results obtained from testing on different classifiers, we selected Bagging algorithm. Bagging is also known as bootstrap aggregating. It is a machine learning ensemble meta-algorithm. Bagging gives each model in the ensemble equal weight in voting. It trains each model in the ensemble with a random subset of the training set. It helps to reduce variance, avoid overfitting, and acquire reliable results.

References

1. Javitt, J.C., Aiello, L.P., Chiang, Y., Ferris 3rd, F.L., Canner, J.K., Greenfield, S.: Preventive eye care in people with diabetes is cost-saving to the federal government. Implications for health-care reform. Diab. Care **17**, 909–917 (1994)
2. Early Treatment Diabetic Retinopathy Study Research Group: Early photocoagulation for diabetic retinopathy. ETDRS report number 9. Ophthalmology **98**, 766–85 (1991)
3. Bresnick, G.H., Mukamel, D.B., Dickinson, J.C., Cole, D.R.: A screening approach to the surveillance of patients with diabetes for the presence of vision-threatening retinopathy. Ophthalmology **107**, 19–24 (2000)
4. Kinyoun, J., Barton, F., Fisher, M., Hubbard, L., Aiello, L., Ferris 3rd, F.: Detection of diabetic macular edema. Ophthalmoscopy versus photography—early treatment diabetic retinopathy study report number 5. The ETDRS Research Group. Ophthalmology **96**(6), 746–750 (1989)
5. Chen, X., Fang, M., Dong, D., Wei, X., Liu, L., Xu, X., Jiang, X., Tian, J., Liu, Z.: A radiomics signature in preoperative predicting degree of tumor differentiation in patients with non-small cell lung cancer. Acad. Radiol. **25**(12), 1548–1555 (2018)
6. Decencière, E., Zhang, X., Cazuguel, G., Lay, B., Cochener, B., Trone, C., Gain, P., Ordonez, R., Massin, P., Erginay, A., Charton, B., Klein, J.-C.: Feedback on a publicly distributed database: the Messidor database. Image Anal. Stereol. **33**, 231–234 (2014)
7. Shakhovska, N.: Advances in Intelligent Systems and Computing. CSIT 2016, September 6–10 Lviv, Ukraine (2016)

8. Karimi, B., Krzy, A.: A novel technique for detecting suspicious lesions in breast ultrasound images. Concurrency Comput. Pract. Exper. **28**, 2237–2260 (2016)
9. Lazar, I., Hajdu, A.: Retinal microaneurysm detection through local rotating cross-section profile analysis. IEEE Trans. Med. Imaging **32**(2), 400–407 (2013)
10. Sinthanayothin, C., Boyce, J.F., Williamson, T.H., Cook, H.L., Mensah, E., Lal, S., Usher, D.: Automated detection of diabetic retinopathy on digital fundus images. Dibetic Med. **19**(2), 105–112 (2002)
11. Ravishankar, S., Jain, A., Mittal, A.: Automated feature extraction for early detection of diabetic retinopathy in fundus images. In: Proceedings/CVPR, IEEE Computer Society Conference on Computer Vision and Pattern Recognition. IEEE Computer Society Conference on Computer Vision and Pattern Recognition, June 2009
12. Yang, W., Wang, S., Jiankun, J., Zheng, G., Valli, C.: Security and accuracy of fingerprint-based biometrics: a review. Symmetry **11**, 141 (2019)
13. Krishnan, M.R., Faust, O.: Automated glaucoma detection using hybrid feature extraction in retinal fundus images. J. Mech. Med. Biol. **13**(01), 1350011 (2013)
14. Haralick, R.M., Shanmugam, K., Dinstein, I.H.: Textural features for image classification. IEEE Trans. Syst. Man Cybern. **SMC-3**, 610–621 (1973)
15. Ha, S., Choi, H., Paeng, J.C., et al.: Radiomics in oncological PET/CT: a methodological overview. Nucl. Med. Mol. Imaging **53**, 14 (2019)

A New Hybrid Adaptive Cuckoo Search-Squirrel Search Algorithm for Brain MR Image Analysis

Sanjay Agrawal, Leena Samantaray, Rutuparna Panda and Lingraj Dora

Abstract This chapter presents a new hybrid adaptive cuckoo search-squirrel search (ACS-SS) algorithm for brain magnetic resonance (MR) image analysis. Thresholding is one of the popular methods utilized for brain image segmentation. Thresholding-based methods are easily implemented. In this context, we present an optimal multilevel thresholding technique for brain MR images using edge magnitude information. The edge magnitude is computed using the gray-level co-occurrence matrix (GLCM) of the brain image slice. The optimum thresholds are found by maximizing the edge magnitude. A new hybrid evolutionary computing technique, namely ACS-SS, is investigated to maximize the edge magnitudes. The proposed scheme is tested with T_2-w brain MR images from Harvard medical education database. The results are compared with cuckoo search (CS), squirrel search (SS), and adaptive cuckoo search (ACS) algorithms. It is witnessed that the findings, using the proposed ACS-SS technique, are superior to the other techniques in terms of qualitative and quantitative measures. The advantages of the proposed technique are as follows: (i) The ACS-SS shows improved fitness function values; (ii) the ACS technique gives speed improvement.

S. Agrawal · R. Panda (✉)
Department of Electronics and Telecommunication Engineering, VSS University of Technology, Burla 768018, India
e-mail: r_ppanda@yahoo.co.in

S. Agrawal
e-mail: agrawals_72@yahoo.com

L. Samantaray
Department of Electronics and Communication Engineering, ABIT, Cuttack, India
e-mail: leena_sam@rediffmail.com

L. Dora
Department of Electrical and Electronics Engineering, VSS University of Technology, Burla 768018, India
e-mail: lingraj02uce157ster@gmail.com

S. Bhattacharyya et al. (eds.), *Hybrid Machine Intelligence for Medical Image Analysis*, Studies in Computational Intelligence 841,
https://doi.org/10.1007/978-981-13-8930-6_5

Keywords Brain MR image analysis · Multilevel thresholding · Hybrid optimization algorithm · Adaptive cuckoo search-squirrel search (ACS-SS) algorithm · Edge magnitude information · Gray-level co-occurrence matrix

1 Introduction

There are many applications of image segmentation from image classification, object identification, to target detection for defense usage. Now, image segmentation in brain MR image analysis has developed as a promising and challenging area of research. It aids in detecting abnormality in any part of the brain using the image acquired from computed tomography (CT) scan, MRI scan, or positron emission tomography (PET) scan. At present, MRI is the most commonly used clinical diagnostic procedures for brain ailments. It provides a good contrast, among the several tissues of the brain. This makes MRI more appropriate for imaging brain, liver, etc., and detecting cancer in contrast to other medical imaging systems like CT or X-rays. One of the popular techniques of image segmentation is thresholding. The accuracy of detecting lesions or disorder in the brain depends on effective thresholding. There are numerous techniques for effective thresholding. A number of thresholding techniques are reported in the literature [1, 2]. Usually, in bi-level thresholding, a threshold value divides an image into two classes. However, with the increase in the number of threshold values, the problem of thresholding becomes difficult. Hence, investigations on the multilevel thresholding problem are even now one of the highest published research works. There are a number of algorithms suggested for bi-level as well as multilevel thresholding based on the image histogram (first-order statistics) [1–4]. We present here an algorithm for multilevel thresholding based on the edge magnitude obtained from the image co-occurrence matrix (second-order statistics) [5]. Over the years, many investigators have proposed several optimal multilevel thresholding procedures, where the optimum thresholds are found by minimizing or maximizing a fitness function as per the requirements of the problem. Several criteria are used for developing the objective function, namely maximizing some measure of separability [6], using fuzzy similarity measure [7], maximizing entropy [3], minimizing Bayesian error [8], and so on. Similarly, some of the aforementioned procedures are being used for brain MR image analysis [9–15].

Maitra and Chatterjee [4] proposed a new optimal multilevel thresholding procedure for brain MR image segmentation. The authors utilized histogram-based criteria for thresholding using bacterial foraging (BF) optimization. They compared their findings with particle swarm optimization (PSO) and stated that BF outdoes PSO. However, a hybridized optimization may give better results. Sathya and Kayalvizhi [16] proposed adaptive bacteria foraging (ABF) for the same problem. However, they employed an adaptive step size to improvise the foraging behavior in the existing BF procedure. They utilized maximization of Kapur's entropy and Otsu's between-class variance as the criteria for the formation of the fitness function. To validate their suggested technique, they compared their findings with genetic algorithm (GA), PSO, and BF. Later, the authors [17] presented an amended bacteria foraging procedure to address the same problem of brain segmentation. They validated their findings with BF, PSO, and GA. The authors have not mentioned any possibility of hybridized optimization approach to the problem on hand. It is observed that the above methods [4, 16, 17] are developed using histogram (first-order statistics) of the images. Shan et al. [18] suggested an automated procedure for whole brain segmentation and volume measurement using T_1-weighted 3D MR images using histogram and morphological operations. The authors have not mentioned any optimization technique in their algorithm. Joliot and Mazoyer [19] proposed an algorithm for 3D reconstruction of the brain from MRI data. The authors used two algorithms of segmentation and interpolation of MRI data. They used morphological operations in segmentation. But they do not mention any optimization technique in their work. Lai and Chang [20] suggested a hierarchical evolutionary algorithm (HEA) based on clustering for medical image segmentation. They represented a hierarchical configuration in the chromosomes that can automatically divide the image into suitable classes. Their method is based on GA, and they compared their work with neural networks.

The authors in [21] presented a hybrid method for information extraction from brain MRI images. They used *K*-means, artificial neural network (KMANN), and GLCM for feature extraction. The authors then created a fuzzy inference system using the extracted features followed by thresholding, morphological operation, and Watershed segmentation for brain tumor detection. It is evident that the authors used a number of methods for their work thereby increasing the complexity. In [22], the identification of tumor is done by segmentation. Initially, the existing methods like expectation—maximization, histogram, and object-based thresholding are analyzed and implemented. But some of the features are missing in all these methods. So a new algorithm is proposed in which all the features from the above methods are fused. The authors in [23] worked for automated identification of tumor from the brain MR images using an intelligent computer-aided method. The authors used a number of approaches to accomplish the task. The brain tumor portion was segmented using K-means clustering. They used the GLCM for feature extraction from the segmented brain tumor portion. Finally, the support vector machine (SVM) classifier was utilized to classify the MR images as normal and abnormal.

But it is believed that a hybrid optimization approach can bring in much improved results [24–28]. In this chapter, we are motivated to investigate a new hybrid optimization technique ACS-SS to perform brain MR image analysis utilizing the edge magnitude information. Our method is derived from the second-order statistics, which is quite different from the existing techniques. The edge magnitude information is derived from the GLCM of the brain image. We then introduce ACS-SS technique to maximize the edge information. This is because optimal multilevel thresholds are obtained by maximizing the edge magnitude information for thresholding the brain MR images.

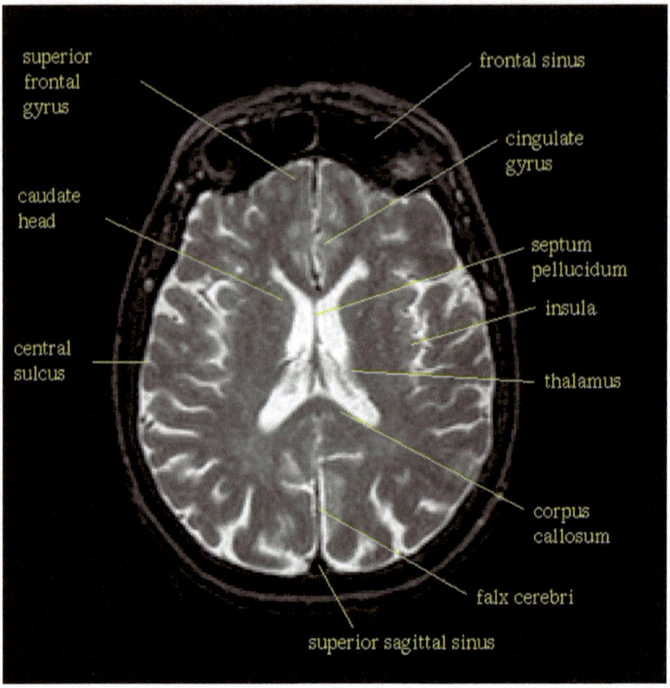

Fig. 1 Example T_2-w structural slice no. 31 of the healthy brain MR image

The brain region has closely spaced intensity levels. So multilevel thresholding may not be easily applied. The edge magnitude information helps in identifying the boundaries in the brain region, because it is embedded in the contrast definition. Further, GLCM considers the neighborhood information. So optimum threshold values using edge magnitude information ease the segmentation of brain image.

Here, we have taken T_2-weighted brain MR image slices from Harvard medical education [29] for the simulation. T_2-w scans represent the changes in the spin–spin (or T_2) relaxation time of various tissues in the brain. Here the fat is distinguished from water. However, it appears blacker than water. Additionally, the cerebrospinal fluid (CSF) will be lighter for brain analysis. It offers superior contrast between brain tissue (dark) and CSF (bright). Certain T_2 structures exhibit more contrast between white matter (darker gray) and gray matter (lighter gray). These images are found most appropriate for brain segmentation problem [30]. Figure 1 displays a sample T_2-w MR image slice of a healthy brain. That is why we have taken T_2-w brain MR samples for simulation.

The remaining of the chapter is structured as follows. The second section briefly explains the idea of edge magnitude information derived from GLCM for multilevel thresholding. The third section presents a brief idea about the different soft computing techniques used. Our suggested technique is given in the fourth section. The fifth section outlines the results and validations. Finally, the sixth section draws the concluding remarks.

2 GLCM and Edge Magnitude Information

The GLCM is a technique of obtaining a second-order statistical feature of an image. The number of rows and columns of the GLCM is same as the number of intensity levels in the image. If an image say I contains L intensity levels from [1,L], then its GLCM—G will be a $L \times L$ matrix. A matrix element $g(i, j)$ of G represents the probability of occurrence of the gray-level pair i and j for a given distance vector D at a given angle θ. A distinct GLCM exists for each such (D, θ) pair. So it is necessary to limit the values of (D, θ) for decreasing the computational complexity. Figure 2 displays the spatial relationship of pixels for computing GLCM for four different angles considering the angular symmetry.

It is stated that there exists a simple relationship for certain pairs of (D, θ) depending on the angular symmetry. This relationship is given as follows:

$$\begin{aligned} G(D, 0°) &= G(D, 180°) \\ G(D, 45°) &= G(D, 225°) \\ G(D, 90°) &= G(D, 270°) \\ G(D, 135°) &= G(D, 315°) \end{aligned} \tag{1}$$

Because of angular symmetry, the information obtained from $G(D, 180°)$, $G(D, 225°)$, $G(D, 270°)$, and $G(D, 315°)$ adds nothing significant to the computation of the GLCM. So researchers have suggested computing a mean GLCM from the four matrices for $\theta = 0°, 45°, 90°, 135°$ at a given distance D. Here, we have computed the GLCM as [31, 32]

$$G_M = \frac{1}{4}\left[G(D, 0°) + G(D, 45°) + G(D, 90°) + G(D, 135°)\right] \tag{2}$$

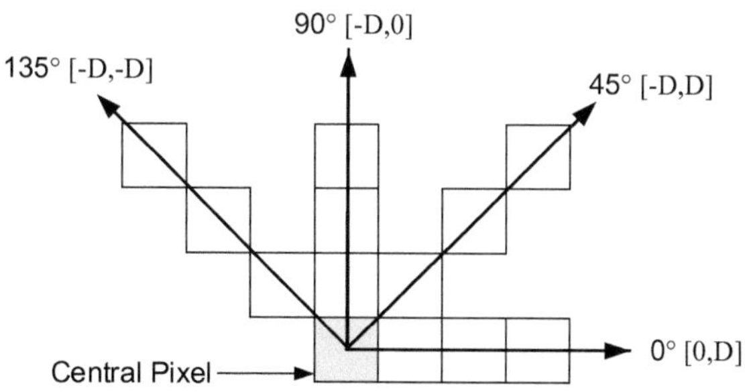

Fig. 2 Spatial relationship of pixels for computing GLCM for different angles

Haralick et al. [33] has extracted many features from GLCM such as homogeneity, contrast, inverse difference moment, correlation, and entropy. We are using here edge magnitude information that is available in the contrast definition and is defined as [34]

$$\text{Contrast} = \sum_{e=0}^{L-1} e^2 \left\{ \sum_{m=0}^{L-1} \sum_{n=0}^{L-1} G_M(m,n) \right\}$$
$$|m-n| = e \tag{3}$$

where e represents the edge magnitude. Note that G_M represents the mean GLCM computed from the four matrices. Mokji and Bakar [31] computed the value of threshold T based on the edge magnitude as

$$T = \frac{1}{\eta} \sum_{m=0}^{L-1-e} \sum_{n=m+e}^{L-1} \frac{m+n}{2} G_M(m,n) \tag{4}$$

where

$$\eta = \sum_{m=0}^{L-1-e} \sum_{n=m+e}^{L-1} G_M(m,n) \tag{5}$$

This value of T is used for bi-level thresholding only. The range of the summation process represents the edge component. Further, the summation range restricts the computation of threshold value within a specified area in the GLCM. This area is identified as $n - m \geq e$ which corresponds to the upper triangle of GLCM due to its symmetry feature as shown in Fig. 3.

The above equations correspond to bi-level thresholding only. The optimal threshold value is computed by maximizing the edge magnitude which is represented as

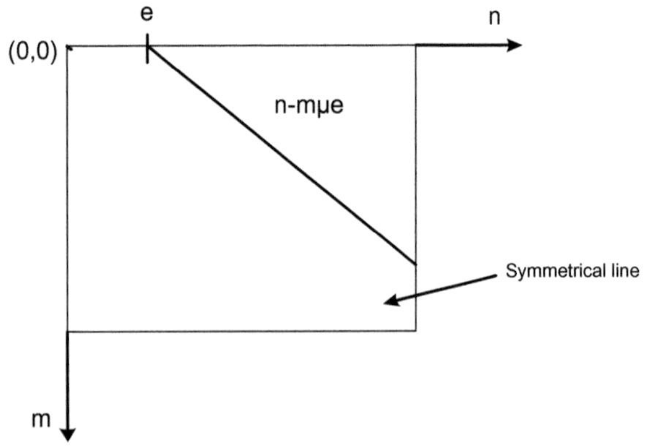

Fig. 3 Computation area in the GLCM for optimum threshold value

$$T_{\text{opt}} = \arg\ \max\{f(e)\} \tag{6}$$

For bi-level thresholding, it is optimum when the summation over e in Eq. (4) is maximized. The optimum threshold values for multilevel thresholding are similarly represented in [5]. The optimal threshold values are computed as in [5]

$$\begin{aligned}
T_{\text{opt1}} &= \arg\ \max\left(\frac{1}{\eta_1}\sum_{m=0}^{e_1}\sum_{n=e_1+1}^{e_2}\left(\tfrac{m+n}{2}\right)G_M(m,n)\right)\\
T_{\text{opt2}} &= \arg\ \max\left(\frac{1}{\eta_2}\sum_{m=e_1+1}^{e_2}\sum_{n=e_2+1}^{e_3}\left(\tfrac{m+n}{2}\right)G_M(m,n)\right)\\
&\vdots\\
T_{\text{opt}(k-1)} &= \arg\ \max\left(\frac{1}{\eta_{k-1}}\sum_{m=e_{k-2}+1}^{e_{k-1}}\sum_{n=e_{k-1}+1}^{L-1}\left(\tfrac{m+n}{2}\right)G_M(m,n)\right)
\end{aligned} \tag{7}$$

where

$$\begin{aligned}
\eta_1 &= \sum_{m=0}^{e_1}\sum_{n=e_1+1}^{e_2} G_M(m,n)\\
\eta_2 &= \sum_{m=e_1+1}^{e_2}\sum_{n=e_2+1}^{e_3} G_M(m,n)\\
&\vdots\\
\eta_{k-1} &= \sum_{m=e_{k-2}+1}^{e_{k-1}}\sum_{n=e_{k-1}+1}^{L-1} G_M(m,n)
\end{aligned}$$

The threshold values are optimum when the addition over edge magnitudes in (7) is maximum. The optimum threshold values are calculated by maximizing the edge magnitude which is represented as

$$\left[T_{\text{opt1}}, T_{\text{opt2}}, \ldots, T_{\text{opt}(k-1)}\right] = \arg\ \max\{f(e_1, e_2, \ldots, e_{k-1})\} \tag{8}$$

subject to

$$0 < T_{\text{opt1}} < T_{\text{opt2}} < \cdots < T_{\text{opt}(k-1)} < L-1$$

3 Soft Computing Techniques

Four different soft computing methods are utilized to find the optimal solutions. Here we have used CS, SS, ACS, and hybrid ACS-SS techniques to obtain the multiple optimal threshold values.

Yang and Deb [35, 36] proposed the CS technique motivated by the parasitic traits of the cuckoo bird. This technique has been widely used as an optimization tool. It has few parameters to tune. In addition, it has in-built constraint-handling mechanism. This makes it computationally efficient and fast. The CS technique is developed on the following assumptions.

1. A cuckoo bird lays a single egg in a time and keeps it in a randomly selected nest.
2. The bestnest with good quality of eggs is considered fit for the next generation.
3. The count of host nests is predetermined. However, the host bird may recognize the cuckoo's egg defined as a probability $p_a \in [0, 1]$. In such case, the host bird may drop the egg or it may leave the nest and make a new nest at a new location.

Here, we have followed a simple representation of the CS algorithm where each egg in a nest corresponds to a solution. Our basic objective is to utilize the new solutions to substitute the weak solution in the nests. Further, a cuckoo bird performs a Levy flight while generating new solutions. The new solution for the next generation is represented as:

$$X_i(g+1) = X_i(g) + \alpha \otimes \text{Levy}(\lambda) \tag{9}$$

where $X_i(g)$ is the current search location of a cuckoo i and $X_i(g+1)$ is the search location of the same cuckoo in the next generation. Here α represents the step size usually considered one, and $\otimes$ represent element-wise multiplication operator similar to that used in PSO. Levy(λ) represents the random walk using the Levy flight. The Levy distribution may be estimated as Levy $\sim u = t^{-\lambda}$ for λ ranging from [1 to 3]. The Levy step applying the popular Mantegna's procedure [37] is represented as

$$\text{Levy} = \frac{u}{|v|^{1/(\lambda-1)}} \tag{10}$$

The authors in [38] investigated the ACS optimization technique. The CS algorithm is made adaptive ignoring the Levy distribution. A new method to control the step size is incorporated to make the CS adaptive. The step size is expressed in terms of the fitness value of individual nests in the search domain for the current generation. That is why the authors have ignored α which is fixed in the conventional CS technique. The adaptive step size is presented as

$$\text{step}_i(g+1) = \left(\frac{1}{g}\right)^{\left|\frac{\text{bestf}(g)-f_i(g)}{\text{bestf}(g)-\text{worstf}(g)}\right|} \tag{11}$$

where g represents the current generation, $f_i(g)$ is the fitness function value of the ith nest in the gth generation, and bestf(g) and worstf(g) represent the best objective function value and the worst fitness function value of the current generation, respectively. The above equation makes the step size adaptive. It decreases with an increase in the number of generations. Now the new solution for the next generation is represented by

$$X_i(g+1) = X_i(g) + \text{randn} \times \text{step}_i(g+1) \tag{12}$$

The authors have claimed ACS to be faster than the CS algorithm as the need to define the step size and initial parameters is eliminated. The details of the ACS algorithm are given in [38].

The squirrel search (SS) algorithm is one of the latest evolutionary computing techniques inspired by the searching manners of southern flying squirrels and effective mode of movement called gliding. The squirrels show a dynamic foraging strategy to optimally use the nutrient resources. The authors have considered the following assumptions while modeling the technique [39].

1. The number of flying squirrels is n, and only one squirrel is sitting on a tree.
2. All the squirrels search for nutrition and uses the existing resources individually.
3. Only three kinds of trees are present in a forest, i.e., normal, oak (acorn nut source), and hickory trees.
4. There are three oak trees (normal nutrient source) and one hickory tree (optimal nutrient source) in the forest region under concern.

The initial position of each squirrel is uniformly distributed in the forest region and is represented using the equation given below:

$$\text{FS}_i = \text{FS}_\text{L} + \text{Rand}(0, 1) \times (\text{FS}_\text{U} - \text{FS}_\text{L}) \tag{13}$$

where FS_L and FS_U are the lower and upper position limits correspondingly of the ith flying squirrel. Rand(0,1) is a uniformly distributed random number $\in [0,1]$.

The quality of the nutrient source explored by a flying squirrel is represented by the fitness value of its location. If a squirrel is on a hickory tree, it is considered as an optimum food source. If it is on an acorn/oak tree, then it is considered as a normal food source. Otherwise, there is no nutrient source and the squirrel is on a normal tree. The fitness value also represents the probability of survival of a squirrel. If a squirrel is having the minimum fitness value, then it is considered to be on the hickory tree, i.e., optimum food source. Based on the fitness values, the subsequent three squirrels are declared to be on the normal food source (i.e., acorn nut trees). This is because it is assumed that there are four food sources only. The squirrels on the acorn nut trees are expected to move to the optimal food source (i.e., hickory nut tree). The rest of the squirrels in the population are assumed to be on normal trees without any food source. Some squirrels are randomly selected and declared to move to the hickory tree provided that they have satisfied their everyday energy needs.

The rest of the squirrels will move to the acorn nut trees for satisfying their everyday energy requirements. The presence of a predator always affects the searching behavior of the squirrels. This is taken care in the model in the update formula using a predefined predator probability (P_{dp}). If the predator is absent, then the flying squirrels move freely and explore the forest efficiently for the optimum food source. The presence of the predator cautions the squirrels, and they utilize small random walk to find a nearby hiding place. The dynamic searching activities of the flying squirrels for optimum nutrition source are modeled as:

$$\mathrm{FS}_{\mathrm{at}}^{g+1} = \begin{cases} \mathrm{FS}_{\mathrm{at}}^{g} + d_g \times G_c \times \left(\mathrm{FS}_{\mathrm{ht}}^{g} - \mathrm{FS}_{\mathrm{at}}^{g}\right); & R_1 \geq P_{\mathrm{dp}} \\ \text{Random location} & \text{otherwise} \end{cases} \quad (14)$$

where d_g is a random flying distance, R_1 is a random figure within [0,1], $\mathrm{FS}_{\mathrm{ht}}$ is the position of the squirrel reaching the optimum food source (hickory nut tree), $\mathrm{FS}_{\mathrm{at}}$ is the position of the squirrel on the acorn nut tree, G_c is the gliding constant which balances the exploitation and the exploration, and g represents the current iteration. Similar equations are modeled for the foraging behavior of the flying squirrels on normal trees as well as when they move to the hickory tree for storing food for the future. The authors claimed that the SS algorithm attains the global best with noteworthy convergence ability as compared to many other techniques.

The synergistic behavior of ACS using better exploitation and exploration capability and the dynamic foraging behavior of SS has motivated us to develop the hybrid ACS-SS algorithm for this application. This will further improve the exploitation and exploration capability of ACS. This is the reason why we have used the hybrid approach of ACS-SS. Further, in this paper constraint-handling mechanism is also incorporated in the proposed method so that the optimum thresholds obtained are well within the defined bounds.

4 Proposed Methodology

Our proposed methodology is shown in Fig. 4. The T_2-weighted brain MR image slice is taken as input from the database. Then GLCM of the brain image is found out using Eq. (2). The edge magnitude is computed using the contrast definition in Eq. (3). The number of thresholds is taken as an input from the user. Accordingly, the objective function presented in Eq. (8) for multilevel thresholding is maximized using all the evolutionary computing techniques.

We propose here the hybrid ACS-SS algorithm to maximize the edge magnitude. The findings are compared with our implementations of ACS, SS, and CS algorithms and reported in this chapter. The ACS algorithm uses the Levy flight feature of cuckoo with an adaptive step size. The SS algorithm uses a very dynamic foraging behavior of the flying squirrels. We introduce here the combination of dynamic foraging behavior and the adaptive step size feature for improved convergence, though at the cost of little additional CPU time.

The proposed ACS-SS algorithm is a hybrid optimization technique. Here, the adaptive step size feature of the ACS algorithm is supplemented by the dynamic foraging behavior of the SS technique. This results in an improved convergence. The adaptive step size helps in preventing the solutions from capturing into local optimum.

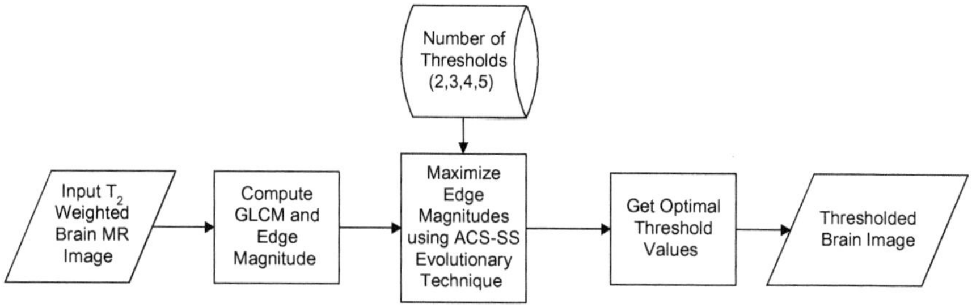

Fig. 4 A schematic block diagram of the suggested technique

A larger step size may also miss the optimal solutions. It is very difficult to get back to that optimal point. A too small step size takes more computing time. Therefore, an adaptive step size is desired. For instance, in the hill climbing problem, smaller steps are adopted as we go higher, which takes more time. Further, the dynamic foraging behavior helps in better exploitation of the food source. The food sources are divided into three levels depending upon the nutrient value. The objective is to reach the source having the highest nutrient value. This foraging strategy results in a better convergence.

The new solution for the next generation in ACS-SS is represented as

$$X_i(g+1) = \begin{cases} X_i(g) + f_g \times F_c \times (\text{bestf}(g) - X_i(g)); & \text{randn} < p_a \\ X_i(g) + \text{randn} \times \text{step}_i(g+1); & \text{otherwise} \end{cases} \tag{15}$$

where f_g is a random flying distance of the cuckoo, F_c is a flying constant to balance the exploitation and exploration, introduced to further enhance the performance of ACS algorithm. The value of randn is between [0,1], and if this value is less than the mutation probability, the cuckoo will take small steps to enhance exploration. If the value of randn is greater than the mutation probability, it will take adaptive steps to enhance the exploration.

The pseudocode for the proposed hybrid ACS-SS algorithm is given below.

Pseudocode of the ACS-SS Algorithm

Set the dimension of search space (d) = the number of thresholds (T);
Initialize number of nests (N) randomly;
Set the mutation probability (p_a), flying constant (F_c), and number of iterations;
Define the fitness function using Eq. (8);
Initially set *generation* $(g) = 1$ and evaluate the objective function (8) for all the nests;
Do {

Compute bestf and worstf of the fitness function values for the current generation,
Compute the adaptive step size using Eq. (11),
Compute the new positions of the nests using Eq. (15),
Compute the fitness function again,
Choose a nest (say *j*) randomly,
if fitness(*j*) > fitness(*i*)
swap the nests *i* and *j*

End
The worst nests are discarded using the mutation probability, and new nests are built at new locations;
Increment $g = g + 1$;
} **while** ($g \leq$ max iterations) or end criteria not satisfied.
Best solutions are ranked and reported.

The optimal thresholds are computed, and the thresholded brain image is obtained by following the reconstruction rule given as:

$$\begin{array}{lll} T & \text{for} & 0 < T \leq T_{\text{opt1}} \\ T_{\text{opt1}} & \text{for} & T_{\text{opt1}} < T \leq T_{\text{opt2}} \\ T_{\text{opt2}} & \text{for} & T_{\text{opt2}} < T < T_{\text{opt3}} \\ T_{\text{opt3}} & \text{for} & T_{\text{opt3}} < T < L - 1 \end{array} \tag{16}$$

where T represents the gray value of the brain MR image. This scheme is presented for three-level thresholding having three optimal threshold values. The same is extended for higher levels of thresholding as well.

5 Results and Validations

The proposed procedure is validated using T_2-w brain MR image slices from [29]. We have considered ten distinct slices in the axial plane for simulation. The simulations are carried out using Intel core i5 processor with 4 GB RAM. The above-mentioned evolutionary computing algorithms are implemented using MATLAB. We have used here six validation indices for comparing the performance among the evolutionary computing techniques.

The parameter setting for ACS-SS and ACS algorithms is shown in Table 1. Table 2 shows the parameter setting for SS and CS algorithms. The parameters are chosen as per the recommendations of the authors in the respective papers.

Table 3 displays the best fitness function values, and Table 4 shows the conforming optimal threshold values. As the fitness function is maximization of the edge

Table 1 Parameter setting for ACS-SS and ACS

ACS-SS	ACS
Dimension of search space = level of threshold Number of nests = 50 Mutation probability = 0.1 Flying constant = 1.9 Number of iterations = 100	Dimension of search space = level of threshold Number of nests = 50 Mutation probability = 0.25 Iterations = 100

Table 2 Parameter setting for SS and CS

SS	CS
Dimension of search space = level of threshold Number of squirrels = 50 Number of nutritious food resources = 4 Hickory nut tree = 1 acorn nut trees = 3 Predator presence probability = 0.1 Gliding constant = 1.9 Number of iterations = 100	Dimension of search space = level of threshold Number of nests = 50 Mutation probability = 0.25 Number of Iterations = 100 Step size (α) = 1 Index (λ) = 1.5

magnitude information, a higher fitness function value is preferred for improved thresholding performance. It is perceived that values obtained utilizing ACS-SS technique is higher than the other techniques. The reason may be the synergistic effect of combining both ACS and SS techniques.

Similarly, Table 5 displays the structured similarity (SSIM) index and feature similarity (FSIM) index [40, 41]. These measures assess the proposed technique's performance. The thresholded image compares to the original image. The more similar is the thresholded image to the original image, the higher will be the value of FSIM and SSIM. It is perceived that both FSIM and SSIM values are better at five-level thresholding. This signifies that image thresholded at five-level thresholding is more alike the original one. Additionally, higher index values are achieved for ACS-SS technique in comparison with the other techniques.

Table 6 shows the comparison of the parameters inverse difference moment (IDM) and contrast for assessing the proposed thresholding technique's performance [42]. IDM is in inverse proportion to contrast. Usually, a large contrast is preferred for a thresholded image at a specific threshold level. We have considered ACS-SS and ACS techniques only for these parameters as the thresholded images are obtained using these techniques only. It is witnessed that the contrast is better in case of ACS-SS method as compared to ACS technique for a given threshold level.

Table 7 compares the techniques utilizing peak signal-to-noise ratio (PSNR) [42] and CPU time. A high PSNR value indicates the presence of more signal information in the thresholded image. A high value of PSNR is desired. It is seen that the PSNR is higher at five-level threshold. Additionally, the PSNR is better for ACS-SS technique in comparison with the other methods for all the stages of thresholding. However,

Table 3 Best cost function values

MR image	*m*	Best cost function values			
		ACS-SS	ACS	SS	CS
Slice-15	2	255.0	253.5	248.9	230.3
	3	315.9	296.5	287.5	227.3
	4	1,154,700.0	1,042,700.0	897,270.0	844,760.0
	5	1,233,500.0	1,145,100.0	1,081,800.0	998,720.0
Slice-25	2	254.9	253.6	252.2	244.1
	3	332.5	288.0	247.0	220.6
	4	1,330,400.0	1,093,100.0	1,045,800.0	655,800.0
	5	2,376,700.0	1,303,800.0	1,232,900.0	1,209,500.0
Slice-35	2	255.9	254.2	235.9	221.5
	3	327.2	312.8	246.7	232.6
	4	1,332,700.0	1,326,100.0	1,255,800.0	217,530.0
	5	1,357,800.0	1,226,200.0	1,210,300.0	1,009,800.0
Slice-45	2	255.2	253.7	244.1	241.6
	3	335.2	322.2	307.3	287.2
	4	1,440,200.0	1,405,500.0	759,000.0	358,520.0
	5	1,507,400.0	1,470,900.0	1,389,400.0	395,920.0
Slice-55	2	255.2	251.4	245.8	241.4
	3	333.0	321.5	295.0	258.4
	4	1,778,500.0	1,595,200.0	1,343,300.0	735,800.0
	5	1,880,800.0	1,751,700.0	1,468,400.0	361,130.0
Slice-65	2	256.5	252.5	250.9	247.7
	3	324.1	307.7	301.9	300.3
	4	1,909,000.0	1,898,800.0	1,869,000.0	821,670.0
	5	1,956,700.0	1,946,000.0	1,870,300.0	450,020.0
Slice-75	2	255.0	254.8	247.1	211.8
	3	340.2	322.6	297.8	286.5
	4	1,888,400.0	1,640,200.0	176,000.0	139,190.0
	5	2,969,300.0	1,834,900.0	1,289,800.0	825,880.0
Slice-85	2	253.8	253.1	252.2	244.8
	3	320.0	266.9	240.5	234.1
	4	1,622,100.0	1,581,100.0	1,529,700.0	388,260.0

(continued)

Table 3 (continued)

MR image	m	Best cost function values			
		ACS-SS	ACS	SS	CS
	5	1,565,100.0	1,561,300.0	1,433,700.0	1,219,200.0
Slice-95	2	253.0	250.2	242.1	210.3
	3	327.1	317.8	287.5	250.8
	4	1,521,700.0	1,463,500.0	445,880.0	358,199.0
	5	1,538,900.0	1,482,900.0	1,085,300.0	998,360.0
Slice-105	2	261.1	259.2	257.9	250.5
	3	322.4	320.9	281.1	245.2
	4	1,347,100.0	1,358,800.0	1,354,500.0	1,142,000.0
	5	2,296,300.0	1,345,900.0	1,324,100.0	1,290,700.0

the ACS technique supersedes all other techniques with respect to CPU time. It takes the minimum time to obtain the thresholded images at all levels. The reason is that ACS method requires very few parameters for tuning.

Figures 5, 6, 7, 8, 9, 10, 11, 12, 13, and 14 display the output images for thresholding levels 2, 3, 4, and 5. Note that the output images for five-level thresholding are more alike the given input images slice. Additionally, the output images using ACS-SS technique are visually better than the ACS technique.

Test Images

The histograms of the sample slice show multimodality, thus making them appropriate for multilevel thresholding. The output thresholded images at level 5 notably show the various regions distinctly as compared to the images thresholded at level 2 or 3.

It is noteworthy to mention here that ACS technique is faster than other approaches. In summary, the suggested ACS-SS technique has shown better results in terms of different performance indices shown in this chapter. The ACS technique is the second contender in this respect. Nevertheless, ACS displays better speed compared to other approaches. The concept of content-based image identification [43] and classification [44] may be explored for the above problem.

Table 4 Optimal threshold values

MR image	*m*	Optimal threshold values			
		ACS-SS	ACS	SS	CS
Slice-15	2	85,222	102,128	114,138	12,160
	3	125,143,167	43,74,161	65,83,170	57,156,228
	4	49,81,164,224	45,68,107,165	29,47,84,129	11,98,210,242
	5	43,83,190,227,243	39,93,196,168,209	48,97,154,169,188	9,58,76,156,233
Slice-25	2	74,151	101,145	103,149	180,218
	3	91,148,227	19,53,147	66,85,151	8,27,48
	4	42,57,84,171	19,51,96,148	46,64,94,152	29,51,111,214
	5	21,59,103,120,142	22,45,119,169,219	36,57,73,130,225	13,152,199,219,235
Slice-35	2	72,170	102,142	102,146	124,200
	3	111,145,156	22,54,160	65,89,157	162,185,232
	4	80,126,139,149	19,53,96,149	41,62,141,205	50,137,166,222
	5	34,53,97,143,192	27,48,91,135,191	29,56,131,179,193	41,99,155,179,220
Slice-45	2	104,189	100,143	104,145	97,116
	3	131,156,234	51,88,162	74,88,164	15,49,179
	4	96,155,177,194	19,52,97,149	68,140,209,243	12,96,145,200
	5	77,96,138,201,226	26,73,104,146,233	13,51,140,197,242	20,69,181,216,247
Slice-55	2	130,205	101,143	100,152	61,78
	3	140,158,176	51,87,162	72,90,160	76,120,229
	4	85,113,127,182	76,92,123,208	21,62,116,174	51,119,178,204
	5	33,76,134,164,189	22,44,106,146,215	35,59,102,153,217	11,86,115,167,207
Slice-65	2	140,186	101,143	117,138	153,197
	3	122,142,165	34,86,157	67,85,160	151,174,187
	4	85,144,201,244	69,128,201,244	44,69,129,189	136,160,184,235
	5	70,102,139,158,236	63,124,190,225,248	7,18,54,115,168	36,66,159,225,241
Slice-75	2	94,220	100,143	110,144	37,186
	3	110,157,223	52,87,162	78,91,166	60,91,215
	4	69,123,174,224	19,52,97,149	34,68,117,185	89,144,212,238
	5	65,97,128,149,199	45,120,165,190,207	28,49,63,97,152	85,112,137,157,174
Slice-85	2	91,143	109,144	109,145	13,45
	3	102,188,249	103,115,132	41,93,143	67,156,212
	4	33,76,157,184	42,94,138,169	38,65,111,172	72,163,207,250
	5	35,54,74,134,175	55,101,154,184,208	46,120,163,185,219	7,19,40,159,245

(continued)

Table 4 (continued)

MR image	*m*	Optimal threshold values			
		ACS-SS	ACS	SS	CS
Slice-95	2	80,165	91,128	86,153	59,187
	3	170,197,215	57,94,157	61,84,151	9,48,173
	4	70,133,159,193	36,79,102,177	57,113,172,207	44,81,90,180
	5	83,122,138,154,198	64,111,164,197,239	16,67,113,175,197	32,169,198,220,234
Slice-105	2	75,138	94,140	84,141	84,131
	3	34,77,121	42,90,140	49,87,138	31,75,99
	4	41,107,167,205	8,78,123,209	9,46,87,138	9,57,112,219
	5	45,114,168,195,233	52,99,155,190,214	8,42,107,165,225	22,49,89,161,206

Table 5 Comparison of evaluation parameters SSIM and FSIM

Test image	*m*	SSIM				FSIM			
		ACS-SS	ACS	SS	CS	ACS-SS	ACS	SS	CS
Slice-15	2	**0.9684**	0.9494	0.9303	0.6252	**0.9921**	0.9891	0.9862	0.9386
	3	**0.9771**	0.9649	0.9375	0.8744	**0.9960**	0.9913	0.9880	0.9710
	4	**0.9961**	0.9744	0.9437	0.8915	**0.9988**	0.9956	0.9897	0.9756
	5	**0.9986**	0.9930	0.9505	0.9212	**0.9995**	0.9982	0.9900	0.9831
Slice-25	2	**0.9737**	0.9644	0.9030	0.6850	**0.9947**	0.9937	0.9841	0.9416
	3	**0.9753**	0.9705	0.9364	0.7944	**0.9954**	0.9939	0.9880	0.9719
	4	**0.9807**	0.9706	0.9471	0.9431	**0.9958**	0.9946	0.9884	0.9882
	5	**0.9901**	0.9900	0.9896	0.9539	**0.9974**	0.9972	0.9910	0.9909
Slice-35	2	**0.9745**	0.9647	0.9464	0.8699	**0.9944**	0.9932	0.9866	0.9801
	3	**0.9859**	0.9739	0.9542	0.9163	**0.9949**	0.9942	0.9889	0.9813
	4	**0.9891**	0.9889	0.9681	0.9284	**0.9975**	0.9972	0.9931	0.9821
	5	**0.9996**	0.9942	0.9882	0.9449	**0.9999**	0.9985	0.9969	0.9886
Slice-45	2	**0.9834**	0.9606	0.9070	0.8111	**0.9936**	0.9922	0.9873	0.9740
	3	**0.9857**	0.9615	0.9530	0.8891	**0.9969**	0.9927	0.9885	0.9784
	4	**0.9862**	0.9713	0.9547	0.8978	**0.9971**	0.9934	0.9897	0.9828
	5	**0.9953**	0.9838	0.9712	0.9697	**0.9992**	0.9961	0.9934	0.9920
Slice-55	2	**0.9673**	0.9581	0.9392	0.8976	**0.9929**	0.9873	0.9871	0.9721
	3	**0.9875**	0.9685	0.9508	0.9103	**0.9962**	0.9897	0.9894	0.9754
	4	**0.9913**	0.9813	0.9635	0.9319	**0.9971**	0.9957	0.9896	0.9853
	5	**0.9978**	0.9859	0.9824	0.9463	**0.9996**	0.9959	0.9935	0.9889
Slice-65	2	**0.9804**	0.9510	0.9399	0.9133	**0.9946**	0.9882	0.9874	0.9820
	3	**0.9967**	0.9738	0.9535	0.9345	**0.9992**	0.9891	0.9876	0.9867

(continued)

Table 5 (continued)

Test image	*m*	SSIM				FSIM			
		ACS-SS	ACS	SS	CS	ACS-SS	ACS	SS	CS
	4	**0.9969**	0.9919	0.9574	0.9574	**0.9994**	0.9984	0.9885	0.9881
	5	**0.9970**	0.9939	0.9798	0.9768	**0.9994**	0.9985	0.9954	0.9942
Slice-75	2	**0.9785**	0.9655	0.9497	0.7315	**0.9917**	0.9886	0.9801	0.9459
	3	**0.9858**	0.9668	0.9507	0.9130	**0.9963**	0.9908	0.9885	0.9828
	4	**0.9894**	0.9814	0.9609	0.9398	**0.9968**	0.9950	0.9896	0.9828
	5	**0.9912**	0.9849	0.9668	0.9541	**0.9969**	0.9955	0.9938	0.9884
Slice-85	2	**0.9628**	0.9452	0.9190	0.7351	**0.9897**	0.9870	0.9810	0.8892
	3	**0.9636**	0.9631	0.9367	0.8144	**0.9907**	0.9887	0.9827	0.9580
	4	**0.9796**	0.9786	0.9447	0.9353	**0.9955**	0.9906	0.9848	0.9652
	5	**0.9825**	0.9820	0.9725	0.9398	**0.9966**	0.9965	0.9907	0.9711
Slice-95	2	**0.9746**	0.9638	0.9551	0.7435	**0.9859**	0.9823	0.9820	0.8977
	3	**0.9826**	0.9665	0.9582	0.8293	**0.9926**	0.9835	0.9824	0.9124
	4	**0.9861**	0.9731	0.9614	0.8753	**0.9939**	0.9856	0.9833	0.9252
	5	**0.9996**	0.9734	0.9692	0.9547	**0.9999**	0.9886	0.9870	0.9766
Slice-105	2	**0.9608**	0.9420	0.9383	0.9308	**0.9956**	0.9916	0.9896	0.9870
	3	**0.9671**	0.9605	0.9392	0.9376	**0.9957**	0.9947	0.9926	0.9881
	4	**0.9685**	0.9613	0.9571	0.9479	**0.9963**	0.9950	0.9935	0.9892
	5	**0.9845**	0.9784	0.9771	0.9666	**0.9971**	0.9963	0.9953	0.9934

Table 6 Comparison of validation parameters IDM and contrast

Test images	*m*	IDM		Contrast	
		ACS-SS	ACS	ACS-SS	ACS
Slice-15	2	**0.0391**	0.0440	**1111.10**	953.41
	3	**0.0370**	0.0694	**1199.40**	609.83
	4	**0.0634**	0.0652	**819.62**	731.28
	5	**0.0703**	0.0789	**788.69**	768.51
Slice-25	2	**0.0383**	0.0464	**1236.60**	882.34
	3	**0.0410**	0.1323	**1134.90**	680.48
	4	**0.0667**	0.1248	**1106.70**	733.61
	5	**0.1176**	0.1219	**1196.40**	834.73
Slice-35	2	**0.0384**	0.0472	**1197.50**	859.10
	3	**0.0368**	0.1187	**1343.50**	688.55
	4	**0.0430**	0.1256	**1187.40**	1150.30
	5	**0.0815**	0.0958	**1099.20**	921.74

(continued)

Table 6 (continued)

Test images	m	IDM		Contrast	
		ACS-SS	ACS	ACS-SS	ACS
Slice-45	2	**0.0365**	0.0365	**1429.00**	1413.30
	3	**0.0329**	0.0598	**1654.80**	767.34
	4	**0.0379**	0.1226	**1476.90**	1441.50
	5	**0.0433**	0.1004	**1495.40**	1250.30
Slice-55	2	**0.0331**	0.0368	**1655.40**	1454.40
	3	**0.0313**	0.0593	**1837.50**	893.15
	4	**0.0399**	0.0429	**1461.60**	1252.30
	5	**0.0856**	0.1128	**1566.30**	1256.70
Slice-65	2	**0.0319**	0.0375	**1873.50**	1528.70
	3	**0.0332**	0.0826	**1848.00**	1051.50
	4	**0.0429**	0.0496	**1339.60**	1152.40
	5	**0.0468**	0.0537	**1470.00**	1108.10
Slice-75	2	**0.0383**	0.0406	**1360.20**	1204.50
	3	**0.0371**	0.0594	**1474.20**	846.53
	4	**0.0495**	0.1233	**1397.10**	1098.90
	5	**0.0494**	0.0705	**1295.00**	1029.20
Slice-85	2	**0.0385**	0.0419	**1375.00**	1217.00
	3	**0.0384**	0.0409	**1387.10**	1224.90
	4	**0.0710**	0.0863	**1063.80**	959.03
	5	**0.0582**	0.0750	**1164.60**	1066.30
Slice-95	2	**0.0435**	0.0445	**1043.90**	961.48
	3	**0.0364**	0.0648	**1306.80**	731.46
	4	**0.0473**	0.0653	**1030.00**	875.03
	5	**0.0483**	0.0604	**1122.90**	959.16
Slice-105	2	**0.0473**	0.0528	**765.24**	657.86
	3	**0.0757**	0.0890	**483.73**	413.37
	4	**0.0793**	0.2656	**2363.50**	583.20
	5	**0.0659**	0.0749	**664.56**	647.46

Bold value indicates the best values

Table 7 Comparison of validation parameters PSNR and CPU time

Test images	*m*	PSNR				Computation time (in s)			
		ACS-SS	ACS	SS	CS	ACS	CS	SS	ACS-SS
Slice-15	2	**24.26**	22.30	21.02	13.91	**2.76**	3.85	5.55	6.11
	3	**25.68**	23.85	21.49	18.55	**2.90**	3.91	6.42	7.06
	4	**32.90**	25.21	21.90	19.15	**3.08**	4.23	6.72	7.39
	5	**37.32**	30.50	22.36	20.54	**3.45**	4.50	7.70	8.47
Slice-25	2	**24.80**	23.60	19.37	14.32	**2.93**	4.19	6.28	6.91
	3	**25.09**	24.24	21.13	16.37	**3.05**	4.36	7.18	7.90
	4	**25.97**	24.36	21.95	21.60	**3.46**	4.53	7.37	8.11
	5	**47.59**	28.75	28.55	22.38	**3.67**	4.68	8.23	9.05
Slice-35	2	**24.80**	23.47	21.67	18.05	**2.87**	4.53	6.38	7.02
	3	**27.13**	24.74	22.31	19.82	**3.06**	4.54	6.99	7.69
	4	**28.13**	28.11	23.78	20.40	**3.48**	4.57	7.00	7.70
	5	**41.73**	30.81	27.85	21.56	**3.54**	4.76	7.09	7.80
Slice-45	2	**26.14**	22.62	19.15	16.27	**2.81**	3.93	6.10	6.71
	3	**26.75**	22.84	21.97	18.43	**2.95**	4.51	6.20	6.82
	4	**27.04**	24.03	22.07	18.58	**3.11**	4.57	6.79	7.47
	5	**31.40**	26.24	23.83	23.62	**3.56**	4.76	8.71	9.58
Slice-55	2	**23.36**	22.29	20.73	18.64	**2.84**	4.43	6.11	6.72
	3	**27.32**	23.49	21.68	18.95	**2.85**	4.50	6.95	7.65
	4	**28.72**	25.66	22.85	20.29	**2.85**	4.51	7.65	8.42
	5	**34.54**	26.77	25.83	21.29	**2.92**	4.79	8.04	8.84
Slice-65	2	**25.07**	21.26	20.46	18.90	**3.02**	4.02	6.69	7.36
	3	**32.46**	23.80	21.47	20.07	**3.23**	4.44	6.93	7.62
	4	**32.69**	28.65	21.87	21.84	**3.42**	4.69	7.21	7.93
	5	**32.87**	29.87	24.79	24.28	**3.43**	4.73	7.26	7.99
Slice-75	2	**24.95**	23.03	21.39	14.57	**2.82**	3.95	6.67	7.34
	3	**26.63**	23.21	21.60	19.16	**3.20**	4.35	7.45	8.20
	4	**27.88**	25.54	22.57	20.74	**3.43**	4.50	7.63	8.39
	5	**28.71**	26.52	23.26	21.90	**3.61**	4.65	7.69	8.46
Slice-85	2	**22.95**	21.29	19.65	14.85	**2.73**	3.88	6.18	6.80
	3	**23.02**	22.96	20.70	16.27	**3.06**	4.19	6.54	7.19
	4	**25.23**	24.97	21.19	20.58	**3.33**	4.54	7.93	8.72
	5	**25.83**	25.73	24.02	20.92	**3.34**	4.63	8.29	9.12
Slice-95	2	**24.64**	23.29	22.40	15.14	**2.90**	3.11	6.45	7.10
	3	**26.21**	23.59	22.70	16.91	**3.17**	4.25	6.61	7.27

(continued)

Table 7 (continued)

Test images	*m*	PSNR				Computation time (in s)			
		ACS-SS	ACS	SS	CS	ACS	CS	SS	ACS-SS
	4	**27.25**	24.40	23.04	18.18	**3.33**	4.35	7.18	7.90
	5	**41.74**	24.56	23.80	22.31	**3.46**	4.48	7.82	8.60
Slice-105	2	**23.22**	21.59	21.26	20.70	**2.84**	3.92	5.86	6.45
	3	**23.91**	23.12	21.31	21.25	**3.20**	4.47	6.57	7.23
	4	**24.07**	23.25	22.81	21.92	**3.32**	4.52	7.28	8.01
	5	**26.84**	25.51	25.26	23.75	**3.54**	4.60	6.57	7.23

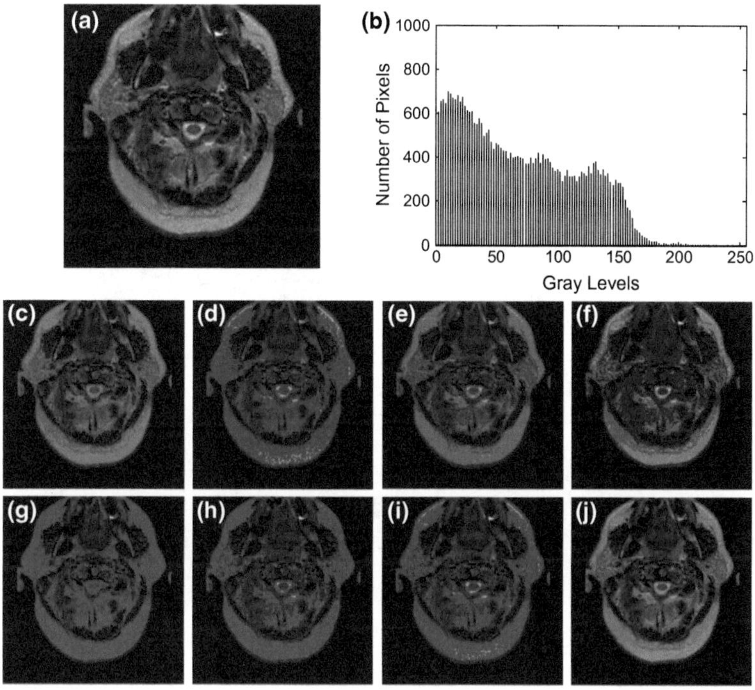

Fig. 5 Input and output thresholded images. **a** Input sample (slice-15), **b** histogram of sample in (**a**), **c** output for level 2 [102,128] using ACS, **d** output for level 3 [43,74,161] using ACS, **e** output for level 4 [45,68,107,165] using ACS, **f** output for level 5 [39,93,196,168,209] using ACS, **g** output for level 2 [85,222] using ACS-SS, **h** output for level 3 [125,143,167] using ACS-SS, **i** output for level 4 [49,81,164,224] using ACS-SS, and **j** output for level 5 [43,83,190,227,243] using ACS-SS

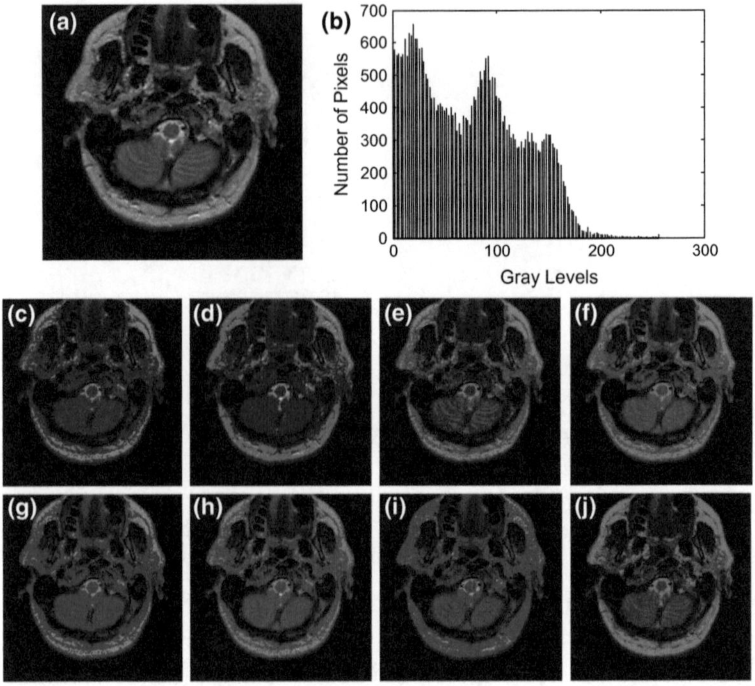

Fig. 6 Input and output thresholded images. **a** Input sample (slice-25), **b** histogram of sample in (**a**), **c** output for level 2 [101,145] using ACS, **d** output for level 3 [19,53,147] using ACS, **e** output for level 4 [19,51,96,148] using ACS, **f** output for level 5 [22,45,119,169,219] using ACS, **g** output for level 2 [74,151] using ACS-SS, **h** output for level 3 [91,148,227] using ACS-SS, **i** output for level 4 [42,57,84,171] using ACS-SS, and **j** output for level 5 [21,59,103,120,142] using ACS-SS

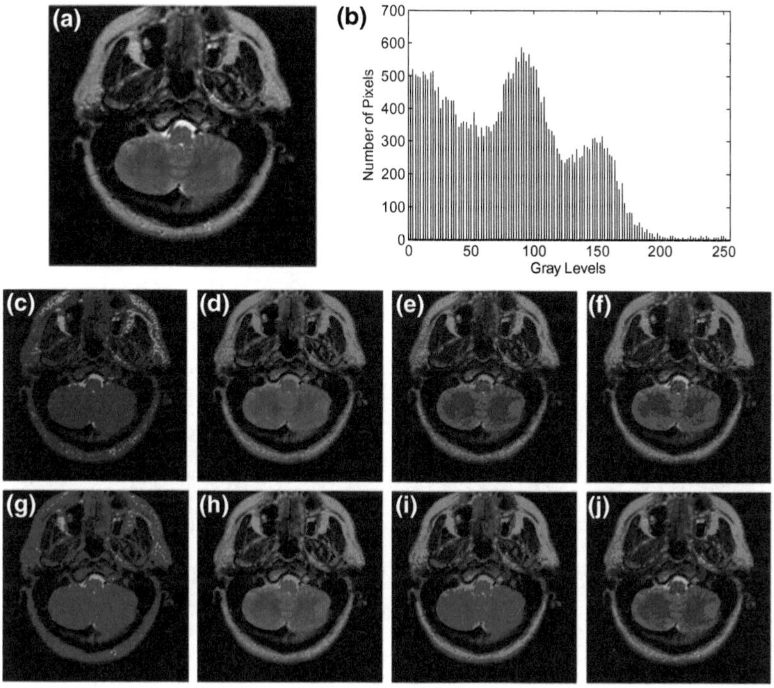

Fig. 7 Input and output thresholded images. **a** Input sample (slice-35), **b** histogram of sample in (**a**), **c** output for level 2 [102,142] using ACS, **d** output for level 3 [22,54,160] using ACS, **e** output for level 4 [19,53,96,149] using ACS, **f** output for level 5 [27,48,91,135,191] using ACS, **g** output for level 2 [72,170] using ACS-SS, **h** output for level 3 [111,145,156] using ACS-SS, **i** output for level 4 [80,126,139,149] using ACS-SS, and **j** output for level 5 [34,53,97,143,192] using ACS-SS

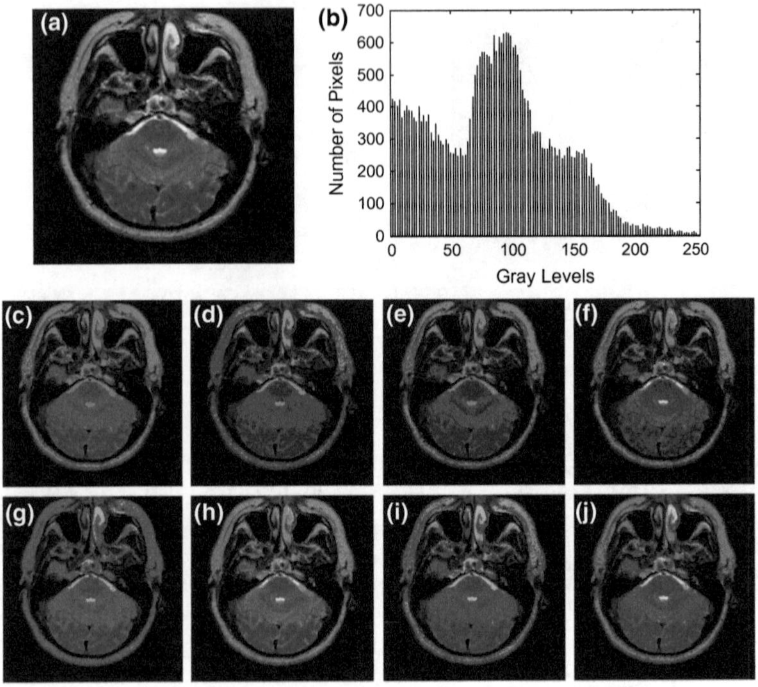

Fig. 8 Input and output thresholded images. **a** Input sample (slice-45), **b** histogram of sample in (**a**), **c** output for level 2 [100,143] using ACS, **d** output for level 3 [51,88,162] using ACS, **e** output for level 4 [19,52,97,149] using ACS, **f** output for level 5 [26,73,104,146,233] using ACS, **g** output for level 2 [104,189] using ACS-SS, **h** output for level 3 [131,156,234] using ACS-SS, **i** output for level 4 [96,155,177,194] using ACS-SS, and **j** output for level 5 [77,96,138,201,226] using ACS-SS

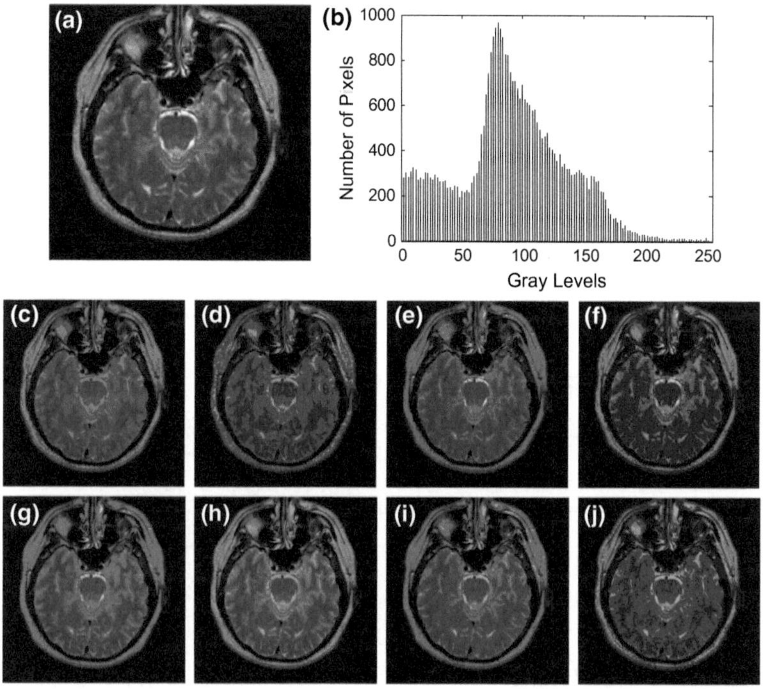

Fig. 9 Input and output thresholded images. **a** Input sample (slice-55), **b** histogram of sample in (**a**), **c** output for level 2 [101,143] using ACS, **d** output for level 3 [51,87,162] using ACS, **e** output for level 4 [76,92,123,208] using ACS, **f** output for level 5 [22,44,106,146,215] using ACS, **g** output for level 2 [130,205] using ACS-SS, **h** output for level 3 [140,158,176] using ACS-SS, **i** output for level 4 [85,113,127,182] using ACS-SS, and **j** output for level 5 [33,76,134,164,189] using ACS-SS

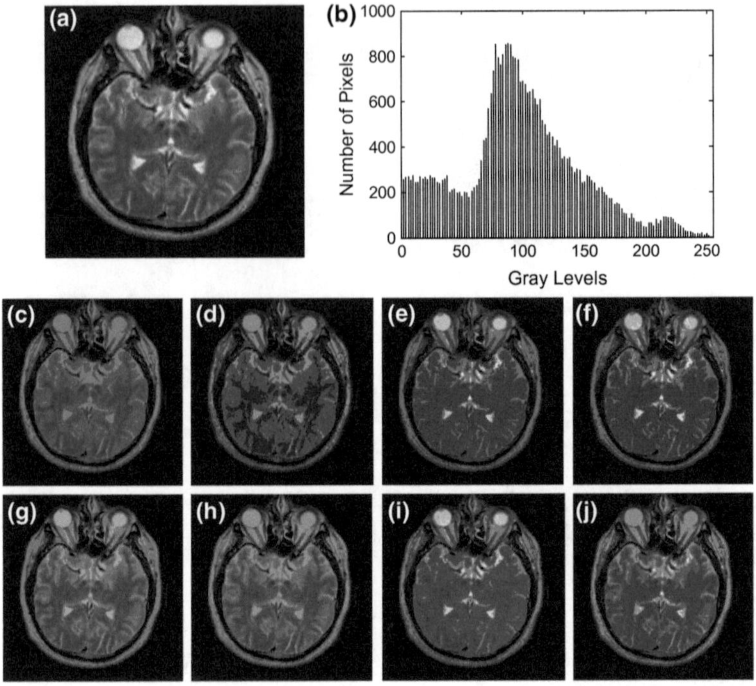

Fig. 10 Input and output thresholded images. **a** Input sample (slice-65), **b** histogram of sample in (**a**), **c** output for level 2 [101,143] using ACS, **d** output for level 3 [34,86,157] using ACS, **e** output for level 4 [69,128,201,244] using ACS, **f** output for level 5 [63,124,190,225,244] using ACS, **g** output for level 2 [140,186] using ACS-SS, **h** output for level 3 [122,142,165] using ACS-SS, **i** output for level 4 [85,144,201,244] using ACS-SS, and **j** output for level 5 [70,102,139,158,236] using ACS-SS

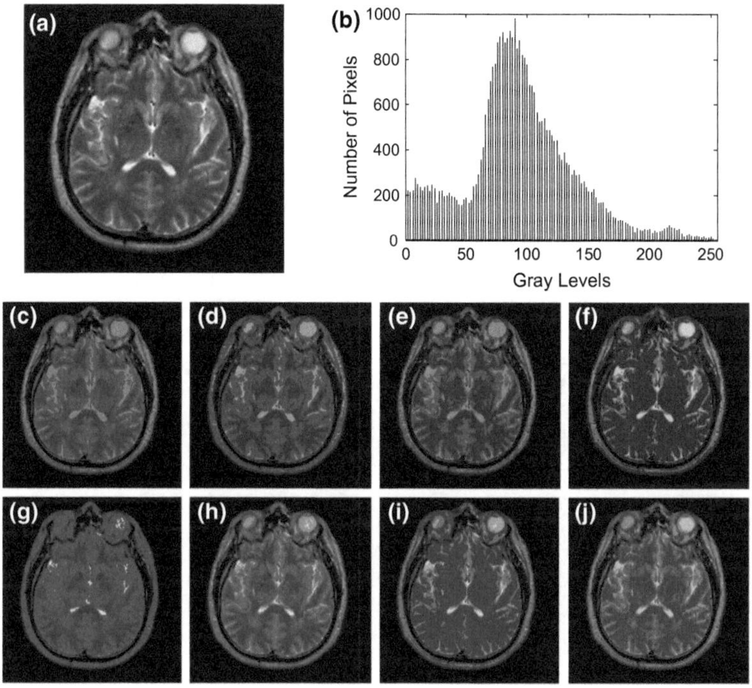

Fig. 11 Input and output thresholded images. **a** Input sample (slice-75), **b** histogram of sample in (**a**), **c** output for level 2 [100,143] using ACS, **d** output for level 3 [52,87,162] using ACS, **e** output for level 4 [19,52,97,149] using ACS, **f** output for level 5 [45,120,165,190,207] using ACS, **g** output for level 2 [94,220] using ACS-SS, **h** output for level 3 [110,157,223] using ACS-SS, **i** output for level 4 [69,123,174,224] using ACS-SS, and **j** output for level 5 [65,97,128,149,199] using ACS-SS

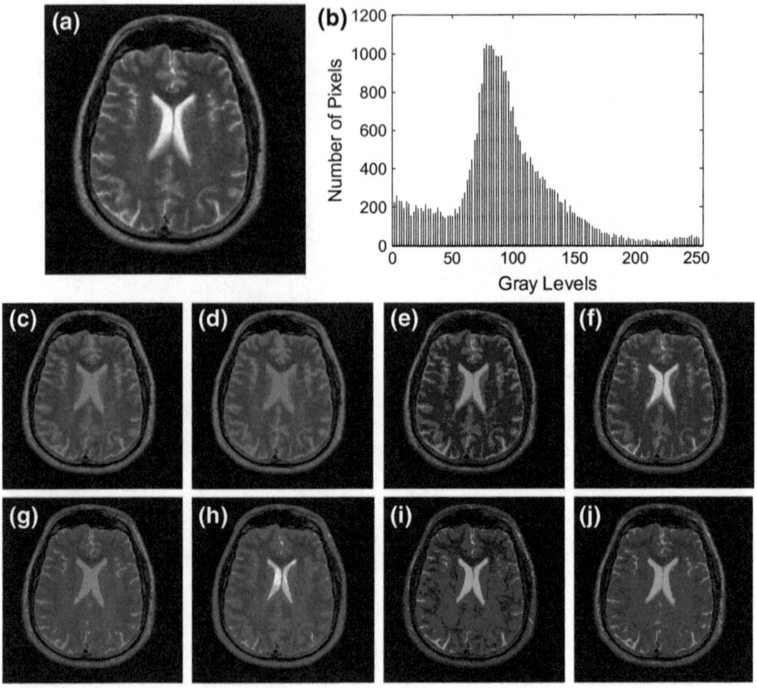

Fig. 12 Input and output thresholded images. **a** Input sample (slice-85), **b** histogram of sample in (**a**), **c** output for level 2 [109,144] using ACS, **d** output for level 3 [103,115,132] using ACS, **e** output for level 4 [42,94,138,169] using ACS, **f** output for level 5 [55,101,154,184,208] using ACS, **g** output for level 2 [91,143] using ACS-SS, **h** output for level 3 [102,188,249] using ACS-SS, **i** output for level 4 [33,76,157,184] using ACS-SS, and **j** output for level 5 [35,54,74,134,175] using ACS-SS

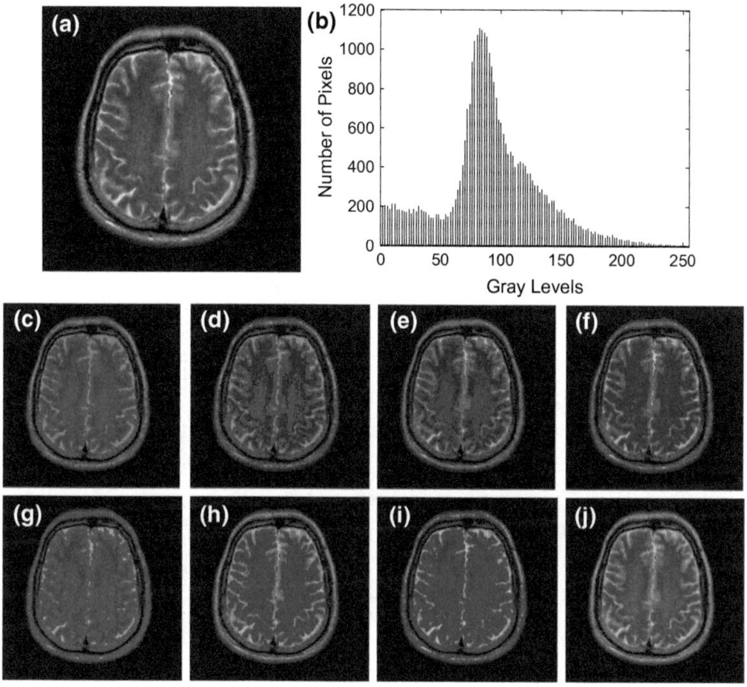

Fig. 13 Input and output thresholded images. **a** Input sample (slice-95), **b** histogram of sample in (**a**), **c** output for level 2 [81,128] using ACS, **d** output for level 3 [57,94,157] using ACS, **e** output for level 4 [36,69,102,177] using ACS, **f** output for level 5 [64,111,164,197,239] using ACS, **g** output for level 2 [80,165] using ACS-SS, **h** output for level 3 [170,197,215] using ACS-SS, **i** output for level 4 [70,133,159,193] using ACS-SS, and **j** output for level 5 [83,122,138,154,198] using ACS-SS

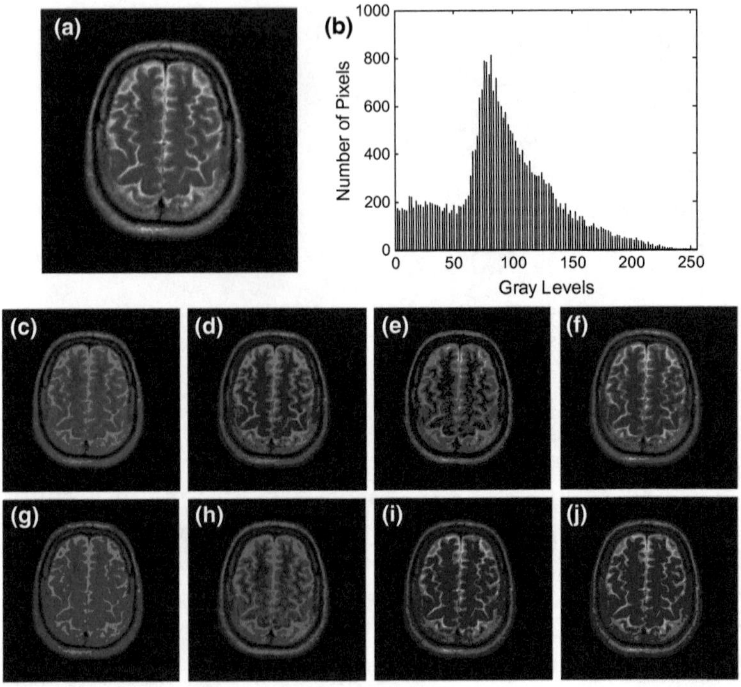

Fig. 14 Input and output thresholded images. **a** Input sample (slice-105), **b** histogram of sample in (**a**), **c** output for level 2 [91,140] using ACS, **d** output for level 3 [42,90,140] using ACS, **e** output for level 4 [8,78,123,209] using ACS, **f** output for level 5 [52,99,155,190,214] using ACS, **g** output for level 2 [75,138] using ACS-SS, **h** output for level 3 [34,77,121] using ACS-SS, **i** output for level 4 [41,107,167,205] using ACS-SS, and **j** output for level 5 [45,114,168,195,233] using ACS-SS

6 Conclusion

In this chapter, we have employed the concept of edge magnitude for brain MR image analysis. Because the brain MR image consists of several tissue regions, such as white matter, gray matter, and cerebrospinal fluid, multilevel thresholding is utilized. Four soft computing techniques are utilized to maximize the edge magnitude to obtain the optimum thresholds. The hybrid ACS-SS technique is investigated for the first time for such application. It is perceived from the validation indices that the ACS-SS method yields better results as compared to the other soft computing tools. This is because the hybridization of squirrel search and the adaptive cuckoo search technique provide us best objective function values. However, the ACS technique is faster (in terms of CPU time) than other techniques. The ACS-SS algorithm is also very effective. Edge magnitude-based technique may set the path for more research in the field of brain MR image analysis.

References

1. Sahoo, P.K., Soltani, S., Wong, A.K.C.: A survey of thresholding techniques. Comput. Vision Graph. Image Process. **41**(2), 233–260 (1988)
2. Sankur, B., Sezgin, M.: Image thresholding techniques: a survey over categories. Pattern Recogn. **34**(2), 1573–1583 (2001)
3. Kapur, J.N., Sahoo, P.K., Wong, A.K.C.: A new method for gray-level picture thresholding using the entropy of the histogram. Comput. Vision Graph. Image Process. **29**(3), 273–285 (1985)
4. Maitra, M., Chatterjee, A.: A novel technique for multilevel optimal magnetic resonance brain image thresholding using bacterial foraging. Measurement **41**, 1124–1134 (2008)
5. Panda, R., Agrawal, S., Bhuyan, S.: Edge magnitude based multilevel thresholding using Cuckoo search technique. Expert Syst. Appl. **40**(18), 7617–7628 (2013)
6. Otsu, N.: A threshold selection method from gray level histograms. Automatica **11**(285–296), 23–27 (1975)
7. Tobias, O.J., Seara, R.: Image segmentation by histogram thresholding using fuzzy sets. IEEE Trans. Image Process. **11**, 1457–1465 (2002)
8. Kittler, J., Illingworth, J.: Minimum error thresholding. Pattern Recogn. **19**(1), 41–47 (1986)
9. Mortazavi, D., Kouzani, A.Z., Soltanian-Zadeh, H.: Segmentation of multiple sclerosis lesions in MR images: a review. Neuroradiology **54**(4), 299–320 (2012)
10. Suzuki, H., Toriwaki, J.I.: Automatic segmentation of head MRI images by knowledge guided thresholding. Comput. Med. Imaging Graph. **15**(4), 233–240 (1991)
11. Zavaljevski, A., Dhawan, A.P., Gaskil, M., Ball, W., Johnson, J.D.: Multi-level adaptive segmentation of multi-parameter MR brain images. Comput. Med. Imaging Graph. **24**(2), 87–98 (2000)
12. Sandhya, G., Kande, G.B.: An efficient approach for the detection of White Matter, Gray Matter, and cerebrospinal fluid from MR images of the brain using an advanced multilevel thresholding. In: 2017 Third International Conference on Advances in Electrical, Electronics, Information, Communication and Bio-Informatics (AEEICB), pp. 422–426. IEEE (2017)
13. Oliva, D., Hinojosa, S., Cuevas, E., Pajares, G., Avalos, O., Gálvez, J.: Cross entropy based thresholding for magnetic resonance brain images using Crow Search Algorithm. Expert Syst. Appl. **79**, 164–180 (2017)

14. Kamathe, R.S., Joshi, K.R.: A novel method based on independent component analysis for brain MR image tissue classification into CSF, WM and GM for atrophy detection in Alzheimer's disease. Biomed. Signal Process. Control **40**, 41–48 (2018)
15. Akkus, Z., Kostandy, P.M., Philbrick, K.A., Erickson, B.J.: Extraction of brain tissue from CT head images using fully convolutional neural networks. In: Medical Imaging 2018: Image Processing, vol. 10574, pp. 1057420. International Society for Optics and Photonics (2018)
16. Sathya, P.D., Kayalvizhi, R.: Optimal segmentation of brain MRI based on adaptive bacterial foraging algorithm. Neurocomputing **74**(14), 2299–2313 (2011)
17. Sathya, P.D., Kayalvizhi, R.: Amended bacterial foraging algorithm for multilevel thresholding of magnetic resonance brain images. Measurement **44**(10), 1828–1848 (2011)
18. Shan, Y., Zu, H., Guang, Z.Y., Liu, J.: Automated histogram-based brain segmentation in T1-weighted three-dimensional magnetic resonance head images. NeuroImage **17**, 1587–1598 (2002)
19. Joliot, M., Mazoyer, B.M.: Three-dimensional segmentation and interpolation of magnetic resonance brain images. IEEE Trans. Med. Imaging **12**(2), 269–277 (1993)
20. Lai, C.C., Chang, C.Y.: A hierarchical evolutionary algorithm for automatic medical image segmentation. Expert Syst. Appl. **36**(1), 248–259 (2009)
21. Sharma, M., Purohit, G.N., Mukherjee, S.: Information retrieves from brain MRI images for tumor detection using hybrid technique K-means and artificial neural network (KMANN). In: Networking Communication and Data Knowledge Engineering, pp. 145–157. Springer, Singapore (2018)
22. Rao, C.R., Kumar, M.N.V.S.S., Rao, G.S.B.: A novel segmentation algorithm for feature extraction of brain MRI tumor. In: Information and Decision Sciences, pp. 455–463. Springer, Singapore (2018)
23. Samanta, A.K., Khan, A.A.: Computer aided diagnostic system for automatic detection of brain tumor through MRI using clustering based segmentation technique and SVM classifier. In: International Conference on Advanced Machine Learning Technologies and Applications, pp. 343–351. Springer, Cham (2018)
24. Ouadfel, S., Taleb-Ahmed, A.: Social spiders optimization and flower pollination algorithm for multilevel image thresholding: a performance study. Expert Syst. Appl. **55**, 566–584 (2016)
25. Mlakar, U., Potočnik, B., Brest, J.: A hybrid differential evolution for optimal multilevel image thresholding. Expert Syst. Appl. **65**, 221–232 (2016)
26. Dehshibi, M.M., Sourizaei, M., Fazlali, M., Talaee, O., Samadyar, H., Shanbehzadeh, J.: A hybrid bio-inspired learning algorithm for image segmentation using multilevel thresholding. Multimedia Tools Appl. **76**(14), 15951–15986 (2017)
27. El Aziz, M.A., Ewees, A.A., Hassanien, A.E.: Whale optimization algorithm and moth-flame optimization for multilevel thresholding image segmentation. Expert Syst. Appl. **83**, 242–256 (2017)
28. Panda, R., Agrawal, S., Samantaray, L., Abraham, A.: An evolutionary gray gradient algorithm for multilevel thresholding of brain MR images using soft computing techniques. Appl. Soft Comput. **50**, 94–108 (2017)
29. Harvard Medical School. [Online]. Available: http://www.med.harvard.edu/AANLIB. Accessed Oct 2018
30. Lim, K.O., Pfefferbaum, A.: Segmentation of MR brain images into cerebrospinal fluid spaces, white and gray matter. J. Comput. Assist. Tomogr. **13**(4), 588–593 (1989)
31. Mokji, M.M., Abu Bakar, S.A.R.: Adaptive thresholding based on co-occurrence matrix edge information. In: First Asia International Conference on Modelling & Simulation, pp. 444–450. IEEE (2007)
32. Chanda, B., Majumder, D.D.: A note on the use of the gray-level co-occurrence matrix in threshold selection. Sig. Process. **15**, 149–167 (1988)
33. Haralick, R.M., Shanmugam, K., Dinstein, I.H.: Textural features for image classification. IEEE Trans. Syst. Man Cybern. **6**, 610–621 (1973)
34. Albregtsen, F.: Statistical Texture Measures Computed from Gray Level Co-occurrence Matrices. Image Processing Laboratory, Department of Informatics, University of Oslo (1995)

35. Yang, X.S., Deb, S.: Cuckoo search via Levy flights. In: World Congress on Nature and Biologically Inspired Computing, pp. 210–214. IEEE (2009)
36. Yang, X.S., Deb, S.: Engineering optimization by Cuckoo search. Int. J. Math. Modeling Numer. Optim. **1**(4), 330–343 (2010)
37. Mantegna, R.N.: Fast, accurate algorithm for numerical simulation of Lévy stable stochastic processes. Phys. Rev. E **49**(4), 4677–4683 (1994)
38. Naik, M.K., Panda, R.: A novel adaptive cuckoo search algorithm for intrinsic discriminant analysis based face recognition. Appl. Soft Comput. **38**, 661–675 (2016)
39. Jain, M., Singh, V., Rani, A.: A novel nature-inspired algorithm for optimization: squirrel search algorithm. Swarm Evol. Comput. (2018)
40. Wang, Z., Bovik, A.C., Sheikh, H.R., Simoncelli, E.P.: Image quality assessment: from error visibility to structural similarity. IEEE Trans. Image Process. **13**(4), 600–612 (2004)
41. Zhang, L., Zhang, L., Mou, X., Zhang, D.: FSIM: a feature similarity index for image quality assessment. IEEE Trans. Image Process. **20**(8), 2378–2386 (2011)
42. Baraldi, A., Parmiggiani, F.: An investigation of the textural characteristics associated with gray level co-occurrence matrix statistical parameters. IEEE Trans. Geosci. Remote Sens. **33**(2), 293–304 (1995)
43. Das, R., Thepade, S., Ghosh, S.: Framework for content-based image identification with standardized multiview features. ETRI J. **38**(1), 174–184 (2016)
44. Thepade, S., Das, R., Ghosh, S.: Decision fusion-based approach for content-based image classification. Int. J. Intell. Comput. Cybern. **10**(3), 310–331 (2017)

Analysis of Human Bone Disorder Using Fuzzy and Possibility Theory

Saikat Maity and Jaya Sil

Abstract X-ray image-based pre-diagnosis modality is the cheapest way of dealing any bone-related problems. Identification of the particular region eliminating bones from the flesh and cartilage in the X-ray image where bone disorder may occur is the objective of this work to facilitate the doctors for proper diagnosis. We employ possibility theory for analyzing the X-ray images to appraise the possibility of the regions having the disorder. The novelty of the work is to facilitate the doctors concentrating only on automatic detection of region of interest (ROI) for correct diagnosis of the patients. The projected technique consists the steps—(i) preprocessing to denoise the X-ray image using fuzzy inference system, (ii) isolation of the bone region from the rest of the X-ray image using Type 1 and Type 2 fuzzy-based edge discovery methods, and (iii) in order to enable the experts for more precise diagnosis, region of interest (ROI) of the bone has been identified using possibility theory. Finally, experimental results of different bone regions using the proposed approach have been demonstrated and compared with the existing methods showing better performance and diagnosis.

Keywords X-ray image · Fuzzy logic · Region of interest · Knowledge acquisition · Possibility theory

1 Introduction

Medical image analysis and study plays a very vital role on the rise of Health Diagnosis System (HDS) [1]. X-ray image-based diagnosis is used long back as conventional method for detecting bone-related disorders. However, the X-ray images either over-

S. Maity (✉)
JIS University, Kolkata 700109, India
e-mail: saikatmaity78@gmail.com

J. Sil
Indian Institute Engineering Science & Technology, Shibpur 711103, India
e-mail: js@cs.iiests.ac.in

S. Bhattacharyya et al. (eds.), *Hybrid Machine Intelligence for Medical Image Analysis*, Studies in Computational Intelligence 841,
https://doi.org/10.1007/978-981-13-8930-6_6

exposed or underexposed due to the mechanism currently followed for capturing the images. The conventional way to recapture the images is, therefore, expensive and not even guaranteed to be noise-free, affecting the results of diagnosis made by the doctors. The purpose of the idea is to develop an intelligent method for cost-effective solution by denoising the X-ray images and obtaining specific bone regions in order to facilitate the doctors to diagnose the disorder more accurately.

Usually, the visual quality of X-ray images is enhanced by removing the additive Gaussian noise and impulse noise present in the images during acquisition. However, the visual quality of the image after noise removal not enables the doctors for taking accurate decision. The noise removal technique proposed in the paper is adaptive in addition to deal the gray-level values of the bone picture, cartilage and flesh regions of the X-ray images which may vary from human to human. Noise-free images are processed to isolate the bone regions and subsequently to the more specific region where there is a possibility of bone disorder. The doctors now can concentrate only on the possible region of interest and apply their knowledge for diagnose taking corrective measures more easily and accurately.

Additive Gaussian noise is frequently characterized by the damaged pixels, which are either set to highest value or get flipped over and quantified by the proportion of pixels which are damaged. Impulse noise changes pixel value by setting zero to the maximum gray-level value and occurs during data transmission or capturing using digital cameras. In digital images, individual pixels may vary unexpectedly in intensity from their neighboring pixel. The noisy pixels may form edges due to change of intensity with the neighbors, therefore, removal of noise sometimes remove important edges and lines. To remove such types of noise, usually low-pass filters are applied to the image, which reduces contrast and noise in the image. However, in case of multiplicative noise the intensity of the noise varies with the signal intensity and difficult to remove from the images using a single noise removal method. The noise removal technique is adaptive because the gray-level values of the bone, cartilage, and flesh regions of the X-ray images may vary from human to human. The adaptive value is known by analyzing X-ray images of different persons and used for developing rules with the help of experts' opinion. Noise-free images are processed to isolate the bone regions and subsequently to the more specific region where there is a possibility of bone disorder. The doctors now can concentrate only on the possible region of interest and apply their knowledge for diagnose and may take corrective measures more easily and accurately. Low-pass filters [2] are frequently utilized as a part of image processing to expel irregular noise from a computerized picture. Lamentably, as noise is truncated from a picture, points of interest in the picture are lost. The objective of truncating noise in a picture is in this way to strike a harmony among commotion and detail, which is suitable for the application. Other than established filters [3], there are distinctive fluffy channels used to expel the noise from the medical images. Fuzzy median filter [4] is outstanding for evacuating drive commotion. Fuzzy detection step utilizes fuzzy incline values in eight directions with a 3×3 window, which shows the level of central pixel as an impulse noise pixel. Gaussian noise reduction filter (GOA) [5] is intended to evacuate Gaussian noise by averaging neighborhood pixels as for a seed pixel and all the while preserving the edge data. The existing methods can able to eliminate a particular type of noise like Gaussian noise, shot noise, or quantization noise but the X-ray image generally consists of multiple type of noise. Therefore, a new adaptive denoising technique is required to eliminate maximum variant of noise from the X-ray images for more accurate diagnosis. A few fuzzy and non-fuzzy filters have been examined for noise lessening [6] from X-ray images.

We suggest a new fuzzy logic-based technique in addition to arithmetical properties of the noisy input X-ray image like mean, standard deviation, maxima and minima, not considered earlier for eliminating the noisy pixels. In order to determine ambiguity in the pixels whether they are noisy or not, fuzzy rules are defined using the statistical parameters represented by fuzzy sets *low*, *medium,* and *high*. The pixels are categorized as noisy or noise free depending on the opinion of the doctors and used as consequent of the fuzzy rules. The noisy pixels are determined by the FLS and eliminated. The denoised image is processed for isolating the bone region from other regions using Type 2 fuzzy technique applied on Prewitt method [7]. The Type 2-based method can handle different types of uncertainty in bone region segmentation, not possible to deal with the existing Type 1-based method. After bone segmentation, we like to concentrate on finding the region of interest (ROI) for focusing the problem lies with the specific bone region. We apply possibility theory to determine the ROI, the unambiguous region where the doctors want to concentrate on starting the treatment.

The ROI has been identified using the proposed possibilistic pixel projection (PPP) [8] approach depending on possibility distribution [9] of the pixels in the segmented regions. The possibility distribution function has been constructed using $P_{r-\pi}$ *Transformation* [9] where every sample (ROI) defines the possibility decision rules for extraction of the possibility pixels. The possibility pixels form a pattern called possibilistic map which has been accepted as knowledge-based expending on the respective confidence index. As the next step Nagao filter [10] is applied on the possibilistic map to obtain maximum possibilistic classification for preserving the pattern contour. The proposed possibilistic theory-based approach detects the defective bone region in absolute terms unlike the other uncertainty-based methods, like fuzzy logic, probability, etc. which measures degree of defectiveness of the region.

While building ROI, the association of each pixel with the region is calculated. This type of methods cannot able to handle uncertainty as well as inaccuracy due to improper and un-uniform transition between normal bone and ROI because of the bad quality of image. We have build up a new concept to find out ROI. The method of developing the ROI is covered with two major points. One, the segmentation method involved Maximum Intensity Projections (MIP) [11]. Multiple MIP projections [12] involves a fusion framework which was used on MRI images. In the paper, a possibilistic knowledge acquisition (PKA) method has been developed to take indecision into explanation in order to avoid the unnecessary segmented areas for consideration.

The idea is arranged as follows: In Sect. 2, we describe the proposed fuzzy adaptive method for denoising the X-ray images; Section 3 proposes segmentation method to isolate the bone region from flesh and cartilages using Type 2 fuzzy edge detection algorithm; Section 4 introduces the notion of possibility theory to choose the ROI of the bone; and experimental results are shown in Sect. 5 while conclusions are arrived at the last section.

2 X-ray Image Denoising Technique

In X-ray imaging, generally, the secondary radiation cannot be blocked due to poor handling of the X-ray machine and the digitization process. Unwanted X-rays always reduce or increase contrast of picture which is very difficult to understand for the doctor. Image processing techniques proved to be efficacious in minimizing the number of acceptable pictures. To maximize the amount of information extracted from the X-ray images and to increment their visual quality, the noise has to be eliminated. Low-pass filters are often utilized in image processing to abstract arbitrary noise from a digital image. Infelicitously, as noise is abstracted from an image, details in the image are withal lost.

In this manuscript, we advise a new fuzzy technique for denoising the X-ray image by determining the crisp range of parameters (mean, standard deviation, maxima, and minima) of signal to noise ratio (SNR) of the input signal. To apply the concept of adaptive threshold is to capture the behavior of the images. Images are divided into blocks within which the data is invariant or have little variance. We use fuzzy sets *low*, *medium*, *high* to represent vagueness in data by measuring the statistical parameters of each block. The three fuzzy levels are sufficient to deal with uncertainty presents in the X-ray images, determined experimentally. We design the rules in consultation with the doctors using different training images where antecedents are statistical parameters with fuzzy values, and the consequent is defined as the range of threshold.

Training and Testing Parameters
Block size(C): The block size has been grouped into two types. One type must be big adequate to hold sufficiently true statistical information to establish the range of thresholds, and other one should be small enough so that we have different thresholds at different spatial locations.

Fuzzy logic rules: Let the fuzzy F build the rule $F_r(r = 1 \ldots n)$ for decisive the range of thresholds $[\lambda_{\min}, \lambda_{\max}]$.

Let x and σ denote mean and standard deviation of the SNR of a given block and the corresponding fuzzy values *low*, *medium,* and *high* are defined in Figs. 1, 2 and 3, respectively. Selecting range of input parameters in such a way so that all pixels of the X-ray image is covered in the rule set and maximum number of noise pixels eliminated.

$$
\begin{aligned}
&\text{Mean\_High}(x) = 0 && x \le 0.5 \\
&\tfrac{x-0.5}{0.1} && 0.5 \le x \le 1.0 \\
&1 && x > 1.0
\end{aligned}
$$

Fig. 1 Graph of 1st function noise

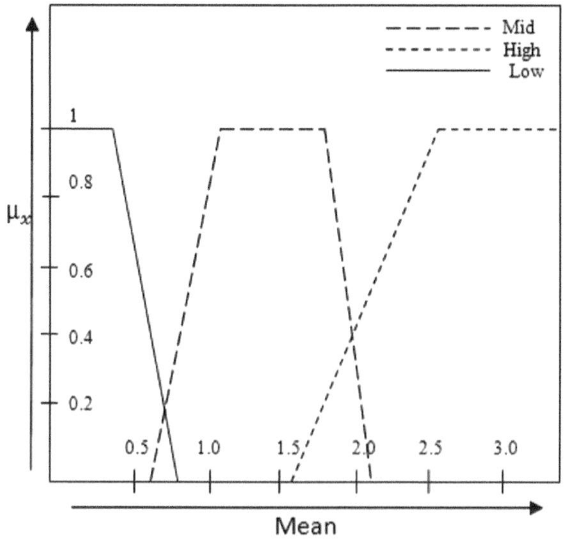

Fig. 2 Graph of 2nd function noise

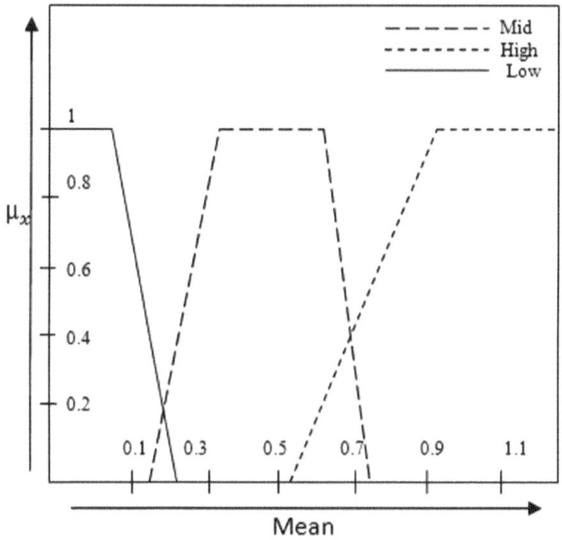

$$
\text{Mean\_mid}(x) = \begin{cases} \frac{x-0.3}{0.1} & 0.3 \le x \le 0.4 \\ & 0.4 \le x \le 0.8 \\ \frac{1.0-x}{0.2} & 0.8 \le x \le 1.0 \\ 0 & x < 0.3, x > 1.0 \end{cases}
$$

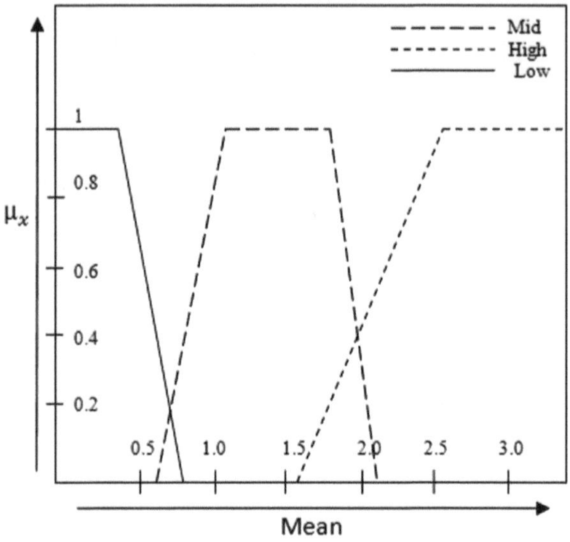

Fig. 3 Graph of 3rd function noise

$$\begin{array}{ll} \text{Mean\_low}(x) = 1 & x \le 0.3 \\ \frac{0.4-x}{0.1} & 0.3 \le x \le 0.4 \\ 0 & x > 1.0 \end{array}$$

2nd Functional Noise gray value range

$$\begin{array}{ll} \text{Mean\_High}(x) = 0 & x \le 0.55 \\ \frac{x-0.55}{0.2} & 0.55 \le x \le 0.75 \\ 1 & x > 0.75 \end{array}$$

$$\begin{array}{ll} \text{Mean\_mid}(x) = \frac{x-0.15}{0.1} & 0.15 \le x \le 0.25 \\ & 0.25 \le x \le 0.55 \\ \frac{0.75-x}{0.2} & 0.55 \le x \le 0.75 \\ 0 & x < 0.15, x > 0.75 \end{array}$$

$$\begin{array}{ll} \text{Mean\_low}(x) = 1 & x \le 0.15 \\ \frac{0.25-x}{0.1} & 0.15 \le x \le 0.25 \\ 0 & x > 0.25 \end{array}$$

3rd Functional Noise gray value range

$$\begin{array}{ll} \text{Mean\_High}(x) = 0 & x \le 0.8 \\ \frac{x-0.8}{0.2} & 0.8 \le x \le 1.0 \\ 1 & x > 1.0 \end{array}$$

$$\text{Mean\_mid}(x) = \frac{x-0.3}{0.1} \quad 0.3 \le x \le 0.4$$
$$0.4 \le x \le 0.8$$
$$\frac{1.0-x}{0.2} \quad 0.8 \le x \le 1.0$$
$$0 \quad x < 0.3, x > 1.0$$

$$\text{Mean\_low}(x) = 1 \quad x \le 0.3$$
$$\frac{0.4-x}{0.1} \quad 0.3 \le x \le 0.4$$
$$0 \quad x > 1.0$$

The rules are listed below considering 27 training images. In one case of the SNR of a block is *low*, it is more to be expected to be intrinsic in nature. We try to maintain the threshold sufficiently high in order to select less number of pixels having SNR below the threshold. If mean of SNR of a block is high, then the eccentric character is used to select pixels.. Similar reasoning holds for other input parameters as well. For example, *mean* = '*low*' and *peak falls* = '*high,*' we shift the threshold range a little to the lower side.

Rule 1:
Mean = '*low*'
Standard deviation = '*low*'
Peak = '*low*' *then* $\lambda_{\text{min}} = \bar{x}$ and $\lambda_{\text{max}} = \bar{x} + 4\sigma$.
Rule 2:
Mean = '*low*'
Standard deviation = '*low*'
Peak = '*Mid*' *then* $\lambda_{\text{min}} = \bar{x} - 5\sigma$ and $\lambda_{\text{max}} = \bar{x} + 4\sigma$.
Rule 3:
Mean = '*low*'
Standard deviation = '*low*'
Peak = '*High*' *then* $\lambda_{\text{min}} = \bar{x} - \sigma$ and $\lambda_{\text{max}} = \bar{x} + 4\sigma$.
Rule 4:
Mean = '*low*'
Standard deviation = '*Mid*'
Peak = '*low*' *then* $\lambda_{\text{min}} = \bar{x}$ and $\lambda_{\text{max}} = \bar{x} + 4\sigma$.
Rule 5:
Mean = '*low*'
Standard deviation = '*Mid*'
Peak = '*Mid*' *then* $\lambda_{\text{min}} = \bar{x} - .25\sigma$ and $\lambda_{\text{max}} = \bar{x} + 4\sigma$.
Rule 6:
Mean = '*low*'
Standard deviation = '*Mid*'
Peak = '*High*' *then* $\lambda_{\text{min}} = \bar{x} - .5\sigma$ and $\lambda_{\text{max}} = \bar{x} + 4\sigma$.
Rule 7:
Mean = '*low*'
Standard deviation = '*High*'
Peak = '*low*' *then* $\lambda_{\text{min}} = \bar{x}$ and $\lambda_{\text{max}} = \bar{x} + 4\sigma$.

Rule 8:
Mean = '*low*'
Standard deviation = '*High*'
Peak = '*Mid*' *then*
$\lambda_{\min} = \bar{x} - 0.05\sigma$ and $\lambda_{\max} = \bar{x} + 0.95\sigma$.
Rule 9:
Mean = '*low*'
Standard deviation = '*High*'
Peak = '*High*' *then* $\lambda_{\min} = \bar{x} - 0.1\sigma$ and $\lambda_{\max} = \bar{x} + 0.09\sigma$.
Rule 10:
Mean = '*Mid*'
Standard deviation = '*low*'
Peak = '*low*' *then* $\lambda_{\min} = \bar{x} - 1.5\sigma$ and $\lambda_{\max} = \bar{x} + 2.5\sigma$.
Rule 11:
Mean = '*Mid*'
Standard deviation = '*low*'
Peak = '*Mid*' *then* $\lambda_{\min} = \bar{x} - 2\sigma$ and $\lambda_{\max} = \bar{x} + 2\sigma$.
Rule 12:
Mean = '*Mid*'
Standard deviation = '*low*'
Peak = '*High*' *then*
$\lambda_{\min} = \bar{x} - 2.5\sigma$ and $\lambda_{\max} = \bar{x} + 1.5\sigma$.
Rule 13:
Mean = '*Mid*'
Standard deviation = '*Mid*'
Peak = '*low*' *then* $\lambda_{\min} = \bar{x} - .75\sigma$ and $\lambda_{\max} = \bar{x} + 1.25\sigma$.
Rule 14:
Mean = '*Mid*'
Standard deviation = '*Mid*'
Peak = '*Mid*' *then* $\lambda_{\min} = \bar{x} - \sigma$ and $\lambda_{\max} = \bar{x} + \sigma$.
Rule 15:
Mean = '*Mid*'
Standard deviation = '*Mid*'
Peak = '*High*'
then $\lambda_{\min} = \bar{x} - 1.25\sigma$ and $\lambda_{\max} = \bar{x} + 0.75\sigma$
Rule 16:
Mean = '*Mid*'
Standard deviation = '*High*'
Peak = '*low*' *then*
$\lambda_{\min} = \bar{x} - 0.45\sigma$ and $\lambda_{\max} = \bar{x} + 0.55\sigma$.
Rule 17:
Mean = '*Mid*'
Standard deviation = '*High*'
Peak = '*Mid*' *then*
$\lambda_{\min} = \bar{x} - 0.5\sigma$ and $\lambda_{\max} = \bar{x} + 0.5\sigma$.

Rule 18:
Mean = '*Mid*'
Standard deviation = '*High*'
Peak = '*High*' *then*
$\lambda_{\min} = \bar{x} - 0.55\sigma$ and $\lambda_{\max} = \bar{x} + 0.45\sigma$.
Rule 19:
Mean = '*High*'
Standard deviation = '*low*'
Peak = '*low*' *then* $\lambda_{\min} = \bar{x} - 3\sigma$ and $\lambda_{\max} = \bar{x} + \sigma$.
Rule 20:
Mean = '*High*'
Standard deviation = '*low*'
Peak = '*Mid*' *then*
$\lambda_{\min} = \bar{x} - 3.5\sigma$ and $\lambda_{\max} = \bar{x} + 0.5\sigma$.
Rule 21:
Mean = '*High*'
Standard deviation = '*low*'
Peak = '*High*' *then* $\lambda_{\min} = \bar{x} - 4\sigma$ and $\lambda_{\max} = \bar{x}$.
Rule 22:
Mean = '*High*'
Standard deviation = '*Mid*'
Peak = '*low*' *then* $\lambda_{\min} = \bar{x} - 1.5\sigma$ and $\lambda_{\max} = \bar{x} + 0.5\sigma$.
Rule 23:
Mean = '*High*'
Standard deviation = '*Mid*'
Peak = '*Mid*' *then*
$\lambda_{\min} = \bar{x} - 1.75\sigma$ and $\lambda_{\max} = \bar{x} + 0.25\sigma$.
Rule 24:
Mean = '*High*'
Standard deviation = '*Mid*'
Peak = '*High*' *then* $\lambda_{\min} = \bar{x} - 2\sigma$ and $\lambda_{\max} = \bar{x}$.
Rule 25:
Mean = '*High*'
Standard deviation = '*High*'
Peak = '*low*' *then* $\lambda_{\min} = \bar{x} - 0.9\sigma$ and $\lambda_{\max} = \bar{x} + 0.1\sigma$.
Rule 26:
Mean = '*High*'
Standard deviation = '*High*'
Peak = '*Mid*' *then*
$\lambda_{\min} = \bar{x} - 0.95\sigma$ and $\lambda_{\max} = \bar{x} + 0.05\sigma$.
Rule 27:
Mean = '*High*'
Standard deviation = '*High*'
Peak = '*High*' *then* $\lambda_{\min} = \bar{x} - \sigma$ and $\lambda_{\max} = \bar{x}$.

Denoising Algorithm

Input: X-ray Image
Begin
X-ray image is separated into number of blocks each of size C.
For each block
Calculate mean, standard deviation, and peak of the SNR
Obtain fuzzy sets low, medium, or high of mean, std. deviation, and peak
Obtain threshold range $[\lambda_{min}, \lambda_{max}]$ *by applying the fuzzy logic rules*
Get Final Denoised image.
Output: Denoised X-ray Image
End.

Here, we have used fuzzy logic to agree on the threshold range variable or rather image specific adaptive so that none of the X-ray images will be missed during the process.

3 Segmentation of X-ray Image

Image Segmentation is a primary footstep utilized in the majority image processing applications to detect specific objects. Recent techniques of edge detection are characterized as fuzzy reasoning quandary [13, 14], to model the gray values of the pixels which often occur in the time of capturing the image. Fuzzy techniques sanction an incipient standpoint to model uncertainties due to the ambiguity of gray values of the pixels present in the images. Miosso and Bauchspiess [15] evaluate the recital of a fuzzy inference system (FIS) in edge detection. The FIS system is robust toward contrast and lighting variations besides evading double edges. More change of the parameters connected with the FIS is till now essential to diminish the ambiguity further for proficiently detecting the non-edge pixels in the X-ray image.

Gradients Calculation
Prewitt operator determines the direction of change of intensity of an image from black to white. It is distinct segregated operators which usually compute and estimate of the gradients at each point as image intensity function producing vertical and horizontal edges in images based on threshold value. Say, D_h and D_v are two images obtained by calculating gradients at each point of the source image in both directions (horizontal and vertical).

In order to identify the edge and non-edge pixels more accurately, experts' knowledge has been utilized in the proposed FIS to frame the rules using Prewitt operator and low-pass filter (LPF). Figure 2 demonstrates how segmentation has been performed in the proposed fuzzy-based system where input is a denoised X-ray image. The first and most important criterion to develop the FIS is to detect the important edges in the denoised X-ray image. The second criterion is to detect the edge

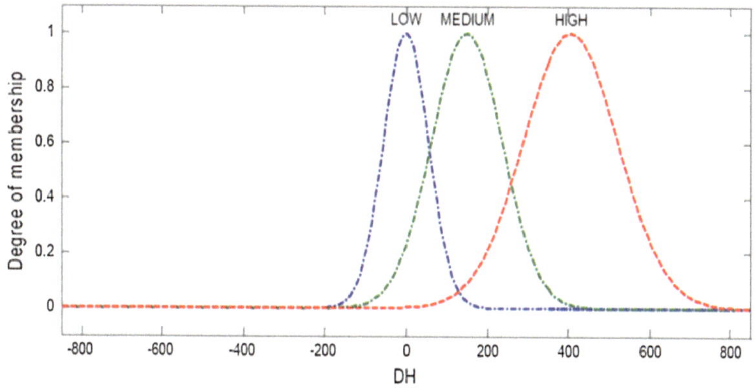

Fig. 4 Membership function for input 1 (D_h)

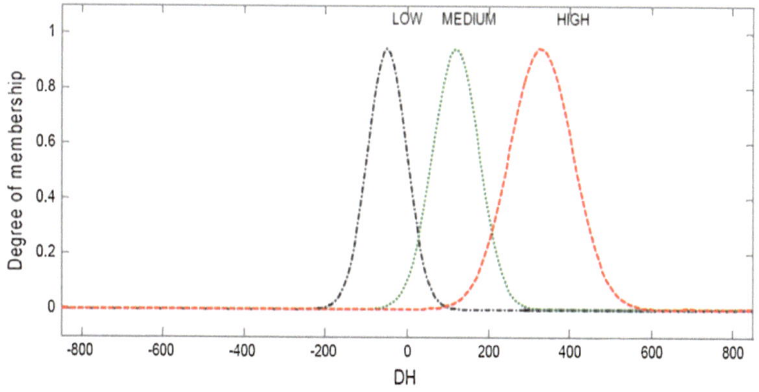

Fig. 5 Membership function for input 2 (D_v)

pixels which are differentiable from the non-edge pixels more accurately. However, gradient information in $X(D_h)$ and $Y(D_v)$ direction do not cover the edges having low-intensity level of the X-ray images. We employ LPF to cover such edges and the edge information obtained in $X(D_{hlpf})$ and $Y(D_{vlpf})$ directions is additionally used to develop the more robust FIS.

Detection of Edge

Mamdani FIS system [16] has been used to detect the edges of denoised image where antecedents and consequences are determined using fuzzified Prewitt operator and LPF. Gradient information is calculated using Prewitt operator and LPF, used as the crisp variables in the rule set. The crisp variables are represented using fuzzy sets to deal ambiguity between the edge and the non-edge pixels in the image and used as antecedents and consequents of the Mamdani rule base. Fuzzy variables and membership functions are Gaussian, shown in Figs. 4, 5, and 6. The fuzzified input–output and rules of the proposed FIS are described below.

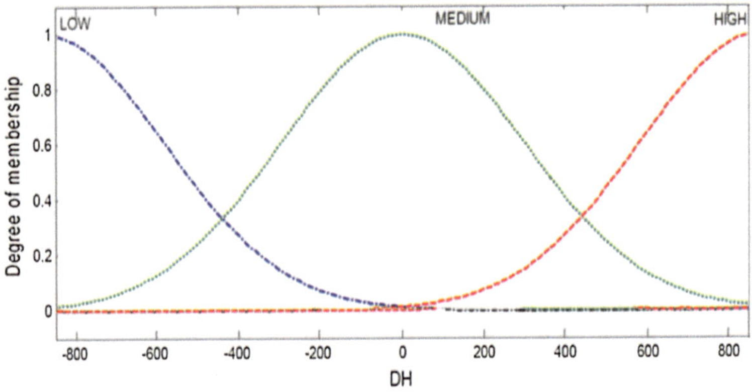

Fig. 6 Membership function for output edges

Where D_h and D_v ranges are

$$D_h = \begin{cases} Low;\ 0.4 \le grey_h < 0.5 \\ high;\ 0.5 \le grey_h \le 1 \end{cases} \tag{1}$$

$$D_v = \begin{cases} Low,\ 0.4 \le grey_v < 0.5 \\ high;\ 0.5 \le grey_v \le 1 \end{cases} \tag{2}$$

$$D_{hlpf} = \begin{cases} Low;\ 0.1 \le grey_{hlpf} < 0.3 \\ high;\ 0.3 \le grey_{hlpf} \le 0.4 \end{cases} \tag{3}$$

$$D_{vlpf} = \begin{cases} Low;\ 0.1 \le grey_{vlpf} < 0.3 \\ high;\ 0.3 \le grey_{vlpf} \le 0.4 \end{cases} \tag{4}$$

Now rules can be defined as:

Rule 1:
$\boldsymbol{D_h}$ = '*low*'; $\boldsymbol{D_v}$ = '*low*'; ***LPF*** = '*Low*'
EDGES = '*Low*'
Rule 2:
$\boldsymbol{D_h}$ = '*low*'; $\boldsymbol{D_v}$ = '*low*'; ***LPF*** = '*High*'
EDGES = '*Low*'
Rule 3:
$\boldsymbol{D_h}$ = '*low*'; $\boldsymbol{D_v}$ = '*High*'; ***LPF*** = '*Low*'
EDGES = '*Low*'
Rule 4:
$\boldsymbol{D_h}$ = '*low*'; $\boldsymbol{D_v}$ = '*High*'; ***LPF*** = '*High*'
EDGES = '*High*'
Rule 5:
$\boldsymbol{D_h}$ = '*High*'; $\boldsymbol{D_v}$ = '*low*'; ***LPF*** = '*Low*'
EDGES = '*Low*'

Rule 6:
$\boldsymbol{D_h}$ = '*High*'; $\boldsymbol{D_v}$ = '*low*'; ***LPF*** = '*High*' *EDGES* = '*High*'
Rule 7:
$\boldsymbol{D_h}$ = '*High*'; $\boldsymbol{D_v}$ = '*High*'; ***LPF*** = '*Low*'
EDGES = '*High*'
Rule 8:
$\boldsymbol{D_h}$ = '*High*'; $\boldsymbol{D_v}$ = '*High*'; ***LPF*** = '*Low*'
EDGES = '*High*'

where

$$Edge = \begin{cases} Low & \left|\langle D_h, D_v, D_{hlpf}, D_{vlpf} = Low\rangle \right. \\ & \langle D_h, D_v, D_{hlpf} = Low, D_{vlpf} = high\rangle \\ & \left\| \begin{pmatrix} 0.5 \le grey_h \le 1 \\ 0.4 \le grey_v < 0.5 \\ 0.1 \le grey_{hlpf} < 0.3 \\ 0.1 \le grey_{vlpf} < 0.3 \end{pmatrix} \begin{pmatrix} 0.4 \le grey_h < 0.5 \\ 0.5 \le grey_v \le 1 \\ 0.1 \le grey_{hlpf} < 0.3 \\ 0.3 \le grey_{vlpf} \le 0.4 \end{pmatrix} \right. \\ & \left\| \begin{pmatrix} 0.4 \le grey_h < 0.5 \\ 0.4 \le grey_v < 0.5 \\ 0.3 \le grey_{hlpf} \le 0.4 \\ 0.1 \le grey_{vlpf} < 0.3 \end{pmatrix} \begin{pmatrix} 0.4 \le grey_h < 0.5 \\ 0.4 \le grey_v < 0.5 \\ 0.3 \le grey_{hlpf} \le 0.4 \\ 0.3 \le grey_{vlpf} \le 0.4 \end{pmatrix} \right. \\ & \left\| \begin{pmatrix} 0.5 \le grey_h \le 1 \\ 0.4 \le grey_v < 0.5 \\ 0.1 \le grey_{hlpf} < 0.3 \\ 0.1 \le grey_{vlpf} < 0.3 \end{pmatrix} \begin{pmatrix} 0.4 \le grey_h < 0.5 \\ 0.5 \le grey_v \le 1 \\ 0.1 \le grey_{hlpf} < 0.3 \\ 0.3 \le grey_{vlpf} \le 0.4 \end{pmatrix} \right. \\ High & \left\| \begin{pmatrix} 0.5 \le grey_h \le 1 \\ 0.5 \le grey_v \le 1 \\ 0.3 \le grey_{hlpf} \le 0.4 \\ 0.3 \le grey_{vlpf} \le 0.4 \end{pmatrix} \begin{pmatrix} 0.4 \le grey_h < 0.5 \\ 0.5 \le grey_v \le 1 \\ 0.3 \le grey_{hlpf} \le 0.4 \\ 0.3 \le grey_{vlpf} \le 0.4 \end{pmatrix} \right. \end{cases} \tag{5}$$

New Approach of Edge Detection Using Type 2 FIS
Type 2 fuzzy approaches cut down the search freedom of the skilled database for the test image and make the algorithm more expeditious. Fuzzy rules are required for each probable relation of the statistical means. We have utilized same bunch of policy are utilized for Type 2 FIS as of Type 1 FIS. For every Type 1 FIS, all the variables are modeled as Type 1 FS but that is not always true for Type 2 FIS. In this accomplishment, the variety of pixel intensities are called as edge pixels. This range can be obtained by Type 2 fuzzy outputs. Type 2 FIS is implemented utilizing Mamdani model using Gaussian-type membership function, shown in Figs. 7, 8, and 9 as footprint of uncertainty (FOU). The MATLAB procedure is given after graphs.

MATLAB Procedure for Type 2 Fuzzy Edge Recognition

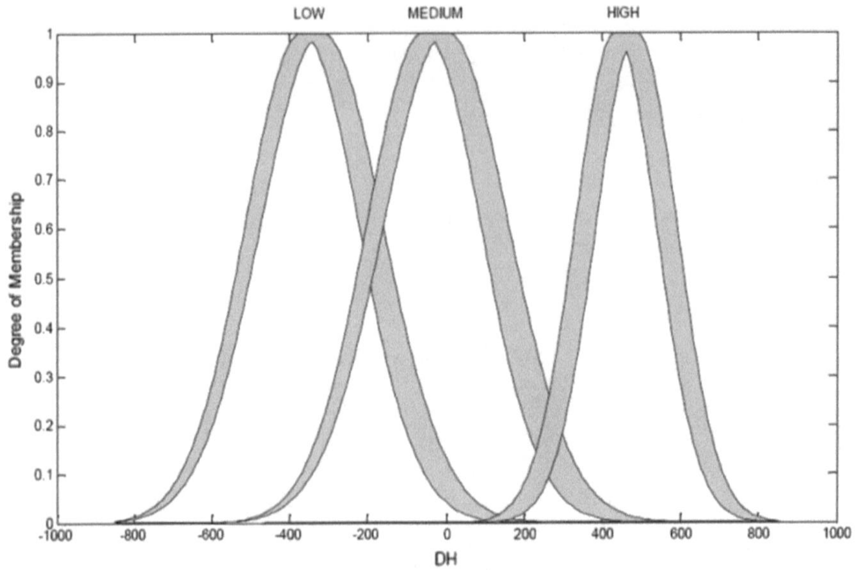

Fig. 7 Type 2 Membership distribution for input D_h

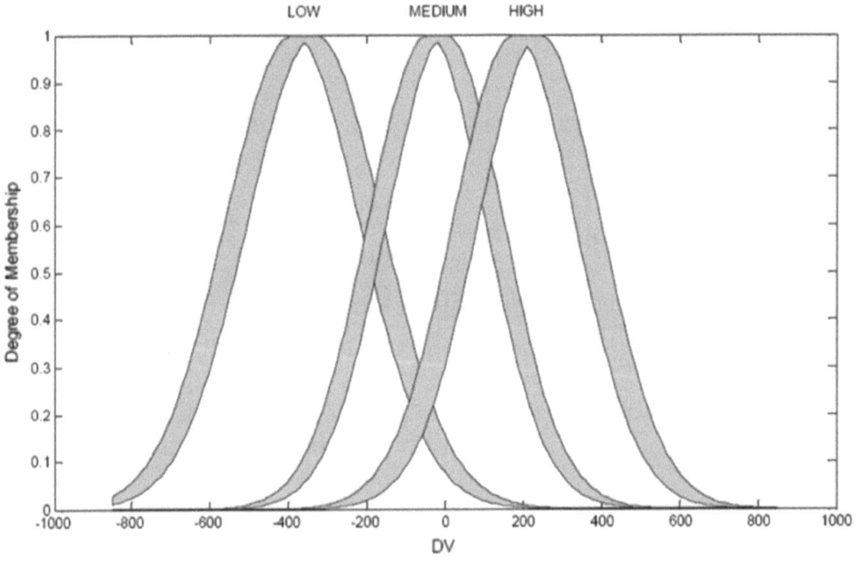

Fig. 8 Type 2 membership distribution for input D_v

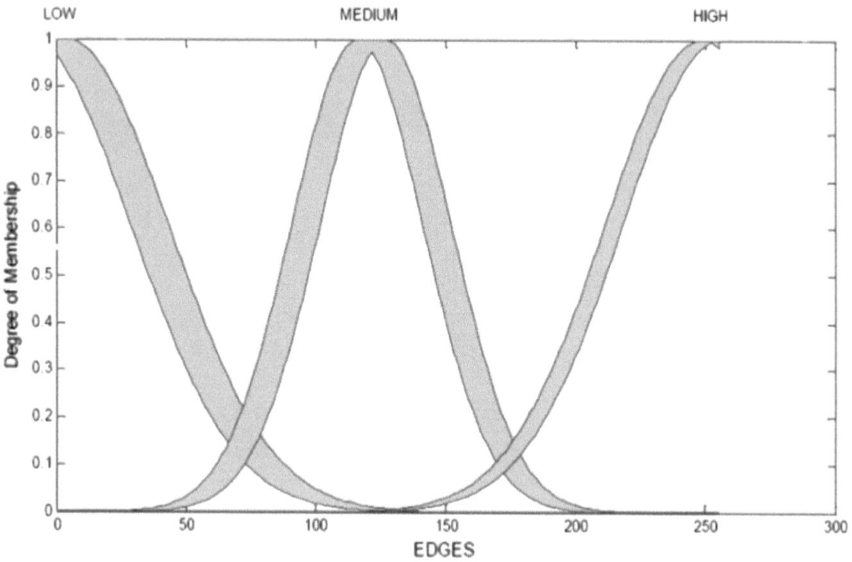

Fig. 9 Type 2 membership function for output edges

Import Preprocessed Image
Calculate X-ray Image Gradient
Using Type 2 fuzzy logic edge-recognition algorithm to split in unvarying province.
Compute the gradient of x*-axis and* y*-axis.*
Create a fuzzy inference system (FIS) for edge detection, named edgeFIS.
Specify FIS Rules
Add rules to make a pixel white if it belongs to a uniform region, make the pixel black.
Evaluate FIS
Evaluate the output of the edge detector for each row of pixels in I using corresponding rows of I_x *and* I_y *as inputs.*
Plot Results.

Region of interest (ROI)
Possibility or plausibility hypothesis is a generally new hypothesis devoted should handle questionable matter in the setting the place the accessible learning may be vague. Possibility theory was invented by Lotfi Zadeh in 1978 assigns belief to a crisp set which measures belongings of a variable in that set depending upon observations. In the work, possibility theory is applied to ensure that some specific regions of a segmented bone image have fracture or disorder. Gray-level values of each segmented bone image are represented by crisp sets. The elements of a crisp set might not provide sufficient knowledge regarding bone disorder, therefore, creating ambiguity in decision-making. Possibility measure has been applied on the subsets of crisp set to take decision with certainty about bone disorder. The result enables

the doctors to concentrate on the specific region for accurate diagnosis. A concept of possibility theory has been briefly described below.

4 Possibility Theory

In this subsection, fundamental of possibility theory including possibility measure and determination of possibility distribution are described.

Element of a crisp set is surrounded by a possibilistic interval, cleared as a possibility measure Ω, necessity measure, N [17]. Let the decision set $\Omega = \{C_1, C_2, \ldots, C_m\}$ having a fixed number of crisp sets on behalf of 'm' mutually limited decisions. Possibility distribution is represented as π, which plots the basic pronouncement from Ω to the interval [0,1], of each element $C_m \in \Omega$. For every event $A \subseteq \Omega$, possibility and necessity measures can be defined by:

$$\prod(A) = \max_{C_m \in A}[\pi(C_m)] \forall m = 1, 2 \ldots M \tag{6}$$

$$N(A) = 1 - \prod(A^c) \min_{C_m \notin A^c}[1 - \pi(C_m)] \tag{7}$$

where A^c represents the reverse of the event A ($A \cup A^c = \Omega$ with $A \cap A^c = \emptyset$). Possibility measure $\prod(A)$ approximates the uniformity about the occurrence of event A using possibility distribution π. $\prod(A) = 0$ means that A is not a possible event while $\prod(A) = 1$, $N(A)$ defined as event A having possibility distribution π. $N(A) = 0$ denotes that the certainty about the occurrence of A is null. On the contrary, $N(A) = 1$ indicates that the occurrence of A is certain. Therefore, uncertainty is related to occurrence of each event $A \subseteq \Omega$ characterized by the interval $[N(A), \prod(A)]$ which allows bounding the probability of occurrence of the event A by the inequality relation; $N(A) \leq P_r(A) \leq \prod(A)$. Decision C_m refers to a given class or category, where events A are composed of a single decision ($A_m = \{C_m\}$, $m = 1, \ldots, M$) of a particular interest. The key issue in applications of possibility theory is determination of possibility distributions [18] and its transformations from probability-to-possibility $P_{(r-\pi)}$ Dubois [19]. The probability-possibility *consistency* principle and the preference preservation principle are given below:

$$P_r(A) < P_r(Bi) \Leftrightarrow \prod(A) \text{ and } P_r(Ai) < P_r(B) \Leftrightarrow \prod(Bi)), \text{ for } \forall Ai, Bi \subseteq \Omega$$

The $P_{(r-\pi)}$ transformation provides satisfactory results in pattern recognition and classification problems [20] and has been adopted in this paper for obtaining region of interest containing bone disorder. The expression of transformation equation is

$$\pi(C_m) = \prod(C_m) = \sum_{m=1}^{M} \min[P_r(C_m), P_r(C_m)]$$

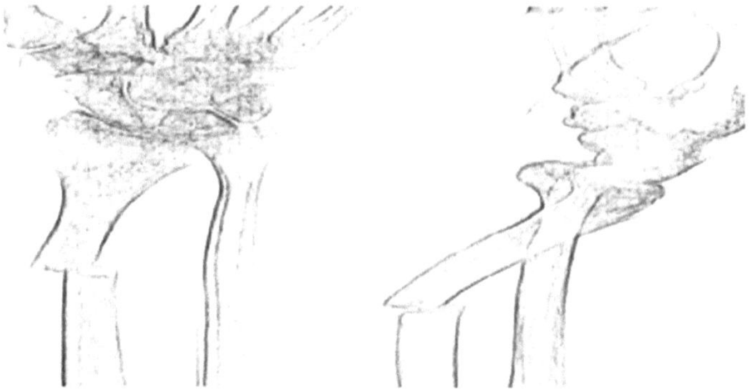

Fig. 10 Exclusive segmented bone images together (Ω_S)

Let $\Omega_S = \{S_1, S_2, \ldots, S_M\}$ are defined as finite and exclusive and exhaustive sets of M mutually exclusive segmented bone images. Ω_S is represented by the interval [0,1] denoting the gray scale value of a set $C_m \in \{S_1, S_2, \ldots, S_M\}$. The value $\pi(C_m)$ represents to what extent it is possible for C_m to be the exclusive and one among Ω_S.

Say, $A \subseteq \Omega_s$ defined using Eq. (8).

$$\prod(A) = \max_{C_m \in A}[\pi(C_m)]$$
$$N(A) = 1 - \prod(A^c) = \min_{C_m \notin A^c}[1 - \pi(C_m)] \tag{8}$$

Events are $\prod(A) = 0$ and $\prod(A) = 1$

Bounding the occurrence of possibility of the event A

$(A_m = (C_m), m = 1, \ldots, M)$, $\prod(.)$ and $N(\cdot)$, thus the expression is,

$$N(A_m) = N(C_m) = 1 - \prod(C_m)^c = \max_{n \neq m} \pi(C_n)] \tag{9}$$

We take an example of ball and socket with hand bone, consisting of nine segments, shown in Fig. 10. For further analysis, we consider the segmented element S_1 only, shown in Fig. 11.

Contrast of the image of Fig. 11 is enhanced, as shown in Fig. 12.

All possible gray-level intensity of pixels present in Fig. 12 is shown in Fig. 13 where different colors represent the intensity of neighboring pixels.

Possibility Decision Rules

In the work, the highest possibility degree of occurrence of the pixels within a region is evaluated to identify the ROI in the segmented bone image. Our aim is to locate the area where the bone disorder exists, called ROI consisting of maximum number of possibility regions. Highest possibility degree of occurrence of the pixels is obtained possibility decision rule, discussed in Aliev [19]. The decision rule R1 is denoted as maximum possibility decision rule having the highest possibility degree of occur-

Fig. 11 Exclusive segmented bone images together (S_i) from (Q_s)

Fig. 12 Enhanced contrast of segmented element from (Ω_S)

rences of pixels. The criteria to define the possibility decision rule are borrowed from two unique similar decision rules:

$R1$, depends on encapsulated possibility distribution π.
$R2$, depends on π.

a. ***Max. Possibility Decision Rule (R1)***

If elementary decision $A_{mo} = (C_{mo}) \subseteq \Omega$ having the maximum possibility degree of occurrence highest possibility degree of occurrence, then

$$(R1): \ Decision \equiv A_{mo} iff \prod(A_{mo}) = \max_{m=1,\ldots M} \prod(A_m) \quad (10)$$

b. ***Max. Confidence Index Decision Rule (R2)***:

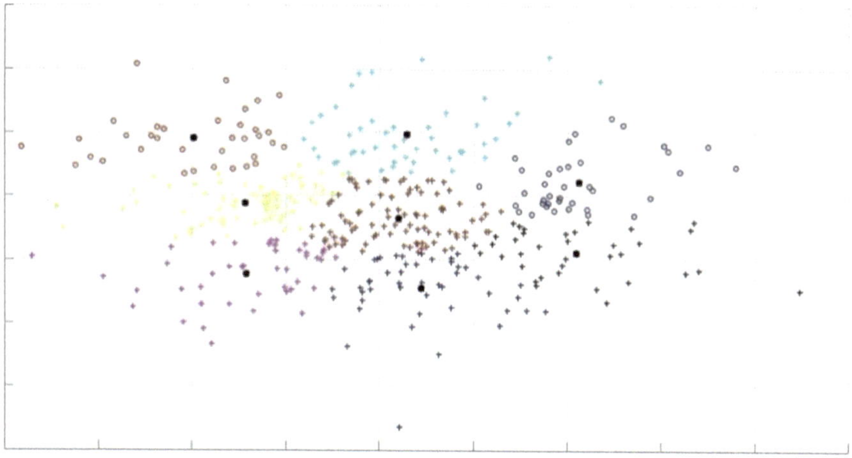

Fig. 13 Possibility of pixels within S_i region

It has been proved [21] that bone disorder region in X-ray image always carries maximum confidence in the decision rule. In the paper, the decision is taken from Max (confidence index Ind), for each event A $\subseteq \Omega$,

$$Ind : 2^{\Omega} \rightarrow [-1, +1]; A \rightarrow Ind(A) = \prod(A) + N(A) - 1$$

For an event A, this index range is [0,1]:

- $Ind(A) = -1 \; iff \prod(A) = N(A) = 0;$
- $Ind(A) = +1 \; iff \prod(A) = N(A) = 1;$

Where events A_m having only one decision $A_m = \{C_m\}$. In this case, $Ind(A_m)$ measures the difference between the possibility measure of the event A_m and the highest possibility degree of all decisions contained in $(A_m)^c$ where $(A_m)^c$ is the complement of A_m in Q and

$$Ind(A_m) = \prod(A_m) + N(A_m) - 1 = \prod(C_m) - \max_{m \neq n} \pi(C_n) \quad (11)$$

Therefore, if elementary decision $A_{mo} = (C_{mo})$ the only event having the highest possibility measure $\pi(C_{mo})$, then A_{mo} is the unique event having a positive confidence index value. All other events may have negative values where we assume $\pi(C_{mo}) > \pi(C_m)$, $\forall \; m = m_o$ and C_{m1}. The decision rule can be formulated by:

$$(R2) :\equiv A_{mo} \text{ if and only if } Ind \; A_{mo} = \max_{m=1,\ldots,M} [Ind(A_m)]$$

The pixels having less possibility of occurrence with respect to possibilistic confidence threshold not required for obtaining the ROI and, therefore, eliminated by

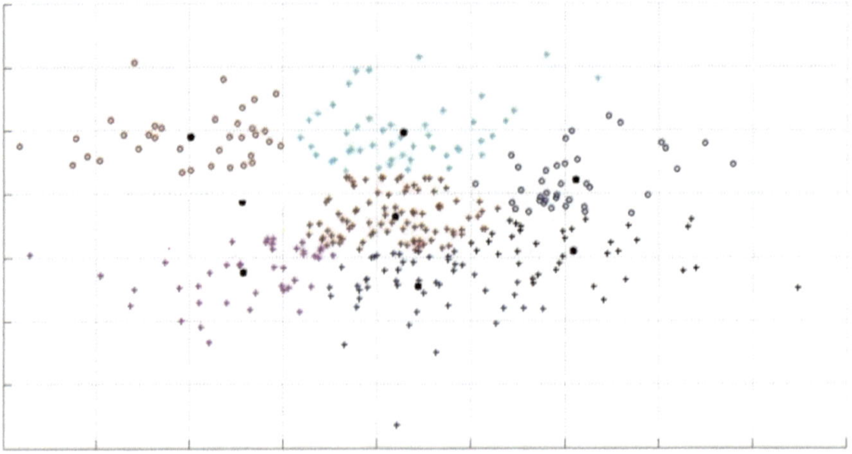

Fig. 14 Possibility of pixels within S_1 after applying rule R_1

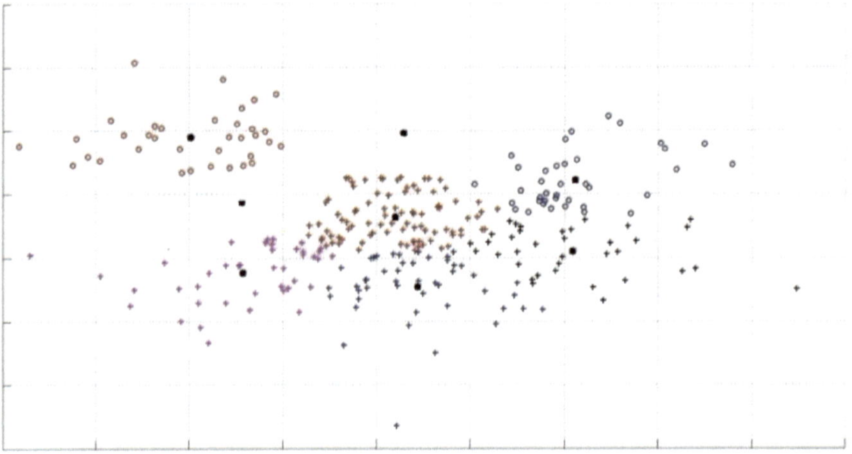

Fig. 15 Possibility of pixels within S_1 after applying R_2

applying the decision rule $R1$, shown in Fig. 14. The pixels having less confidence index (bone disorder region always carry maximum confidence) are eliminated further by applying the decision rule $R2$, shown in Fig. 15.

It reveals the strength of the uncertainty measure for conclusion construction. The decision rule associated with $R2$ will be much acceptable conclusion simply when the index value $Ind\ A_{mo}$ crosses a tuned threshold named *possibilistic confidence threshold*. Applying Rules $R1$ and $R2$ on the elements of the decision set of segmented bone images, we obtain regions R_{roi} where all pixels have highest possibility degree associated with the elementary decisions. Now we finally concentrate on finding the ROI from the R_{roi} using possibilistic pixel seed extraction procedure.

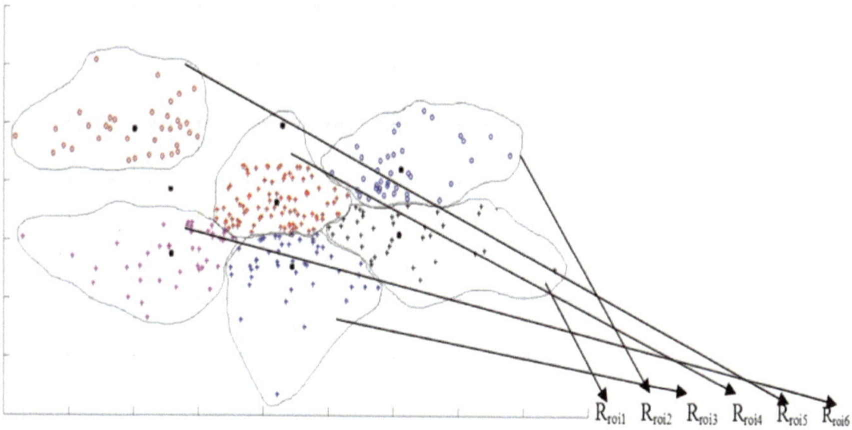

Fig. 16 Possible regions

Possibilistic Pixel Seed Extraction

In this paper using possibility distribution semantic knowledge has been extracted from the regions R_{roi} by analyzing the pixels called possibilistic seeds [22]. Each pixel P_0 in region R_{roi} can be compared with spatial possibility distribution [23] which provides a spatial analysis of points of interest, amorphous on the place of M classes say, $Q = \{Cq_1, Cq_2, \ldots, Cq_M\}$.

$$\bar{\pi} P_o = \bar{\pi} P_o(Cq_1), \bar{\pi} P_o(Cq_1), \bar{\pi} P_o(Cq_3), \ldots, \bar{\pi} P_o(Cq_M) \quad (12)$$

where $\bar{\pi} P_o(Cq_M)$ for $m = 1, 2, \ldots, M$ describes the possibility degree of the pixel P_o in different classes. A knowledge base is built by obtaining the possibility map using the spatial possibility distribution $\bar{\pi} P_o$. $I\bar{n}d$ is computed for every class defined by (13).

$$I\bar{n}d(A_M) = \bar{\pi} P_o(Cq_m) - \max_{n \neq m} \bar{\pi} P_o(Cq_n) \quad (13)$$

where $A_m = (Cq_m) = 1, \ldots, M, P_o$ is considered as Cq_m—possibilistic seed pixel if $I\bar{n}d(A_m) \in C_m$, cross threshold $S = 0.5$. Without errors the threshold $S = 0.5$. For low discrimination complexity, threshold $S = 0.5$. P_o was considered as Cq_m (possibilistic seed) if maximum confidence index value, where $I\bar{n}d(A_m) \in C_m$, across the value of $S = 0.5$. We use two operators, $\emptyset < 1_{\text{mean}}$ and I_{Po}, on each possibilistic map, $PM_{\tau,C_m}(P_o)$. Pixel $P_o \approx C_m$—possibilistic seed if and only if both having threshold $S = 0.5$. In Fig. 16, six different regions (R_{roi1}, R_{roi2}, R_{roi3}, R_{roi4}, R_{roi5}, R_{roi6}) obtained from Fig. 16 are shown and marked manually based on the range of gray-level value, different for different regions.

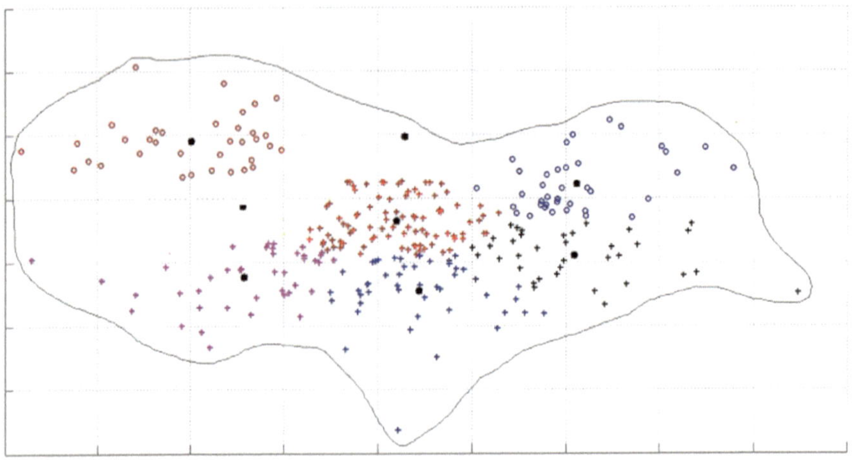

Fig. 17 Possibilistic map $PM_{\tau,C_m}(P_o)$ from R_{roi}

The possibilistic pixel projection considering the possibilistic seed extraction from the image region R_{roi} now required for the distribution of the possibilistic map $PM_{\tau,C_m}(P_o)$. The distribution of the possibilistic map is defined below:

$\pi C_m : D \rightarrow [0, 1]$ and $x(P) \rightarrow \pi C_m(x(P))$, where $D = x(P)$ of the pixel P. For each class C_m, $\pi C_m(x(P))$ associates each pixel $P \in R_{\text{roi}}$, observed through a feature $x(P) \in D$ with a possibility degree of belonging to the class C_m, $m = 1, \ldots, M$. In Fig. 17, possibilistic map is shown constructed manually with the help of regions ($R_{\text{roi1}}, R_{\text{roi2}}, R_{\text{roi3}}, R_{\text{roi4}}, R_{\text{roi5}}, R_{\text{roi6}}$). The pixels in the possibilistic map are used to refine S_1 after applying to Nagao filter.

An input R_{roi}, the application of these M possibility distributions $\pi C_m(\cdot)$, $m = 1, \ldots, M$, on the image R_{roi} called possibilistic maps $PM_{\tau,C_m}(P_o)$ and written as:

$$PM_{\tau,C_m}(P_o) = \pi C_m(x(P(i, j))) \tag{14}$$

Possibilistic Knowledge Acquisition (PKA)

The possibilistic maps help to enrich the current knowledge representation. The knowledge diffusion principle always helps to find out the neighborhood pixel gray value extraction from a specific point pixel gray value. It also helps to segment the image which consists of homogeneous regions. Possibilistic knowledge acquisition step (PKA) is conducted after creating the possibilistic map. Possibilistic knowledge acquisition (PKA) is conducted after the construction of the possibilistic map. In Fig. 18, horizontal and vertical cliques of Nagao filter [24] are shown. These filters are used to refine the image which produces relatively low error rates. Figure 19 shows the S_1 region obtained from the possibilistic map where it shows the zoom of a section of S_1 and other part shows the zoomed image after applying the Nagao filter. The zoom affect is used as the picture is pity big, and it is very difficult to visualize even the changes cannot be shown in a small area according to eye perception.

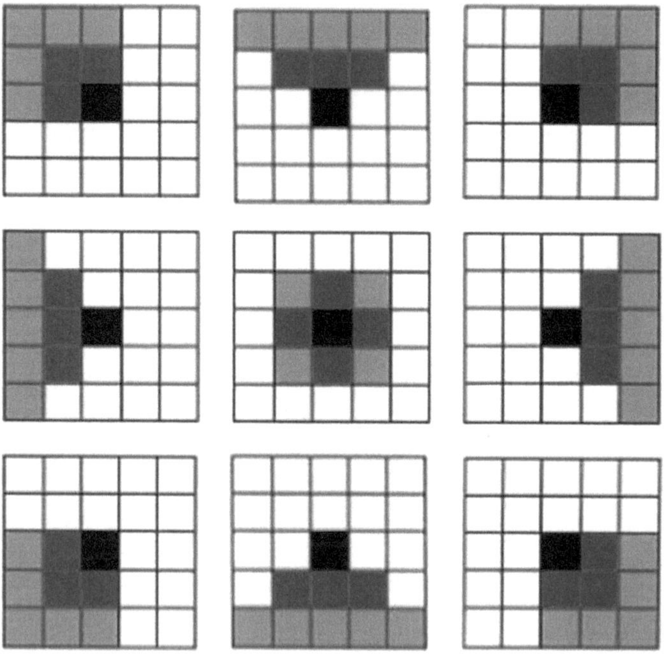

Fig. 18 Nagao filter diagonal, vertical and horizontal cliques $f = \{f_1, f_2, f_3, \ldots, f_9\}$

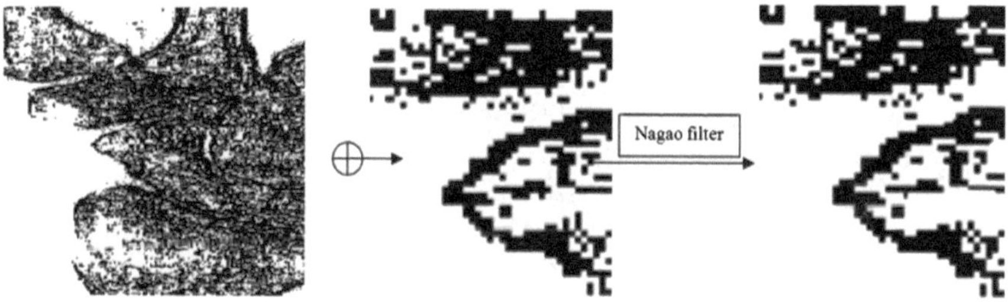

Fig. 19 S_1 region from possibilistic map with zoom in section after applying Nagao filter

Nagao filter has been used for contours. This is constructed as 'cliques' $f = \{f_1, f_2, f_3, \ldots, f_9\}$. For every possibilistic map, $PM_{\tau,C_m}, (P_o)$ is at last in the refining process. Winning clique fg has the maximum possibilistic mean value. Nagao filter at iteration $n + 1$ as: $PM_{\tau,C_m}(P_o, n + 1)$. $x(i, j) = (i, j)^{\text{th}}$ pixel gray value. After the process, we get the final segmented image region which is the prime concern of the doctors to deal with bone disorder. Figure 18 shows different steps of inspecting the ROI created from the segmented bone image S1 and provided to the doctor for Diagnosis.

Six different stages show how the application works after selection manually done by any doctor for precise diagnosis of disorder shown in Fig. 20.

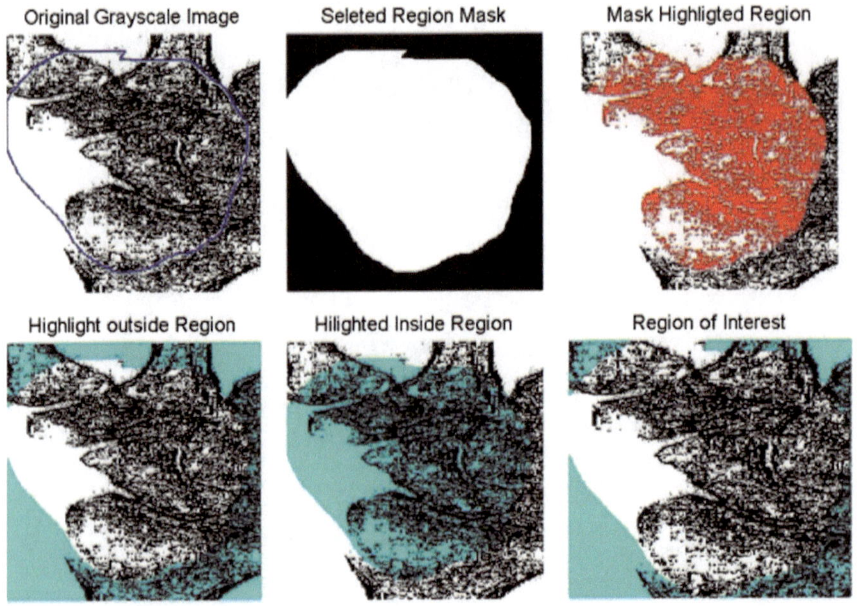

Fig. 20 Step-by-step process of building region of interest

5 Results and Discussion

In the paper, the proposed algorithms are implemented using MATLAB R2010a in a standalone personal computer (Intel Quad Core 2.2 Ghz, processor with 10 GB of 1666 MHz. FSB RAM and 2 GB of DDR5 Accelerated Graphics Port using Windows 10 with 64 Bit Graphical User interface platform), where the execution time is not more than 10 s as calculated while the ROI selection takes less than 5 min for a well-trained physician. Training X-ray images are collected from the Mission Hospital, Durgapur while the test images from different hospitals and internet medical X-ray image databank. The X-ray raw images are converted into JPEG format and scaled 20% of the total image. Each step of the proposed method has been illustrated with six sets of images: (a) collar bone disorder, (b) foot disorder, (c) neck disorder, (d) leg bone disorder, (e) hand and leg foot disorder, and (f) Tersometatarsal disorder, shown in Fig. 21. As the possibility theory part sections have been subdivided into several sections that is why instead of taking nine part of every segmented image of X-ray we have considered a single picture for it. Each and every part of the proposed approach will justify.

- Robustness of our approach in accordance with theoretical approach.
- Accuracy of our approach taken from datasets.
- Clinical assessment of our idea has been shown according to a clinical dataset and simulated datasets.

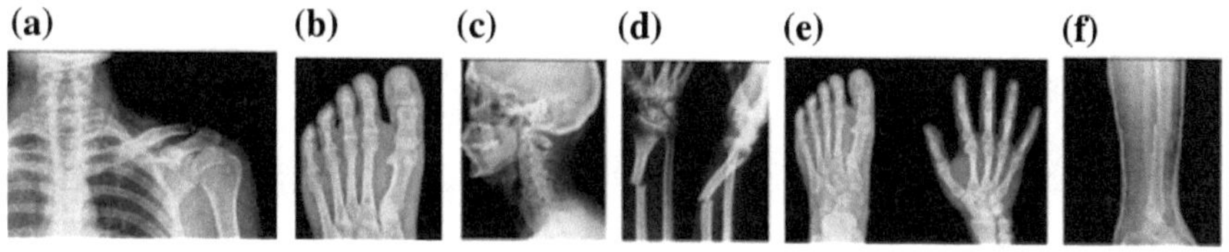

Fig. 21 Set of Input Images, **a** Collar bone disorder, **b** foot disorder, **c** neck disorder, **d** leg bone disorder, **e** hand and leg foot disorder, and **f** Tersometatarsal disorder

Table 1 Calculative values using NFA method

NFA					
Image Name	PSNR	UIQI	EME	PCC	MAE
(a) Collar bone	17.22	0.45	13.21	167034.63	5.44
(b) Feet	17.45	0.3	6.25	142032.21	5.05
(c) Neck	17.12	0.19	7.5	2040233.12	6.02
(d) Leg	16.82	0.22	11.42	336020.38	6.04
(e) Hand and leg	16.55	0.24	5.26	295323.8	6.22
(f) Tersometatarsal	16.48	0.22	3.22	232060.93	6.12

Table 2 Calculative values using Median method

Image name	Median				
	PSNR	UIQI	EME	PCC	MAE
(a) Collar bone	17.12	0.51	13.22	167088.63	6.33
(b) Feet	17.48	0.3	6.23	142092.21	6.23
(c) Neck	17.35	0.19	7.23	2040255.12	6.12
(d) Leg	17	0.22	11.42	336016.38	6.22
(e) Hand and leg	16.98	0.24	5.56	295368.8	6.21
(f) Tersometatarsal	16.96	0.23	3.39	232064.93	6.12

The changes of peak values in the histograms before and after denoising imply elimination of noise which is shown in Fig. 22. Moreover, the images show smoothness in the contrast of the image after denoising. The bone regions are segmented using the proposed Type 2 fuzzy segmentation algorithm to isolate the flesh and cartilage from the bone. Image quality is measured with the help of measuring quality indices, given in Tables 1 2, and 3, respectively. Here, we measure the quality of improvement by peak signal-to-noise ratio (PSNR) [25], universal image quality index (UIQI) [26], enhancement measurement error (EME) [27], Pearson correlation coefficient (PCC) [28], and mean absolute error (MAE) [29]. The comparisons between Type 1 and Type 2 Fuzzy segmentation are given in Fig. 22.

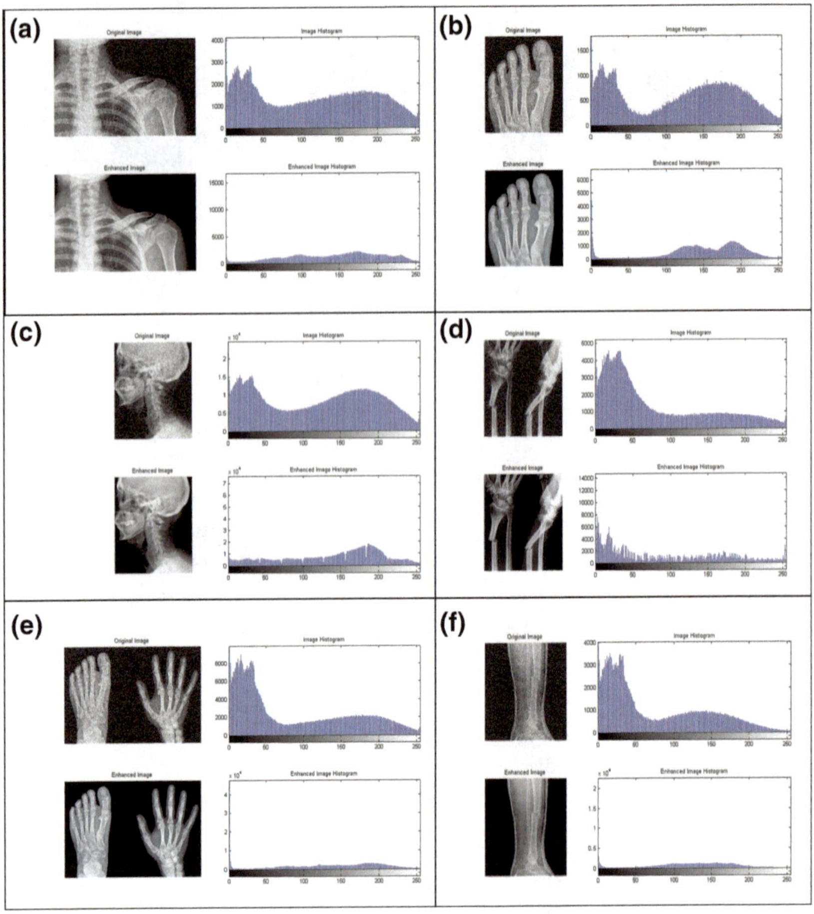

Fig. 22 Histograms of the denoised images after using fuzzy adaptive algorithm

Table 3 Calculative values using proposed method

Image name	Proposed method				
	PSNR	UIQI	EME	PCC	MAE
(a) Collar bone	17.48	0.55	13.05	167083.63	6.46
(b) Feet	17.55	0.31	6.87	142094.21	6.32
(c) Neck	17.35	0.2	7.54	2040258.12	6.36
(d) Leg	17	0.28	11.48	336017.38	6.41
(e) Hand and leg	16.98	0.26	5.52	295371.8	6.36
(f) Tersometatarsal	16.96	0.24	3.84	232062.93	6.42

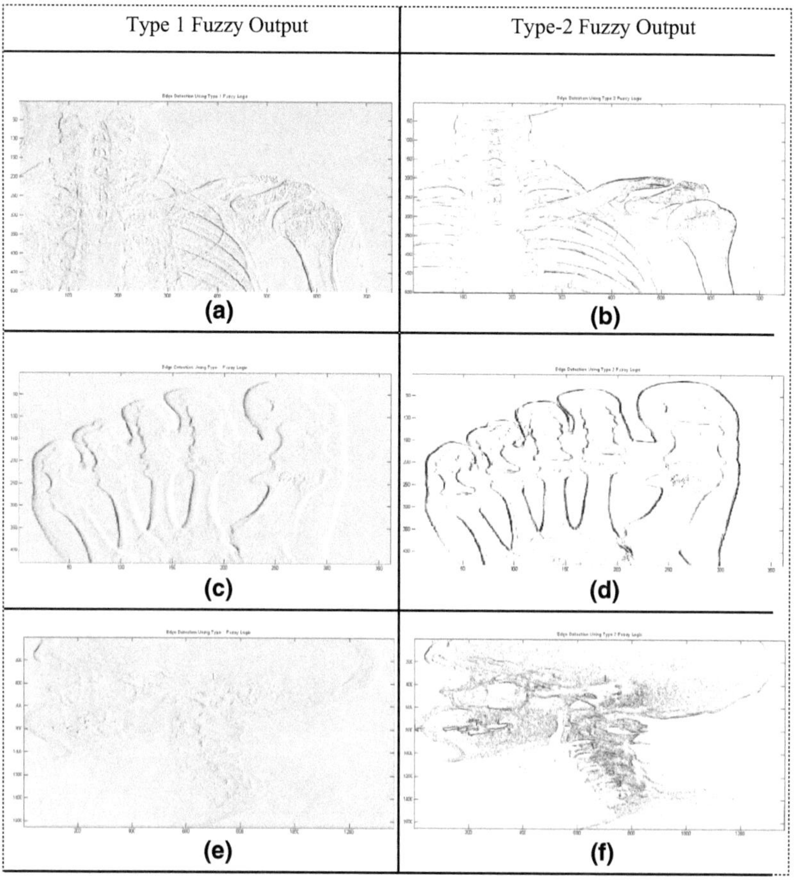

Fig. 23 Segmented bone regions

Type 2 image is more prominent, and edges are well marked than that of Type 1 images which ease to eliminate the flesh and cartilage from a bone. Segmented bone regions for six sets of images are shown in Fig. 22. Figure 23 shows the classification rate for the use of Prewitt operator [30] for the same six sets of images. The upper area smoothness shows the uniformity of the contrast in the image whereas the other color signifies the contrast distribution of the image after using Prewitt operator (Fig. 24).

Figure 25 is the possibilistic projection map (PPM), the less possible points are denoted by color cyan and yellow which has been eliminated from the PPM.

The possibilistic density of the image consists of actual problem shown in Fig. 26.

After building the PPM, Nagao filter is applied for eliminating the unnecessary points which are not part of the bone region. The final image after filtering has been shown on which a doctor can easily mark the region for diagnosis. Comparative outputs of other filters are also shown in Fig. 27 with OTSU are FCM for six sets of images.

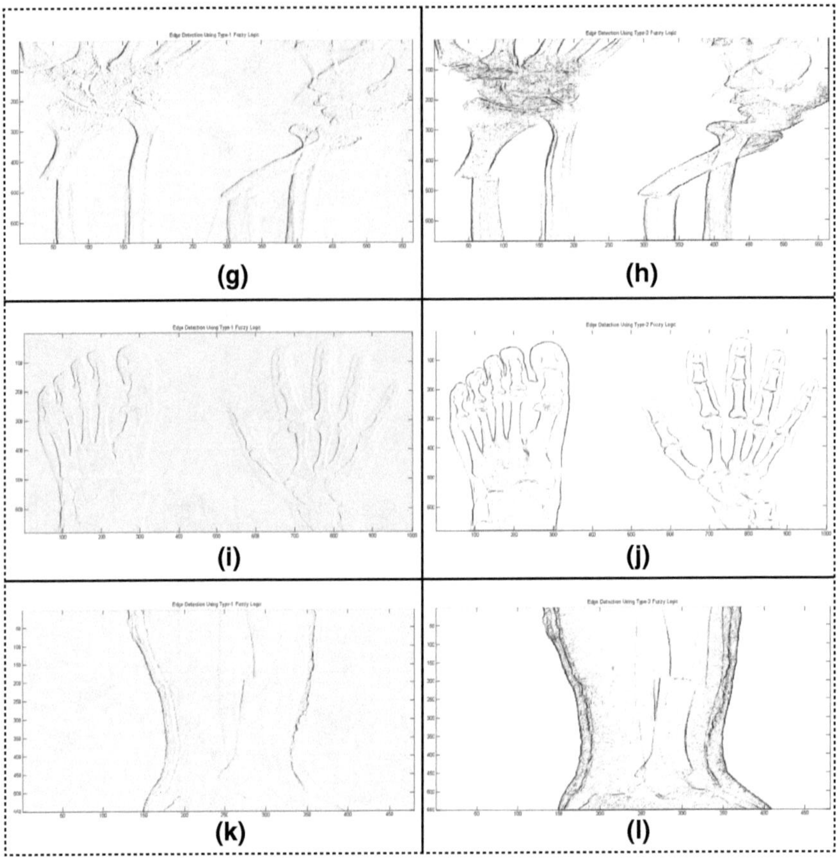

Fig. 23 (continued)

We develop the method using possibility theory is quite helpful for the doctors to take certain decisions. The robustness of our approach is quite good with respect to other manual uncertain approaches. In the result section for different evaluation indexes like PSNR, MSE, UIQI, EME, PCC, and MAE has been calculated and compared for different image sets. Comparison with different approaches, such as NFA, Median has been demonstrated of our idea. The idea behind the paper is of interest for a new approach of possibility theory. The marking of the region has been programmed, and it acts as a GUI interface for the doctor where doctor can able to select the region of interest by clicking and dragging of mouse. The clear region has finally shown and marked as final ROI as shown in Fig. 28.

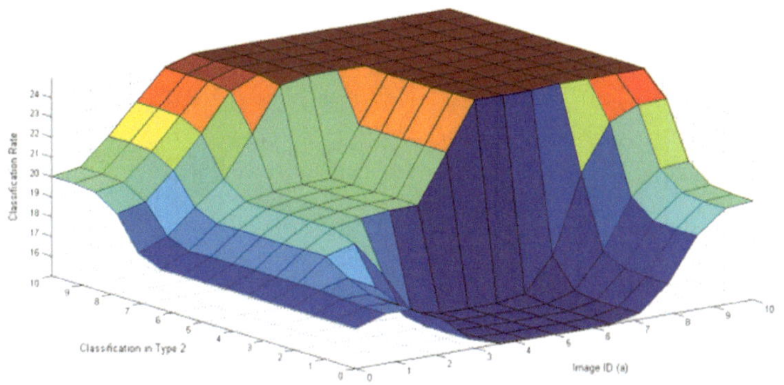

(a) Classification rate using Prewitt Operator on Image set

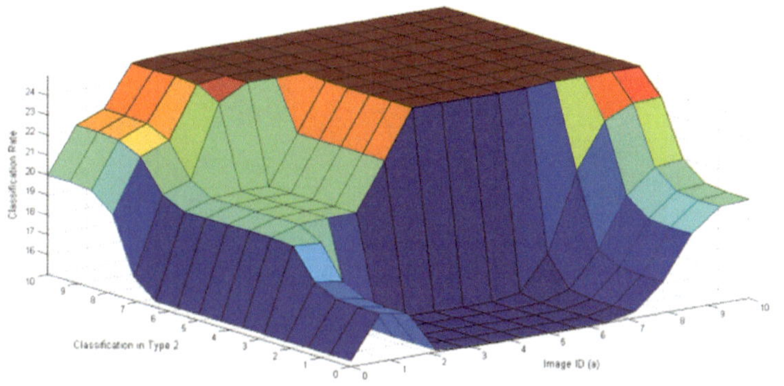

(b) Classification rate using Prewitt Operator on Image set

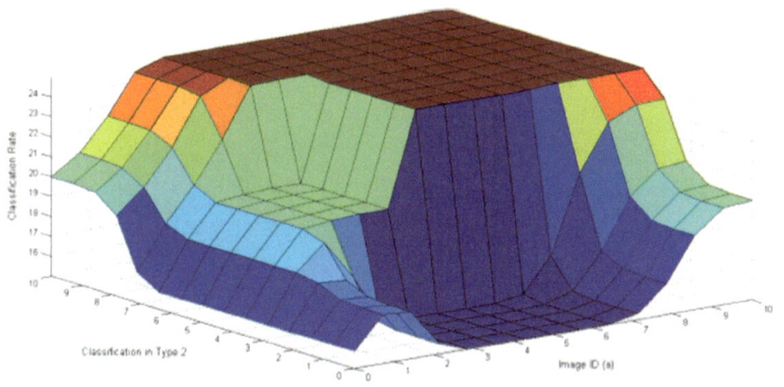

(c) Classification rate using Prewitt Operator on Image set

Fig. 24 Classification rate for the use of Prewitt operator [31]

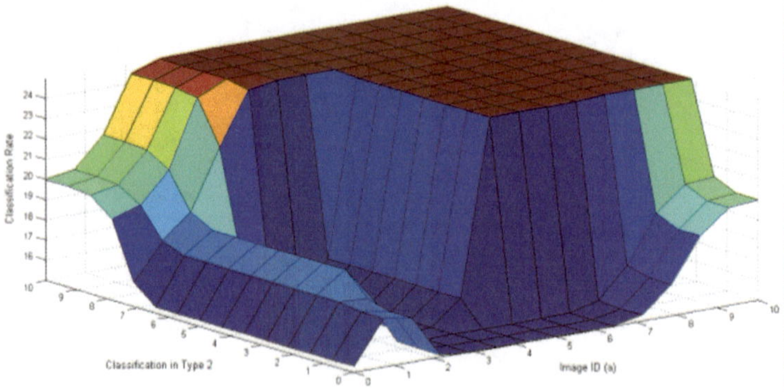

(d) Classification rate using Prewitt Operator on Image set

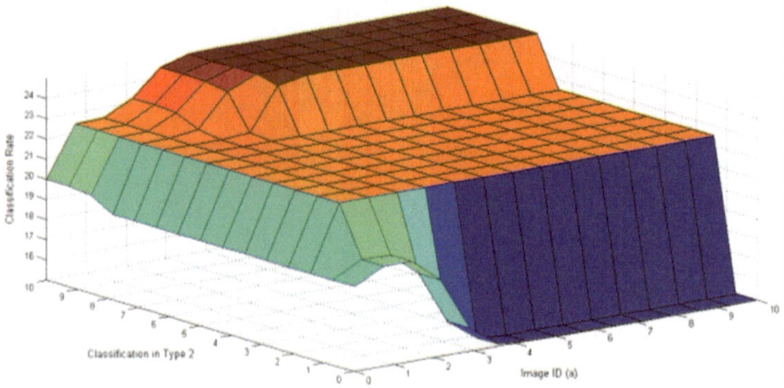

(e) Classification rate using Prewitt Operator on Image set

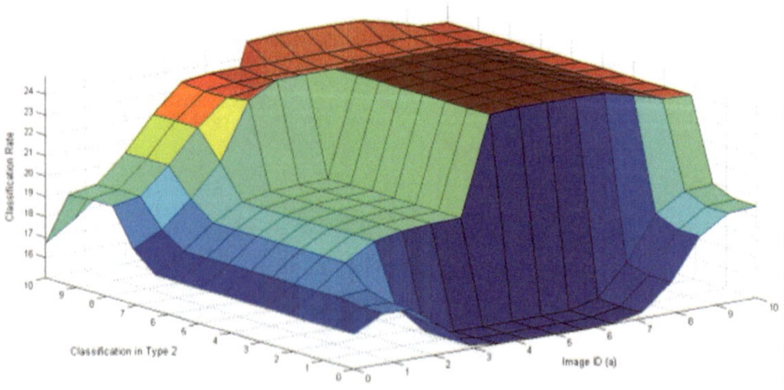

(f) Classification rate using Prewitt Operator on Image set

Fig. 24 (continued)

Enhanced Contrast Image

Possibilistic Map

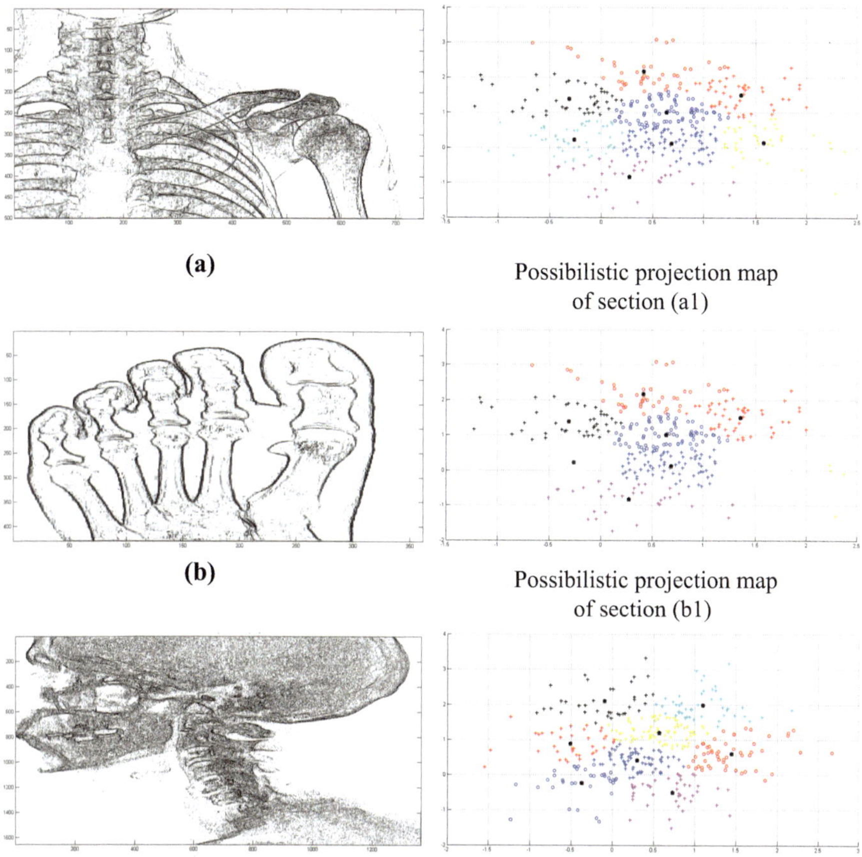

(a)

Possibilistic projection map of section (a1)

(b)

Possibilistic projection map of section (b1)

(c)

Possibilistic projection map of section (c1)

Fig. 25 Possibilistic projection map formation

6 Conclusion

The fuzzy and possibility theories are the most relative choice in the proposed work where we found very less deficiency related to medical diagnosis. In order to finding out ROI from X-ray image ROI and modeling uncertainty, we use possibilistic theory in the urbanized system. The proposed system provides enhanced competence and presentation over the conventional processing system. Starting from segmentation part to detection of ROI shows the effectiveness of the proposed method. Performance has been measured for each step. The proposals are sufficient enough to confirm the stoutness of the proposed system and minimize diverse errors occurred in imperfect

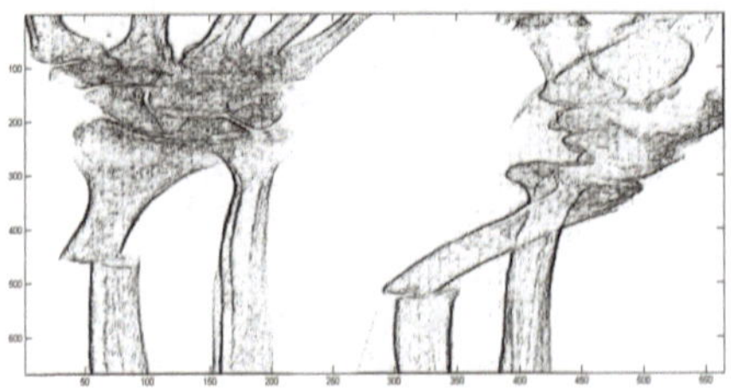

(d)

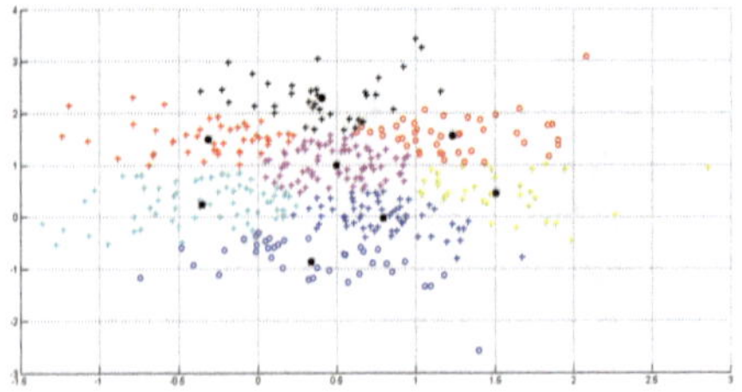

Possibilistic projection map of section (d1)

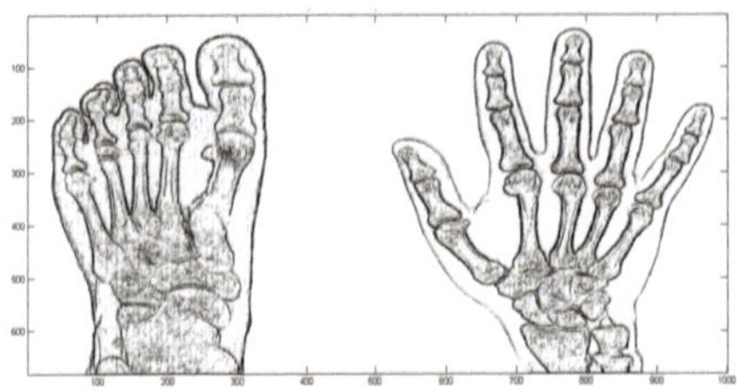

(e)

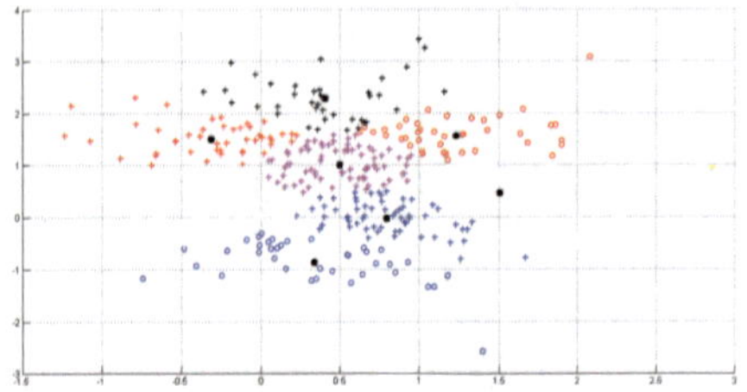

Possibilistic projection map of section (e1)

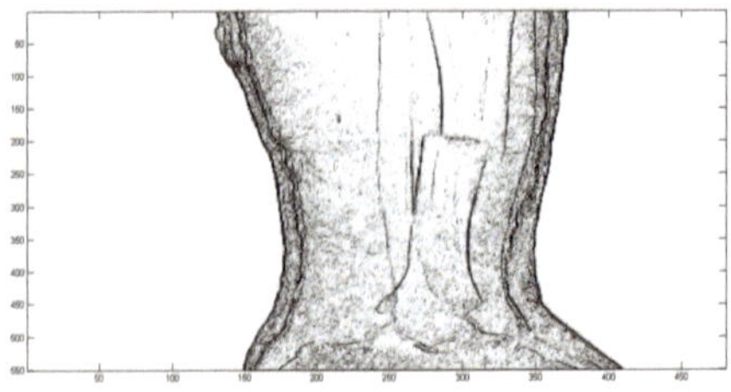

(f)

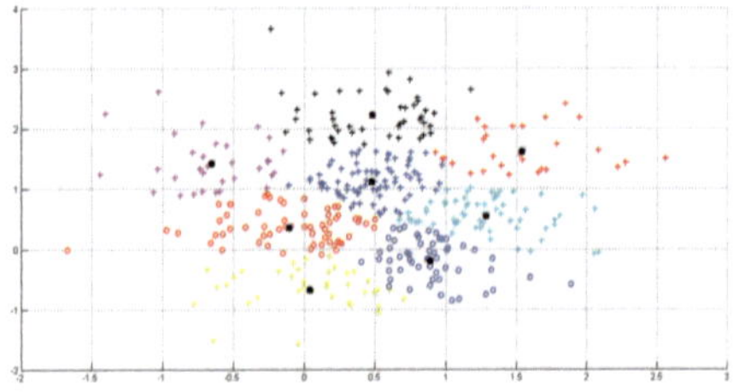

Possibilistic projection map of section (f1)

Fig. 25 (continued)

X-ray images. In future, we will try the proposed system to verify the other medical images like Functional Magnetic Resonance of Image (fMRI) and Positron Emission Tomography (PET) which are very widely used and most acceptable test image in the field of medical diagnosis. The GUI interface is supposed to be a very useful tool for any doctor to select ROI manually which can also be implemented in other medical images too . Also, we can use this system with other probability-possibility

Fig. 26 Possibility density graphs for six sets of images after building ROI

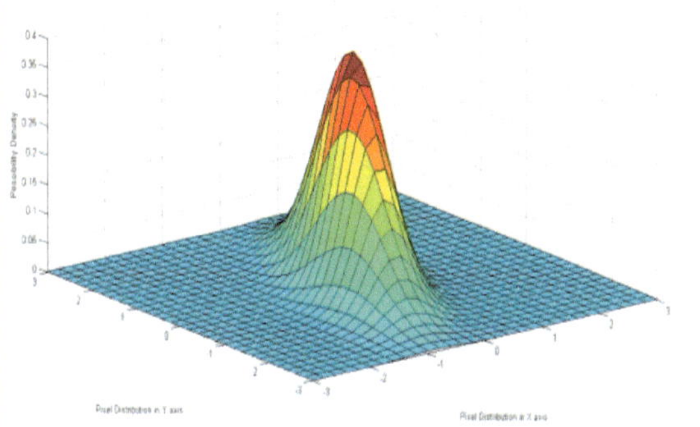

(a) Possibility Density graph for

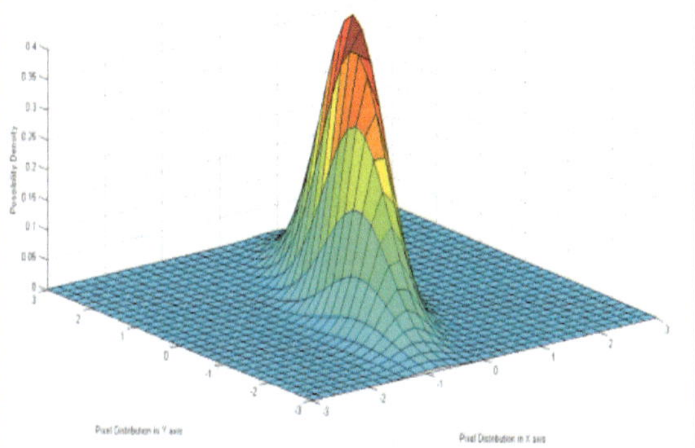

(b) Possibility Density graph for

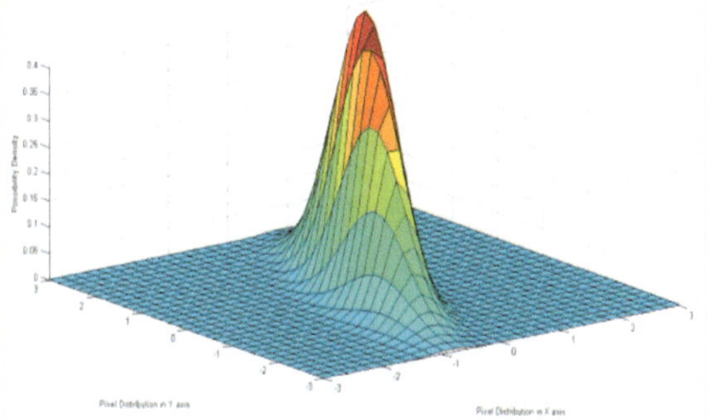

(c) Possibility Density graph for

Fig. 26 (continued)

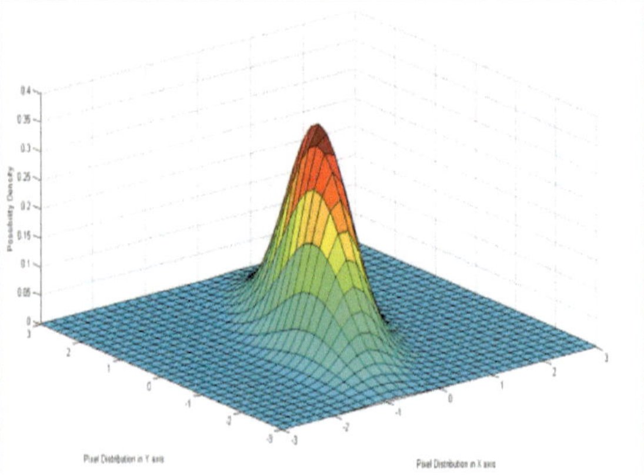

(d) Possibility Density graph for

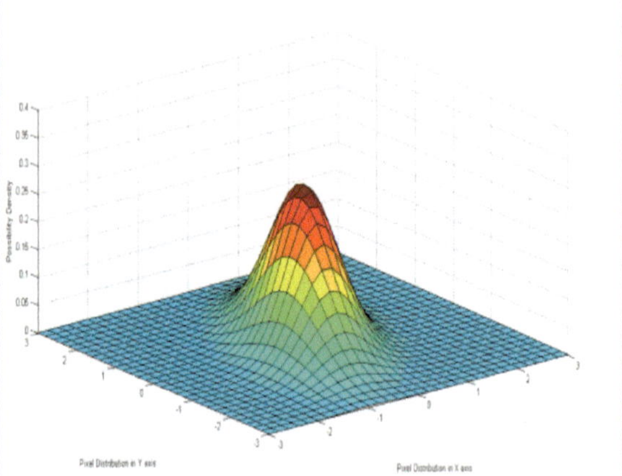

(e) Possibility Density graph for

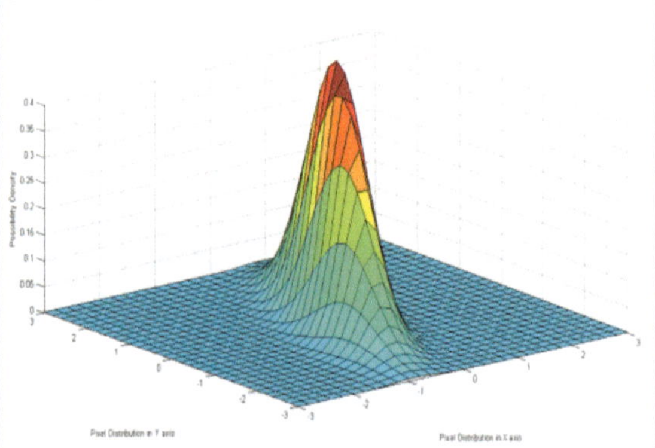

(f) Possibility Density graph for

Fig. 27 Comparison of Nagao filter with other filters

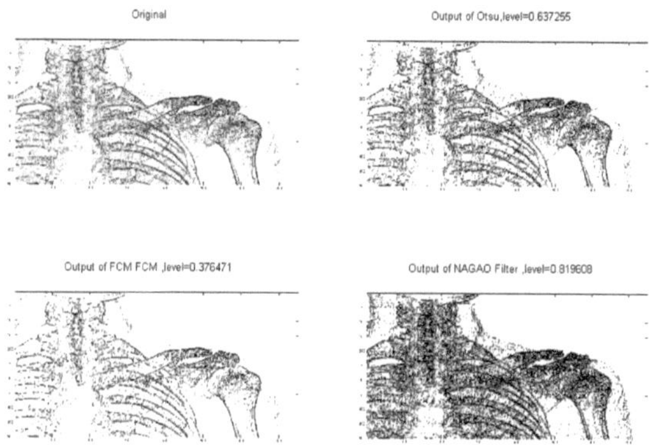

(a) Different Filter output comparative study for

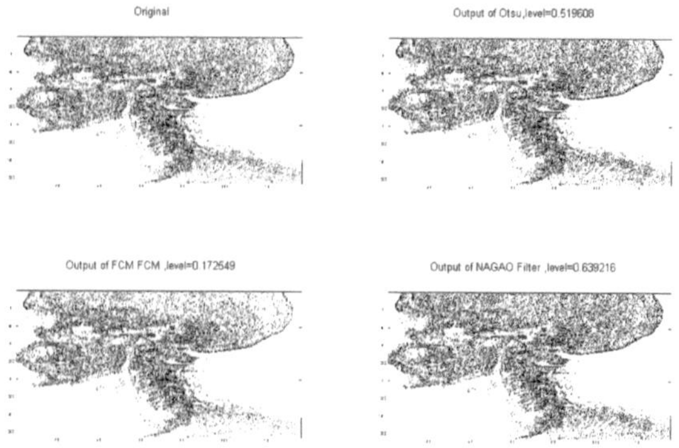

(c) Different Filter output comparative study for

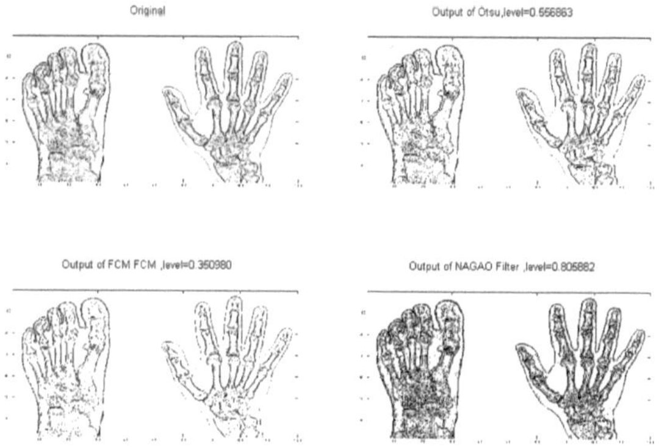

(e) Different Filter output comparative study for

Fig. 27 (continued)

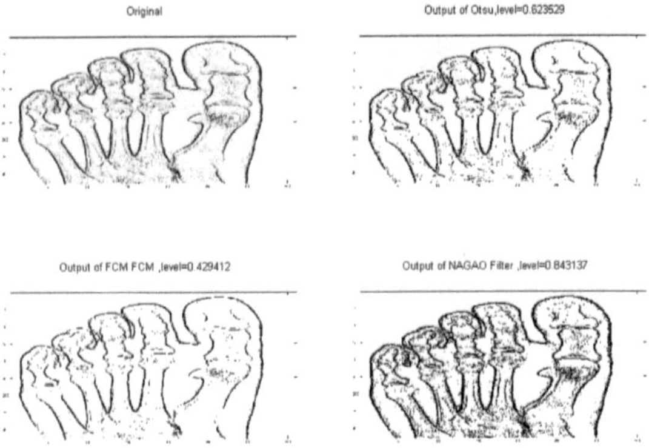

(b) Different Filter output comparative study for

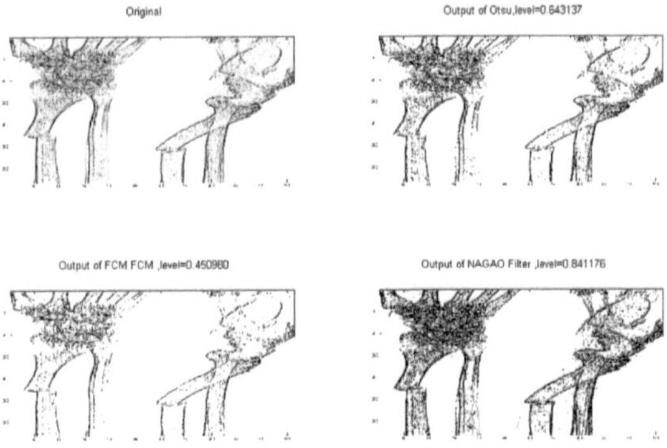

(d) Different Filter output comparative study for

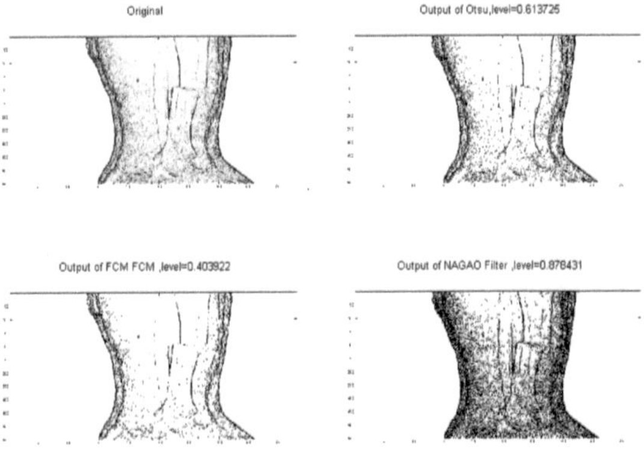

(f) Different Filter output comparative study for

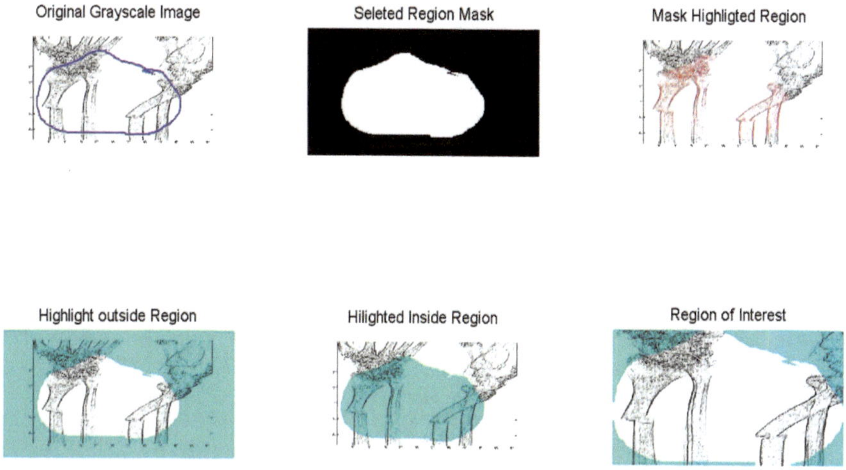

Fig. 28 Marked ROI by doctor

transformation for very small size of PET images too.

Acknowledgements I would like to thank the management of Mission Hospital for giving me the opportunity to work and share the digital images for my research work. I also would like to thank Dr. Jaya Sil of IIEST for her continuous guidance and support.

References

1. Bustince, H., Burillo, P.: Mathematical analysis of interval-valued fuzzy relations: application to approximate reasoning. Fuzzy Sets Syst. **113**, 205–219 (2000)
2. Chaneau, J.L., Gunaratne, M., Altschaeffl, A.G.: An application of type-2 sets to decision making in engineering. In: Bezdek, J. (ed.) Analysis of Fuzzy Information, vol. II: Artificial Intelligence and Decision Systems. CRC Press, Boca Raton, FL (1987)
3. Bounhas, M., Mellouli, K., Prade, H., Serrurier, M.: From Bayesian classifiers to possibilistic classifiers for numerical data. In: Deshpande, A., Hunter, A. (eds.) SUM 2010, LNAI 6379, pp. 112–125, Springer, Heidelberg (2010)
4. Dempster, A.P.: Upper and lower probabilities induced by a multivalued mapping. Ann. Math. Stat. **38**(2), 325–339 (1967)
5. Operations in a fuzzy-valued logic. Inform. Control **43**, 224–240 (1979)
6. Fuzzy Sets and Systems: Theory and Applications. Academic, New York (1980)
7. Dubois, D., Prade, H.: Possibility theory as a basis for preference propagation in automated reasoning. In: Proceedings of the IEEE International Conference on Fuzzy Systems, pp. 821–832. San Diego (1992)
8. Dubois, D., Prade, H.: On possibility/probability transformations. Fuzzy Logic, 103–112 (1993)
9. Izumi, K., Tanaka, H., Asai, K.: Resolution of composite fuzzy relational equations of type 2. Trans. Inst. Electr. Common. Engineers Japan, pt. D **J66D**, 1107–1113 (1983) (in Japanese)
10. John, R.I.: Type-2 inferencing and community transport scheduling. In: Proceedings of the 4th European Congress Intelligent Techniques Soft Computing, pp. 1369–1372. Aachen, Germany (1996)

11. Farbiz, F., Menhaj, M.B., Motamedi, S.A.: Edge preserving image filtering based on fuzz logic. In: Proceedings of the 6th EUFIT Conference, pp. 1417–1421 (1998)
12. Kwan, H.K., Cai, Y.: Fuzzy filters for image filtering. In: Proceedings of Circuits and Systems (MWSCAS-2002) (2002)
13. Tolt, G., Kalaykov, I.: Fuzzy-similarity-based image noise cancellation. In: Lecture Notes in Computer Science, vol. 2275, pp. 408–413 (2002)
14. Balster, E.J., Zheng, Y.F., Ewing, R.L.: Feature-based wavelet shrinkage algorithm for image denoising. IEEE Trans. Image Proc. **14**(3), 2024–2039 (2005)
15. Type-2 fuzzy logic systems: Type-reduction. In: Proceedings of the IEEE Conference on Systems, Man and Cybernetics, pp. 2046–2051 (1998)
16. Vertan, C., Buzuloiu, V.: Fuzzy nonlinear filtering of Colour images. In: Fuzzy Techniques in Image Processing (Vol. 52 of Studies in Fuzz. and Soft Comp.), pp. 248–264 (2000)
17. Vermandel, M., Betrouni, N., Taschner, C., Vasseur, C., Rousseau, J.: From MIP image to MRA segmentation using fuzzy set theory. Comput. Med. Imag. Graph. **31**, 128–140 (2007)
18. Xu, H., Zhu, G., Peng, H., Wang, D.: Adaptive fuzzy switching filter for images corrupted by impulse noise. In: Pattern Recognition Letters, vol. 25, pp. 1657–1663 (2004)
19. Schulte, S., De Witte, V., Nachtegael, M., Mélange, T., Kerre, E.E.: A new fuzzy additive noise reduction method. In: Lecture Notes in Computer Science, vol. 4633 (Proc. of ICIAR 2007), pp. 12–23 (2007)
20. Lukac, R.: Adaptive vector median filtering. In: Pattern Recognition Letters, vol. 24, pp. 1889–1899 (2003)
21. Romberg, J.K., Choi, H., Baraniuk, R.G.: Bayesian tree-structured image modelling using wavelet-domain hidden Markov models. IEEE Trans. Image Proc. **10**(7), 1056–1068 (2001)
22. Dabov, K., Foi, A., Katkovnik, V., Egiazarian, K.: Image denoising with block-matching and 3D filtering. In: Proceedings of SPIE Electronic Imaging, pp. 354–365 (2006)
23. Jentzen, W., Freudenberg, L., Eising, E.G., Heinze, M., Brandau, W., Bockisch, A.: Segmentation of PET volumes by iterative image thresholding. J. Nucl. Med. **48**(1), 108–114 (2007)
24. Alsahwa, B., Solaiman, B., Almouahed, S., Bossé, É., Guériot, D.: Iterative refinement of possibility distributions by learning for pixel-based classification. IEEE Trans. Image Process. **25**(8), 3533–3545
25. Zadeh, L.A.: Fuzzy sets as the basis for a theory of possibility. Fuzzy Sets Syst. **1**, 3–28 (1978)
26. Maity, S., Sil, J.: Feature extraction of bone scintigraphy for diagnosis of disease—CSNT 2012, IEEE Explore, pp. 179–183. https://doi.org/10.1109/csnt.2012.46
27. Bezdek, J.C.: Pattern recognition with fuzzy objective function algorithms. Plenum Press, New York (1981)
28. Vial, S., Gibon, D., Vasseur, C., Rousseau, J.: Volume delineation by fusion of fuzzy sets obtained from multiplanar tomographic images. IEEE Trans. Med. Imag. **20**, 1362–1372 (2001)
29. Schaefer, A., Kremp, S., Hellwig, D., Rube, C., Kirsch, C.M., Nestle, U.: A contrast-oriented algorithm for FDG-PET-based delineation of tumour volumes for the radiotherapy of lung cancer: derivation from phantom measurements and validation in patient data. Eur. J. Nucl. Med. Mol. Imag. **35**(11), 1989–1999 (2008)
30. Zadeh, L.A.: Fuzzy sets. Inf. Control **8**, 338–353 (1965)
31. Christensen, G.E., Johnson, H.J.: Consistent image registration. IEEE Trans. Med. Imag. **20**(7), 568–582 (2001)

Segmentation of Anomalies in Abdomen CT Images by Convolution Neural Network and Classification by Fuzzy Support Vector Machine

S. N. Kumar, A. Lenin Fred, H. Ajay Kumar, P. Sebastin Varghese and Salga Ann Jacob

Abstract The application of computer-aided algorithms for disease diagnosis and treatment is gaining prominent in the past years, and the role of machine learning algorithms is inevitable. This chapter focuses on the segmentation of liver and anomalies like tumor and cyst from abdomen CT images using deep learning convolution neural network (DLCNN) and classification of tumor stages by fuzzy support vector machine (FSVM). The segmentation result of DLCNN outperforms the backpropagation neural network, group method data handling neural network, and decision tree algorithm. The FSVM-based tumor classification results were superior when compared with classical SVM. This chapter focuses on the following: (i) machine learning algorithms for classification and segmentation of medical images, (ii) role of DLCNN in medical image segmentation, (iii) role of FSVM in anomalies classification, and (iv) validation of segmentation and classification results by performance metrics. The simulation results are generated in Matlab 2015a and Java and validated on real-time abdomen CT images.

S. N. Kumar · H. Ajay Kumar
School of Electronics and Communication Engineering, Mar Ephraem College of Engineering and Technology, Elavuvilai, Kanyakumari, India
e-mail: appu123kumar@gmail.com

H. Ajay Kumar
e-mail: ajayhakkumar@gmail.com

A. Lenin Fred (✉)
School of Computer Science and Engineering, Mar Ephraem College of Engineering and Technology, Elavuvilai, Kanyakumari, India
e-mail: leninfred.a@gmail.com

P. Sebastin Varghese
Metro Scans and Research Laboratory, Trivandrum, India
e-mail: sebastin464@gmail.com

S. A. Jacob
Department of Electronics and Communication Engineering, Mar Baselios College of Engineering and Technology, Nalanchira, Thiruvananthapuram, India
e-mail: annjacob34@gmail.com

S. Bhattacharyya et al. (eds.), *Hybrid Machine Intelligence for Medical Image Analysis*, Studies in Computational Intelligence 841,
https://doi.org/10.1007/978-981-13-8930-6_7

Keywords Segmentation · Machine learning · Convolution neural network · Support vector machine · Fuzzy support vector machine

1 Introduction

Artificial intelligence refers to the usage of computers for performing an application in a smart manner like humans, analysis of data, and taking decisions. Machine learning is termed as a subset of artificial intelligence that uses statistical measures for improving accuracy in the decision making. Deep learning is termed as the subset of machine learning that uses neural network; automatic extraction of features makes it an efficient one while comparing with the classical approaches that require a feature extraction algorithm [1]. Machine learning algorithms are further classified into three categories: supervised, unsupervised, and reinforced. Supervised learning relies on the training of data; unsupervised learning discovers solutions based on similarity or difference with no labeled data, and reinforced learning depends on the feedback or rewards.

The artificial neural network (ANN) gains more importance in computer vision and pattern recognition for many real-time applications [2]. The two main phases of neural network architecture are training and testing. The classical neural network architectures heavily depend on feature extraction algorithms. The widely used feature extraction techniques are Haralick, GLCM, LBP, STFT, wavelet, etc. [3, 4]. Based on topology, neural networks are classified into the feedforward neural networks and recurrent neural networks [5, 6]. The feedforward neural network (FFNN) has no feedback loops, and the output does not depend upon the past input. Some of the widely used FFNNs in medical image processing are multilayer perceptron (MLP), self-organized map (SOM), and radial basic function (RBF) algorithms. In general, FFNNs are classified into a single-layer and multilayer neural networks [5]. The feedback neural network is also called as recurrent neural network (RNN). It is a typical case of FFNNs with one or more hidden layer and at least a feedback loop. Some of the widely used RNNs are Hopfield, cellular, convolution, and pulse-coupled neural networks [7].

The supervised classifiers [8] like logistic regression, support vector machine, decision trees, and unsupervised classifiers [9] like *K*-means, mean shift, and fuzzy *C* means (FCM) clustering also gain importance in medical image processing for the analysis of anomalies. In this work, the segmentation of liver and tumor from abdomen CT images was performed by DLCNN, and tumor classification was performed by FSVM. Section 2 discusses the widely used machine learning algorithms for anatomical organ extraction and abnormalities classification; the inferences about the machine learning algorithms are also presented. Section 3 describes the role of DLCNN in the medical image segmentation, and Sect. 4 describes FSVM for the classification of tumor stages. Section 5 highlights the results and discussion with performance analysis, and conclusion is highlighted in Sect. 6.

2 Related Works in Machine Learning for Segmentation and Classification of Medical Images

The ANN is basically a data processing architecture that mimics the operation of the human brain. The basic building block of ANN is neuron, and the learning of ANN is similar to the humans' learning from examples. The important aspect in the supervised algorithm is training, and a set of standard data is used for training and analyzed in the testing phase.

2.1 Learning Vector Quantization

The learning vector quantization (LVQ) architecture comprises two layers, competitive layer and linear layer, depicted in Fig. 1a. The learning takes place in competitive layer for input vector classification. The linear layer transfers the classes generated by the competitive layer into target classes [10]. The competitive layer classes are termed as subclasses, and the linear layer classes are termed as target classes. Subclasses are always larger than the target classes [11]. The input vector is accommodated into a class based on the smallest value of Euclidean distance (smallest Euclidean distance value indicates the degree of similarity between the code vector and the input vector).

2.2 Multilayer Perceptron

The multilayer perceptron (MLP) is a supervised neural network comprising of the input layer, hidden layer, and output layer; the training is done by error backpropagation [12]. In MLP, the selection of hidden layer neurons is crucial, and the less number of neurons decreases the complexity of the network and prevents overfitting of the network. MLP is widely used for pattern classification [13]. A typical MLP is depicted in Fig. 1b. The number of neurons in the input layer depends on the features extracted, and the neurons in the hidden layer are 50% of the neurons in the input layer. During the training phase, the error value is computed by comparing with the actual output, and it is backpropagated through the network. The backpropagation algorithm adjusts the weight values until the error value is minimized. The local minima problem in classical MLP was solved by optimized MLP that uses dynamic learning rate; performance metrics reveal that CAD system aids the physicians for the early detection of breast cancer [14].

The MLP when coupled with particle swarm optimization gains its importance in the handling of medical data for decision making related to high-risk pregnancy [15]. The machine learning algorithm finds its application in image reconstruction; MLP was proposed for reconstruction of the subsampled data using k-space subsampling patterns with a drastic decrease in aliasing artifacts [16]. The watershed segmenta-

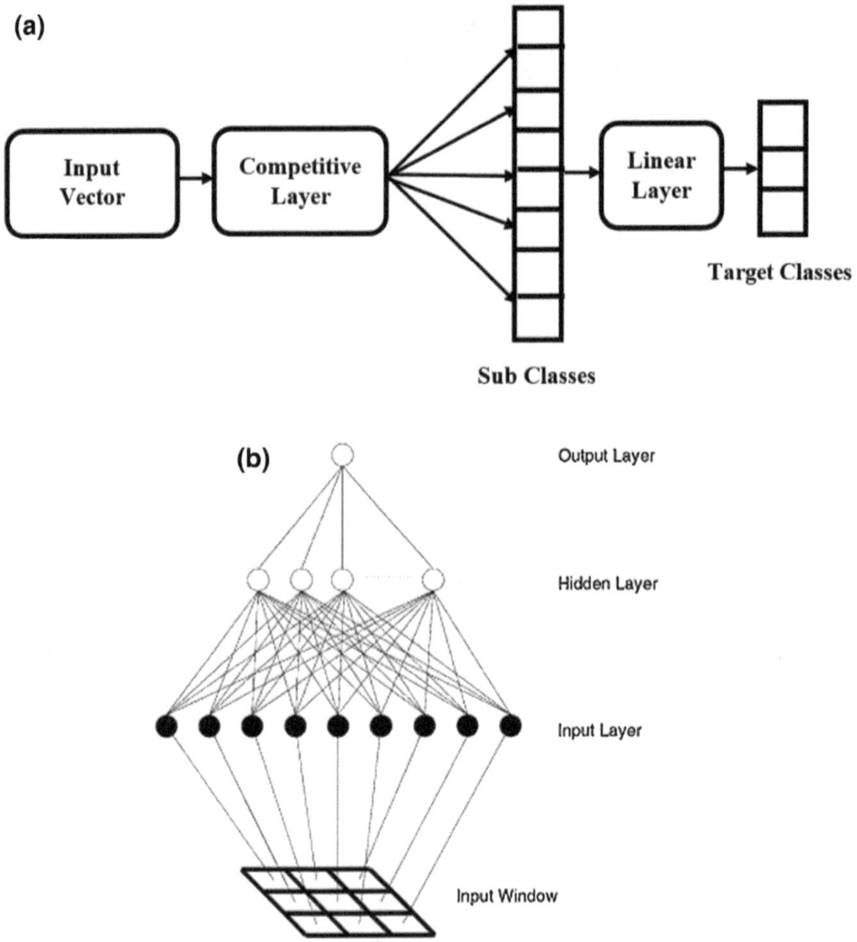

Fig. 1 **a** LVQ architecture and **b** BPNN architecture

tion with MLP generates good results for brain stroke classification in MR images [17]. The MLP along with an active contour model generates efficient result for segmentation of anatomical structures in MR images. The left and right lung boundaries were clearly traced in MR images of the torso with good precision, recall, and effectiveness rate [18]. The MLP was trained with the liver segmentation result obtained from FCM with morphological filtering algorithm. The trained MLP was used for the further segmentation of adjacent slices; successful results were produced in spite of the gray-level matching of adjacent organs and tumors [19].

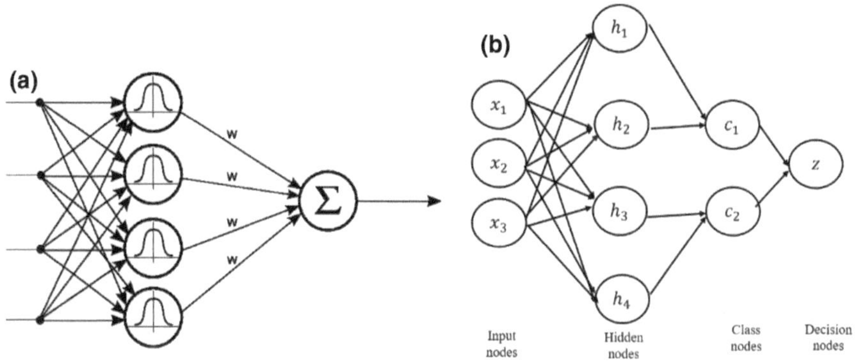

Fig. 2 **a** RBF neural network and **b** probabilistic neural network

2.3 Radial Basis Function Neural Network

The output of each node in radial basis function neural network (RBFNN) is determined by enforcing a confidence measure to all the hidden layer units. A node is selected for the estimation of output in the output layer when a hidden layer neuron unit obeys the confidence measure. The generalized RBFNN is depicted in Fig. 2a.

The computation time of RBFNN is minimized and has been successfully applied to a variety of visual processing and analysis problems. The RBFNN comprises three categories of neurons; input layer neurons feed the input data to the hidden layer. The hidden layer comprises nonlinear neurons having Gaussian transfer function, and a linear neuron determines weighted sum at the output layer.

The brain tumor classification in MR images was done by RBFNN and BPNN, and GLCM was used for feature extraction. The ROC characteristics and convergence rate indicate that RBFNN is having efficient accuracy than BPNN [20]. The self-generating hierarchical RBFNN gains its importance in transfer function initialization for volume rendering in CT/MR images. The complexity is reduced in transfer function initialization and fully automates the volume rendering process [21]. The RBFNN was coupled with the FCM algorithm for the segmentation of brain tumor in MR images. The dice coefficient indicates the efficiency of the hybrid algorithm (FCM-RBFNN) when compared with variants of the FCM algorithm that are sensitive to noise [22]. The RBFNN along with FCM was used for the classification of pulmonary fibrosis on microscopic images of lungs. Better results are produced when compared with the SVM classifier [23].

2.4 Probabilistic Neural Network

The probabilistic neural network (PNN) is a FFNN, and weighted PNN generates efficient results than classical PNN. In the first layer, the Euclidean distance is determined between the input vector and training input vector, and it is an indication of closeness to the training vector. Each input vector is correlated with one of the *K* classes. The elements are then passed into the second layer that consists of a radial transfer function. The resultant product terms are then passed into the pattern layer or summation layer where the grouping takes place into classes based on the training. The PNN classifier has good accuracy, less training requirement and small retraining period, and highly robust to weight changes. The classical PNN is depicted in Fig. 2b.

The simple network architecture and fast learning speed make PNN attractive for disorder analysis; the PNN when coupled with fast discrete curvelet transform was proposed for the automatic brain anomalies detection [24]. In [25], authors performed a comparative analysis of classifiers for classification of fatty and cirrhosis liver in CT images. The ROC analysis reveals that PNN outperforms BPNN and LVQ algorithms. The modified PNN was found to be robust for brain tissue segmentation in MR images. The weighted PNN takes into account the partial volume effects, and the performance is better than *K*-means, PNN, and expectation maximization algorithms [26]. PNN was found to be effective for brain tumor classification of MR images. Feature extraction was performed by principal component analysis; computation time was low, and accuracy was good when compared with the classical neural networks [27].

2.5 Hopfield Neural Network

The Hopfield neural network (HNN) was introduced by Hopfield (1982) and comprises a set of neurons with connections between each unit. The main feature of HNN is that it is an unsupervised neural network with symmetrical weights. With respect to image segmentation, the image is depicted as a 2D network comprising a grid of $N \times M$ neurons. Each neuron is represented as a point $(\boldsymbol{i}, \boldsymbol{k})$, where $\mathbf{1} \leq \boldsymbol{i} \leq \boldsymbol{N}$ **and** $\mathbf{1} \leq \boldsymbol{k} \leq \boldsymbol{M}$ For organ segmentation, the grid comprising $N \times M$ neurons is depicted in Fig. 3a, and pixels are grouped into respective classes thereby generating a region of interest.

The output of each neuron is related to the current state as follows:

$$V_{ik} = f(U_{ik}) \tag{1}$$

where $f(U_{ik})$ is the maximum evolution function. The state of the neuron is updated as follows:

$$U_{ik} = \sum_{\boldsymbol{j=1}}^{N}\sum_{\boldsymbol{l=1}}^{M} \boldsymbol{W}_{\boldsymbol{ijkl}}\boldsymbol{V}_{\boldsymbol{jl}} + \boldsymbol{I}_{\boldsymbol{ik}} \tag{2}$$

The HNN minimized energy function is expressed as follows:

(a)

Pixel number	Class number					
	1	2	3	4	..	M
1	0	1	0	0	..	0
2	0	0	0	1	..	0
3	0	0	0	1	..	0
..	0	0	0	0	..	0
..	..	..	..	..	..	..
N	1	0	0	0	..	0

(b)

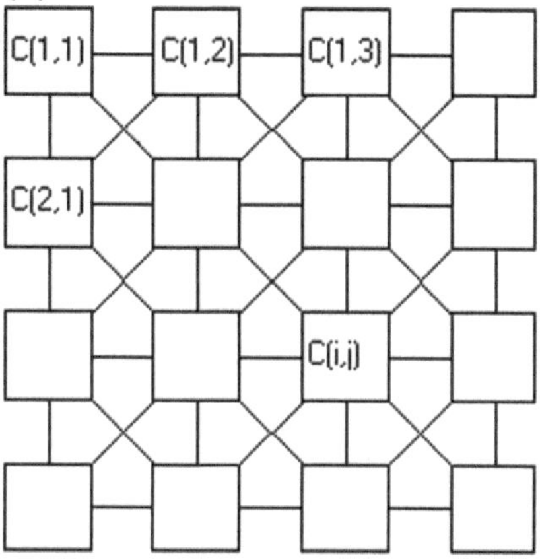

Fig. 3 **a** HNN topology for segmentation and **b** topology of cellular neural network

$$E = \frac{1}{2}\sum_{j=1}^{N}\sum_{k=1}^{M}\sum_{j=1}^{N}\sum_{l=1}^{N} W_{ijkl}V_{ik}I_{jl} - \sum_{i=1}^{N}\sum_{l=1}^{M} V_{ik}I_{ik} \tag{3}$$

2.6 Cellular Neural Network

The cellular neural network (CNN) comprises the spatial arrangement of locally coupled cells, where each cell is a dynamic system and output is depicted as an electrical circuit. There is a local interconnection between each cell, and the processing unit communicates with the cells in the neighborhood within a specific area. The parallelism is highly utilized in the cellular neural network. The topology of a typical cellular neural network is depicted in Fig. 3b. The discrete cellular neural network (DTCNN) with genetic algorithm was an effective tool for the radiologist to locate the small pulmonary nodules in CT images of lungs; it can detect the nodules missed by human intervention [28]. The CNN algorithm was found to be efficient for the automatic detection of hippocampal sclerosis on MR images of the brain; the classification accuracy was good for normal and abnormal tissues [29].

2.7 Convolution Neural Network

It is a multilayer perceptron in which each layer comprises multiple planes and requires little pre-processing of input images. The automatic feature detection and shared weights make it suitable for massive parallelism applications. The typical convolution neural network is depicted in Fig. 4.

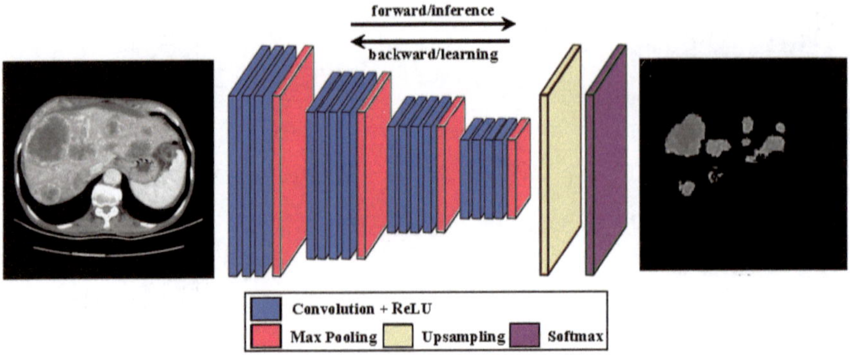

Fig. 4 Convolution neural network

It can be depicted in the form of a directed graph in which nodes represent the pixels and the edges represent the filters. Stochastic gradient descent learning algorithm is used, and the segmentation problem with respect to CNN is to split the nodes of the graph into disjoint sets that minimize the cost function (graph cut: Max Flow/Min Cut). The network will be trained with the input images along with the ground truth generated by the experts. The advanced fuzzy CNN (AFCNN) was able to accurately detect the boundary of the liver in CT images; AFCNN results were superior when compared with fuzzy CNN (FCNN) [30]. The CNN was found to be efficient for the cardiac MR image segmentation, and the proposed deep learning architecture was based on policy gradient reinforcement learning [31].

3 Inferences About Machine Learning Algorithms for Medical Image Segmentation and Classification from the Related Works

The choice of a neural network depends upon the applications, and there is no universal network for all applications. The feedforward networks were found to be efficient in medical image processing when input/output vectors are accurate in training under supervised condition. The RBF networks are effective in the analysis of 3D medical structures; the size and topology are flexible, making them applicable for a variety of real-time problems. The HNN is flexible in setting the weights between neurons to generate desired set of patterns, potentially vital in medical image processing for tumor classification, where the output is derived from partial or similar patterns during training. In pattern recognition and segmentation models, SOM was found to be efficient since it can automatically organize the neuron layers based on the nature of inputs. The feedback neural networks are widely used for medical image segmentation; HNN is an efficient one and needs to minimize an energy function during the convergence of the network.

In image compression, SOM gains its importance in vector quantization due to its topology-preserving nature. The time required for applying the trained neural network to solve a problem is considerably small when compared with the classical image processing techniques. The computation time incurred in the training of the neural network is high for a supervised network. The choice of a neural network for an application is not simple, and in many scenarios, trial–and-error method is adopted. The motivation behind the use of DLCNN and FSVM for segmentation and classifications is as follows. The DLCNN operates by the process of convolution, a point-wise multiplication operation of small matrices of weights across the entire image to create corresponding neural maps. Usually, the size of convolution mask is 3 × 3, 5 × 5, and 7 × 7. The technique of matrix convolution is repeated until the new hidden deep layers of neural maps are generated. The optimization of the neural network by loss estimation and backpropagation are then performed similar to classical ANN. In DLCNN, for feature extraction, there is no need for a specific algorithm.

The gray-level co-occurrence matrix (GLCM), gray-level difference matrix (GLDM), local binary pattern (LBP), and wavelet are some of the widely used feature extraction techniques used along with classical neural network algorithms. The automatic feature extraction takes place in DLCNN, and the extracted features rely on the number of convolution layers. The DLCNN generates superior results in terms of segmentation accuracy when compared with classical neural networks like BPN and GMDH. The parameter tuning is difficult in GMDH, and computation complexity is also high for the training process. The classical SVM was not found to be satisfactory for large datasets for classification problem. In fuzzy SVM, the kernel function is based on the fuzzy membership function that is integrated with the SVM classifier for the attainments of better accuracy (Table 1).

Table 1 Neural network architecture characteristics

S. No.	Neural network architectures	Features
1	MLP	Supervised learning and it is a part of general regression testing (GRT) regression modules. Applicable for classification and regression for any *N*-dimensional signal. Nonlinear mapping: mapping of *N*-dimensional input signal into *M*-dimensional output signal. Possibility to get stuck in local minima and convergence rate is slow. The setting of number of neurons in the hidden layer is crucial, low value will cause underfitting, and high value will cause overfitting in the network
2	BPNN	Three-layer network: input, hidden, and output layers; the number of neurons in hidden layer depends upon the complexity of the application. Learning takes place through a two-stage process: forward propagation and backpropagation of error. It is applicable for linear as well as nonlinear classification; even multiclass classification is possible. The training period is high and suffers from the local minima problem

(continued)

Table 1 (continued)

S. No.	Neural network architectures	Features
3	HNN	Unsupervised learning and comprises a single layer of neuron. The classification is done using associative pattern that tries to determine memorized pattern closely matching with input. Too many input patterns will generate spurious outputs, and strong similar patterns can generate an error in the output. The advantage is that in many cases, the data item can be retrieved, even when incomplete or corrupt information is presented
4	SOM	Unsupervised learning and in general, neurons are spatially arranged. Training is done by inputs rather like MLP, where training takes by input output pairs. Simple architecture and easy to train; however, needs full data vectors for training with respect to each input sample. Computation time is high especially as the input dimension increases
5	RBFNN	Three-layer network: Input is simply a fan-out layer, hidden layer performs nonlinear mapping of inputs, and output layer performs a weighted sum of inputs with Gaussian output function. It trains quicker than MLP; the hidden layer is easy to interpret. Computation time involved in testing phase is high and not much applicable in higher-dimension problems. Training is easy and not susceptible to stuck in local minima
6	Cellular neural network	A 2D rectangular array comprising analogical nonlinear processors named cells. It can be easily extended without modifying the entire network. It gives excellent results for noise detection, connected component detection, and thinning, and even it can be used as an associative memory. It is good for complex real-time applications that require massive parallelism. The hardware implementation of CNN involves high power consumption
7	Deep learning convolution neural network	The DLCNN comprises convolutional layers, pooling layers, fully connected layers, and terminates with a softmax layer for the generation of outputs. The computation complexity is high, and for complex applications, good GPU is required

4 Deep Learning Convolution Neural Network for Medical Image Segmentation

The DLCNN can process 2D image data as well as speech signals. The evolution of deep learning architecture started in 1965; however, nowadays, it gains prominence for real-world applications like image classification, video classification, natural language processing, visual tracking, and speech recognition. The segmentation role is vital for the analysis of the region of interest (ROI), and from the perspective of medical image processing, the ROI corresponds to anatomical organs or anomalies like tumor and cyst. In general, the segmentation algorithms are classified into semi-automatic and fully automatic techniques. The CNN is one of the widely used deep learning architectures, and the proposed segmentation algorithm is a fully automatic algorithm and its efficiency relies on the training.

The CNN structure mimics the biological structure of animal visual cortex. The CNN gains a prominent role in image and video processing, natural language processing due to the minimal pre-processing required. Prior to segmentation, the pre-processing was performed by min-max normalization. The CNN comprises multiple convolution layers preceded by one or more fully connected layer. The convolution layer is the basic building block of CNN, and it comprises Gaussian kernel masks. The filter mask is convolved with the height and width of the image and generates a feature map or activation map. The convolution layer is followed by rectified linear units (ReLU) layer, and here in this layer, the ReLU function is applied to the feature map. The pooling is a key feature of CNN in which the downsampling is performed. The pooling dimension size is predefined and based on that, the input feature map image was subdivided into rectangles; the maximum value in the sub-block is the resultant value. The fully connected layer is the last layer, and it uses hyperbolic sigmoid as the activation function.

The four main components in the convolution neural network are as follows:

1. Convolution layer
2. ReLU layer
3. Pooling or subsampling layer
4. Fully connected layer.

4.1 Convolution Layer

The choice of feature extraction algorithm is a tedious work, and it changes with respect to applications. The deep learning architecture is independent of feature extraction algorithms. The number of convolution layers and pooling layers changes with respect to application, and kernel size also plays a vital role. The convolution layer is termed as the key layer in CNN. The parameter of the convolution layer is the filter size, and the response is determined by convolving the filter with the input image. The spatial relationship between the pixels is maintained in convolution neural network, and the response is called the convolved feature. The number of filters and kernel size is predefined for each layer. The number of filters, the more image features yet extracted.

The size of the feature map is governed by three parameters:

Depth: It corresponds to the number of filters used for the convolution operation.
Stride: It represents the pixel count when the filter kernel is moved across the image matrix when the stride value is 1; the filter is moved one pixel at a time. The larger the slide, smaller feature maps are produced.
Zero padding: It applies filter to border pixels of the input image matrix. The zero padding helps to control the size of feature maps. When zero padding is incorporated, it is called wide convolution, and not using zero padding is called narrow convolution.

The feature map derived from the convolution layer is mathematically represented as follows:

$$f_{ij}^{k} = \tan h\Big(\big(w^{k} * y\big)_{ij} + b_{k}\Big) \quad (4)$$

For a given layer, the $\boldsymbol{k}$**th** filter is represented as $\mathbf{f^k}$ weights ($\boldsymbol{w}^k$); bias values ($\boldsymbol{b}_k$); **tan** $\boldsymbol{h}$ represents the hyperbolic function, and $*$ represents the convolution operation.

4.2 ReLU Layer

After convolution, an additional operation termed as ReLU is done. ReLU represents rectified linear unit, and it is a nonlinear operation. ReLU is a pixel-wise operation and replaces the negative pixels in the feature map by zero. The sample input image is represented in Fig. 5, and the pixel values are only 0 and 1. The 3 × 3 mask along the convolved image and the convolved feature value is also depicted above. The 3 × 3 mask is also called as a filter, kernel, or feature detector, and the convolved image is formed by moving the mask over the image and dot product is computed. The resultant dot product value is termed as convolved feature or activation map or the feature map.

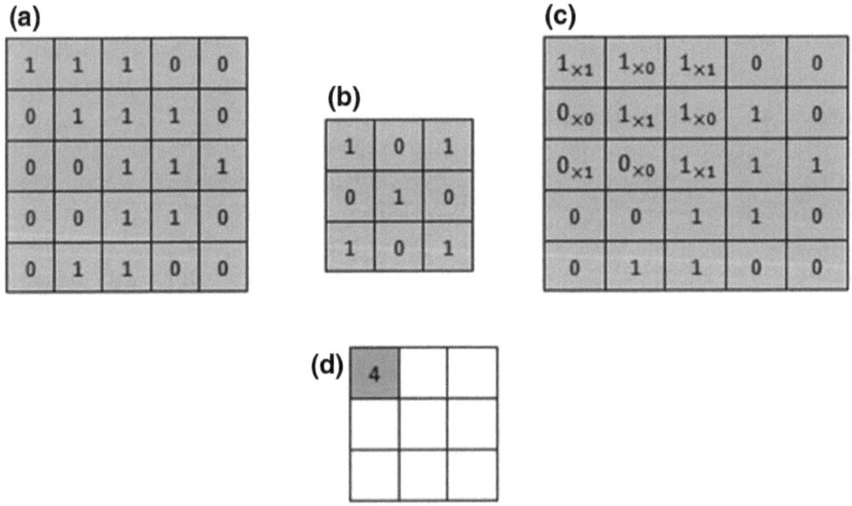

Fig. 5 **a** Sample input image, **b** 3 × 3 mask, **c** convolved image, and **d** convolved feature value

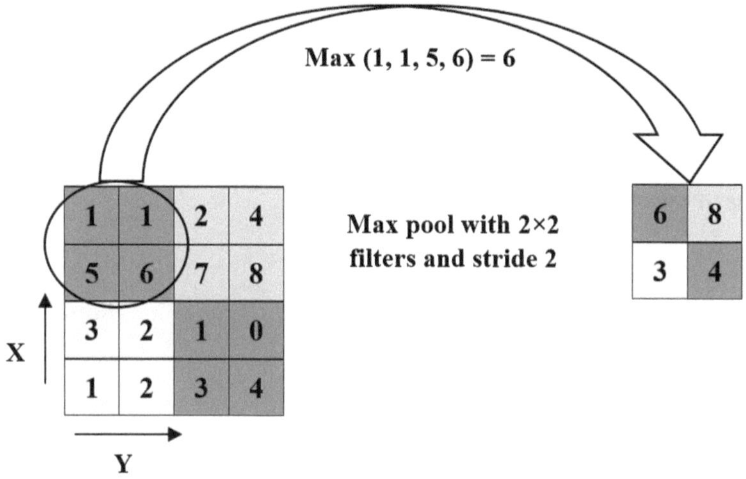

Fig. 6 Pooling operation in convolution neural network

4.3 *Pooling Layer*

In this layer, subsampling or downsampling is done, thereby the dimension of the feature map is preserved. The pooling operation can be of different types such as max, average, and sum. The max pooling is used in many applications. In Max pooling, a spatial neighborhood ($n \times n$ window size) is defined, and the maximum value from the rectified feature map with the kernel is selected. However, in the case of average pooling, the average of the entire elements in the kernel is selected. The pooling operation minimizes the feature dimension and thereby reduces the computation complexity and is depicted in Fig. 6.

4.4 *Fully Connected Layer*

The fully connected layer employs softmax function, and each neuron in the prior layer is connected to the neuron in the foregoing layer.

The training processes of convolution neural network are summarized as follows:

1. Initialization of parameters: number of filters, filter size, weight values.
2. The input image when applied, feature map is created; nonlinear, pooling operations are done. The output probabilities are determined for each class.
3. Estimate the total error at the output layer.

$$\text{Error} = \sum \frac{1}{2}(\text{target probability} - \text{output probability})^2$$

4. The weight values are adjusted for the minimization of error. The backpropagation algorithm is employed to determine the gradient of the error with respect to the weights in the network.
5. For all input images, steps 2–4 are repeated.

In the testing phase, based on the query input image, resultant output is produced based on the training done.

5 Fuzzy Support Vector Machine for Tumor Classification

Machine learning algorithm plays a vital role in tumor classification and detection. This research work proposes SVM and fuzzy SVM for the tumor classification in abdomen CT images. Prior to classification, feature extraction was performed by first-order and second-order statistics. The fuzzy SVM was found to generate efficient classification results when compared with the classical SVM algorithm.

5.1 Support Vector Machine (SVM)

The support vector machine (SVM) is based on the statistical learning theory and is widely used for binary classification. The classical SVM is based on the grouping of input patterns into two different classes.

Let '*T*' represents the set of labeled training points.

$$(X_1, Y_1)\ldots(X_n, Y_n) \tag{5}$$

Each training point $\boldsymbol{X}_i \in \boldsymbol{R}^N$ belongs to either of two classes and is given a label $\boldsymbol{Y}_i \in \{-\mathbf{1}, \mathbf{1}\}$.

For real-time problems, the estimation of appropriate hyperplane corresponding to an input space is difficult. To solve this, the input data items are mapped into a higher-dimensional feature space, and optimal hyperplane is searched for the input feature vectors.

Let $z = \emptyset(\boldsymbol{x})$ represents the features of input data with a mapping $\emptyset$ from $\mathbf{R}^N$ to a feature space z. The objective is to find a hyperplane.

$$WZ + a = 0 \tag{6}$$

The hyperplane is defined by the pair $(\boldsymbol{W}, \boldsymbol{a})$; the input data points are grouped based on the function.

$$f(X_i) = \mathrm{sign}(WZ_i + a) = \begin{cases} 1 \text{ if } Y_i = 1 \\ -1 \text{ if } Y_i = -1 \end{cases} \tag{7}$$

where $\boldsymbol{W} \in \boldsymbol{Z}$ and $a \in R$.

The set 'T' is said to be linearly separable if the following conditions are satisfied.

$$\begin{cases} (WZ_i + a) \geq 1 \text{ if } Y_i = 1 \\ (WZ_i + a) \leq -1 \text{ if } Y_i = -1 \end{cases} \quad I = 1, 2, \ldots n. \tag{8}$$

An optimum hyperplane can be estimated for linearly separable set (T) for grouping the data into two classes. In some cases, the input data may not be linearly separable; the above expression (8) is modified as follows by the inclusion of a negative variable.

$$Y_i(WZ_i + a) \geq 1 - \varphi_i \quad i = 1, 2, \ldots n \tag{9}$$

The nonzero $\boldsymbol{\varphi}_i$ in (9) represents for which the points $\boldsymbol{X}_i$ that do not satisfy (8). The term $\sum_{i=1}^{n} \boldsymbol{\varphi}_i$ is considered as misclassification rate. The optimal hyperplane is considered as the solution to the problem

$$\begin{aligned} &\text{minimize } \frac{1}{2} W \cdot W + C \sum_{i=1}^{n} \varphi_i \\ &\text{Subject to } \begin{cases} Y_i(WZ_i + a) \geq 1 - \varphi_i & i = 1, 2, \ldots n \\ \varphi_i \geq 0 & i = 1, 2, \ldots n \end{cases} \end{aligned} \tag{10}$$

where 'C' is a constant termed as regularization parameter.

The searching of the optimal hyperplane is a QP problem and can be solved by Lagrangian multipliers.

$$\begin{aligned} &\text{maximize } W(\beta) = \sum_{i=1}^{n} \beta_i - \frac{1}{2} \sum_{i=1}^{n} \sum_{j=1}^{n} \beta_i \beta_j Y_i Y_j Z_i Z_j \\ &\text{subject to } \sum_{i=1}^{n} Y_i \beta_i = 0 \quad 0 \leq \beta_i \leq C, \quad i = 1, 2, 3, \ldots, n \end{aligned} \tag{11}$$

where $\boldsymbol{\beta} = \boldsymbol{\beta}_1, \boldsymbol{\beta}_2, \ldots \boldsymbol{\beta}_n$ is the vector of non-negative Lagrangian multipliers.

The solution of Eq. (11) was determined by Kuhn–Tucker theorem.

The solution $\bar{\boldsymbol{\beta}}_n$ of (11) satisfies

$$\bar{\beta}_n \left(Y_i \left(\overline{W} Z_i + \bar{a} \right) - 1 + \varphi_i \right) = 0; \quad i = 1, 2, 3, \ldots n \tag{12}$$

$$(C - \beta_i)\varphi_i = 0, \quad i = 1, 2, 3, \ldots n \tag{13}$$

The nonzero values of $\bar{\boldsymbol{\beta}}_n$ in (12) represent the solution that satisfies the constraints in (9). The point $\boldsymbol{X}_i$ subjected to the constraint $\boldsymbol{0} < \bar{\boldsymbol{\beta}}_n < C$ is called a support vector. The support vector is of two types in a non-separable case.

Case (i): $\mathbf{0} < \bar{\boldsymbol{\beta}}_{\iota} < C$, the support vector satisfies the condition

$$\boldsymbol{Y}_{\mathrm{i}}\left(\overline{\boldsymbol{W}}\boldsymbol{Z}_{i} + \bar{\mathbf{a}}\right) = \mathbf{1}, \boldsymbol{\varepsilon} = \mathbf{0}$$

Case (ii) $\bar{\boldsymbol{\beta}}_{\iota} < C, \boldsymbol{\varphi}_{i} \neq \mathbf{0}$, the support vector does not satisfy (9).
For the construction of the optimal hyperplane $\bar{w}z + \bar{a}$, the condition is

$$\bar{w} = \sum_{i=1}^{n} \bar{\beta}_{\iota} Y_i Z_i \tag{14}$$

where the scalar factor is $\bar{\boldsymbol{a}}$ determined from the Kuhn–Tucker conditions. The generalized form of the decision function is as follows:

$$f(x) = \operatorname{sign}(WZ + a) = \operatorname{sign}\left(\sum_{i=1}^{n} \bar{\beta}_{\iota} Y_i Z_i Z + a\right) \tag{15}$$

The polynomial kernel of degree d is used to construct an SVM classification.

$$K\left(X_i, X_j\right) = \left(1 + X_i, X_j\right)^d \tag{16}$$

The nonlinear separating hyperplane is found as the solution of maximization.

$$\text{Subject to } W(\beta) = \sum_{i=1}^{n} \bar{\beta}_{\iota} - \frac{1}{2}\sum_{i=1}^{n}\sum_{j=1}^{n} \beta_i \beta_j Y_i Y_j K\left(X_i, X_j\right) \tag{17}$$

The decision function is expressed as follows:

$$f(X) = \operatorname{sign}(WZ + a) = \operatorname{sign}\left(\sum_{j=1}^{n} \beta_i Y_i Y_j K\left(X_i, X_j\right) + a\right) \tag{18}$$

The efficiency of the statistical learning theory leads to a concept called VC dimension, the maximum number of points that can be separated in all possible ways using the functions of the given class. The occurrences of the inner product are replaced by the kernel function (K) defined as follows:

$$K\left(X_i, X_j\right) = \emptyset(X_i)\emptyset\left(X_j\right) \tag{19}$$

Mercer's theorem states that conversely, the mapping exists for a kernel.
The RBF kernel is expressed as follows:

$$K\left(X_i, X_j\right) = \exp\left(-g\left\|X_i - X_j\right\|^2\right) \tag{20}$$

The penalty parameter (C) and the kernel parameter (Y) are properly tuned for classification. The optimal values of C and Y determine the boundary complexity and thus improve the classification performance. For parameter optimization, cross-validation or grid search is employed. Cross-validation is a technique by which the n samples are divided into $n - 1$ training set and one testing set. Each training set is checked, and cross-validation accuracy determines the percentage of correctly classified data. The grid search technique searches for the optimal pair of parameters that generate the highest classification accuracy.

5.2 *Fuzzy Support Vector Machine*

The fuzzy membership is incorporated in the classical SVM for the formulation of fuzzy SVM. The classical SVM groups the data into two distinct classes. In the case of real-time scenario, each data member can have a degree of belongingness in multiple classes. The fuzzy membership value determines the degree of closeness for a class with respect to each data member. Fuzzy SVM is simply an extension of SVM. In SVM, label is assigned on the basis of hyperplane and kernel function, while in fuzzy SVM, membership in terms of probability is also determined for each sample to be fall in each class.

The main objective of SVM is that a data point belongs to anyone of the class. In real-time applications, a data point may have 80% preference in one class and 20% in another class. In terms of fuzzy membership, a set of labeled training set 'T' is expressed as follows:

$$(X_1, Y_1, F_1) \ldots (X_n, Y_n, F_n) \tag{21}$$

The fuzzy membership $\boldsymbol{\sigma \leq F_i \leq 1}$ with $\boldsymbol{i = 1, 2, \ldots l}$ and $\sigma > 0$, the term $\boldsymbol{F_i \psi_i}$ is a measure of error in FSVM.

The formulation of the optimal hyperplane is considered as the solution to

$$\begin{aligned} &\text{minimize } \frac{1}{2} WW + C \sum_{i=1}^{l} F_i \psi_i \quad i = 1, 2, 3, \ldots .l \\ &\text{subject to } Y_i(WZ_i + a) \geq 1 - \psi_i, \quad i = 1, 2, 3, \ldots l \\ &\psi_i \geq 0, \quad i = 1, 2, 3, \ldots l \end{aligned} \tag{22}$$

The above optimization problem is solved by Lagrangian multipliers.

$$L(W, a, \psi, \alpha, \beta) = \frac{1}{2} WW + L \sum_{i=1}^{l} F_i \psi_i$$

$$-\sum_{i=1}^{l}\alpha_i(Y_i(WZ_i+a)-(1+\psi_1))-\sum_{i=1}^{l}\beta_i\psi_i \quad (23)$$

The objective is to determine the saddle point of $\boldsymbol{L(W, a, \psi, \alpha, \beta)}$. The parameters should satisfy the following conditions

$$\frac{\partial L(W,a,\psi,\alpha,\beta)}{\partial w}=W-\sum_{i=1}^{l}\alpha_i Y_i Z_i=0 \quad (24)$$

$$\frac{\partial L(W,a,\psi,\alpha,\beta)}{\partial a}=-\sum_{i=1}^{l}\alpha_i Y_i=0 \quad (25)$$

$$\frac{\partial L(W,a,\psi,\alpha,\beta)}{\partial \psi_i}=F_iC-\alpha_i-\beta_i=0 \quad (26)$$

The optimization problem (22) can be reformulated based on the above conditions. Maximize

$$W(\alpha)=\sum_{i=1}^{l}\alpha_i-\frac{1}{2}\sum_{i=1}^{l}\sum_{j=1}^{l}\alpha_i a_j Y_i Y_j K\left(X_i,X_j\right)$$

Subject to

$$\sum_{i=1}^{l}Y_i\alpha_i=0 \quad 0\le\alpha_i\le F_iC \quad i=1,2,3,\ldots l \quad (27)$$

The Kuhn–Tucker condition is expressed as follows:

$$\bar{\alpha}_t\left(Y_i\left(\overline{W}Z_i+\bar{a}\right)-1+\bar{\psi}_l\right)=0 \quad i=1,2,3,\ldots l \quad (28)$$

$$(F_iC-\bar{a}_l)\bar{\psi}_l=0 \quad i=1,2,3,\ldots l \quad (29)$$

The data points $\boldsymbol{X_i}$ with $\bar{\boldsymbol{\alpha}}_\iota > 0$ are termed as support vector. There are two cases. Case (i) $\mathbf{0} < \bar{\boldsymbol{\alpha}}_\iota < \boldsymbol{F_i C}$, the support vector lies on the margin of the hyperplane. Case (ii) $\bar{\boldsymbol{\alpha}}_\iota < \boldsymbol{F_i C}$, the support vector is misclassified.

The parameter 'C' regulates SVM, the trade-off between the maximization margin, and amount of misclassifications. Higher the value of 'C', lesser the misclassification with a narrow margin. Lower the value of 'C', misclassification increases with a wider margin. In FSVM, along with the parameter C, F_i can also be tuned. Lower value of F_i indicates that the data point is less important for training. The selection of appropriate fuzzy membership function is also vital. Firstly, the lower bound of fuzzy membership must be defined, and main property of the data is analyzed; a connectivity is made between the property and fuzzy membership.

The linear fuzzy membership is expressed as follows:

$$F_i = f(t_i) = at_i + b \tag{30}$$

$$F_i = f(t_i) = \frac{1-\sigma}{t_i - t_i}t_i + \frac{t_i\sigma - t_i}{t_i - t_i} \tag{31}$$

The FSVM and SVM algorithms are initially trained with normal, benign, and malignant cases, and then testing is performed.

6 Results and Discussion

The DLCNN algorithm for segmentation is developed in Matlab 2015a, and for comparative analysis of segmentation results, BPNN, GMDH neural network, and decision tree classifier are used. The tumor classification algorithms, support vector machine and fuzzy support vector machine, are developed in Java. The system specifications are as follows: 64-bit operating system with Intel core i3 processor @3.3 GHz and 4 GB RAM. The abdomen CT images are acquired from Metro Scans and Research Laboratory, Trivandrum.

For the evaluation of segmentation results, the metrics are used, and here in this research work, supervised segmentation is employed. The supervised segmentation requires a gold standard image and is generated by the physicians. The gold standard image (G) and machine-generated result (M) are used for the formulation of metrics. The Hausdorff distance measures the distance between the machine-generated result and the gold standard result. The Hausdorff distance ($\boldsymbol{HD}$) is expressed below:

$$\text{HD} = \max(H_{MG}, H_{GM}) \tag{32}$$

where

$$H_{MG} = \max\left(d_i^{Mg}\right); \quad i = 1, 2, 3, \ldots n_M$$

$$H_{GM} = \max\left(d_i^{Gm}\right); \quad i = 1, 2, 3, \ldots n_G$$

where $\boldsymbol{d}_i^{Mg}$ is the minimum distance for the ith surface voxel in M to the set of surface voxels in G.

The Jaccard coefficient (JC) is expressed as follows:

$$\text{JC} = \frac{|\text{TP}|}{|\text{TP}| + |\text{FN}| + |\text{FP}|} \tag{33}$$

The expression for Rand index (RI) is represented below:

$$\mathrm{RI} = \frac{|\mathrm{TP}| + |\mathrm{TN}|}{|\mathrm{TP}| + |\mathrm{FN}| + |\mathrm{TN}| + |\mathrm{FP}|} \tag{34}$$

The JC determines the amount of spatial overlap between the gold standard image and machine-generated result. The value of JC is in the range 0–1. The value of '1' indicates perfect overlap, and lower the value, closer to '0' indicates poor matching. The RI represents the consistency of pixels in the gold standard and machine-generated results. The value of '1' represents the perfect similarity between M and G, and the value of '0' represents the poor similarity between M and G.

Global consistency error (GCE) forces all local refinements in the same direction, and it is expressed as follows:

$$\mathrm{GCE}(M, G) = \frac{1}{N}\min\left\{\sum_i \mathrm{LRE}(M, G, x_i) \sum_i \mathrm{LRE}(G, M, x_i)\right\} \tag{35}$$

where LRE is local refinement error.

Lower the value of GCE, closer to 0 indicates the efficiency of the segmentation algorithm and as the value of GCE increases, poor the quality of the segmentation algorithm.

The variation of information (VOI) is expressed as follows:

$$\mathrm{VOI}(M, G) = H(M) + H(G) + I(M, G) \tag{36}$$

Lower the value of VOI, better is the segmentation result. The root mean square error (RMSE) is expressed as follows:

$$\mathrm{RMSE} = \sqrt{\frac{1}{N}\sum_{i=1}^{n} e_i^2} \tag{37}$$

where e is the error between the ground truth and machine-generated result.

The success and error rates are expressed as follows:

$$\mathrm{Sensitivity} = \frac{|\mathrm{TP}|}{|\mathrm{TP}| + |\mathrm{FN}|} \tag{38}$$

$$\mathrm{Speicify} = \frac{|\mathrm{TN}|}{|\mathrm{TN}| + |F|} \tag{39}$$

$$\mathrm{Precision} = \frac{|\mathrm{TP}|}{|\mathrm{TP}| + |\mathrm{FP}|} \tag{40}$$

The BPNN is a classical neural network algorithm, and in GMDH neural network, automatic organization of architecture takes place based on the prediction error coefficient and generates better results than BPNN [32]. GMDH neural network is characterized by using heuristic self-organization and is used for modeling

Table 2 Features of machine learning algorithms employed for segmentation and classification

Machine learning algorithms	Feature extraction employed	Parameters
Backpropagation neural network	Gray-level co-occurrence	No of hidden neurons: 20
	Matrix (GLCM) and first-order statistics (FOS)	Training ratio: 0.7
		Validation ratio: 15
		Testing ratio: 0.15
		Neighborhood size for feature extraction: 7 × 7
GMDH	GLCM and FOS	Automatic selection of hidden layer neurons based on Akaike's information criteria
		Neighborhood size for feature extraction: 7 × 7
Decision tree	FOS, GLCM, and local binary pattern (LBP)	Number of clusters: 6
		Fuzziness value: 3
		No of iterations: 20
		Neighborhood size for feature extraction: 7 × 7
Convolution neural network	–	The filter size of convolution layer: 5
		Number of filters in convolution layer: 12
		Pooling dimension: 2
Support vector machine	First-order statistics (FOS) and second-order statistics (SOS)	Regularization parameter, $C = 20$; kernel = RBF
Fuzzy support vector machine	First-order statistics (FOS) and second-order statistics (SOS)	Regularization parameter, $C = 200$; kernel = RBF

nonlinear and complex systems [33]. The decision tree is a flow diagram like a tree structure comprising root, where each internal node represents a test on an attribute, branches represent the outcome of the test, and leaves represent classes (Table 2).

In the plots below, ID1–ID8 depict the datasets, and the *Y*-axis represents the performance metrics values of the segmentation algorithms.

The liver segmentation results are depicted in Figs. 7 and 8. The performance metrics' plots for liver segmentation are depicted in Figs. 11, 12, 13, 14, 15, and 16. The sensitivity, specificity, and precision plots for liver segmentation are depicted in Figs. 23, 24, and 25. The performance evaluation reveals the efficiency of CNN, when compared with the BPNN, GMDH, and decision tree algorithms. The liver tumor segmentation results are depicted in Figs. 9 and 10. The performance metrics' plots for liver tumor segmentation are depicted in Figs. 17, 18, 19, 20, 21, and 22. The

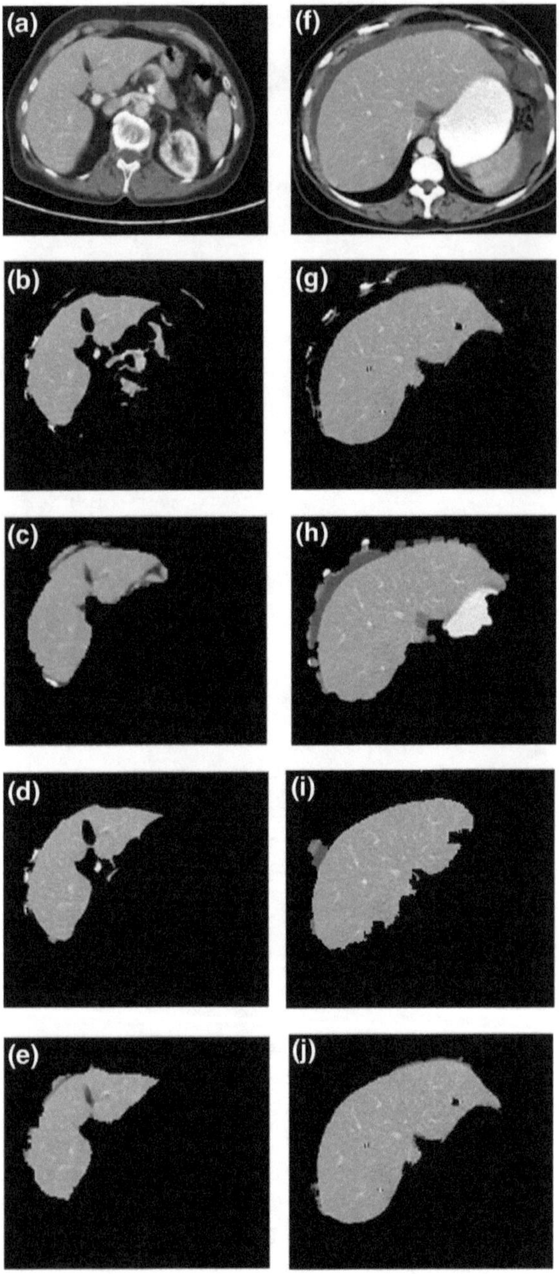

Fig. 7 ID1 and ID2 liver segmentation results: The first row depicts the input images, the second row depicts the BPNN results, the third row depicts the GMDH results, the fourth row depicts the decision tree results, and the fifth row depicts the CNN results

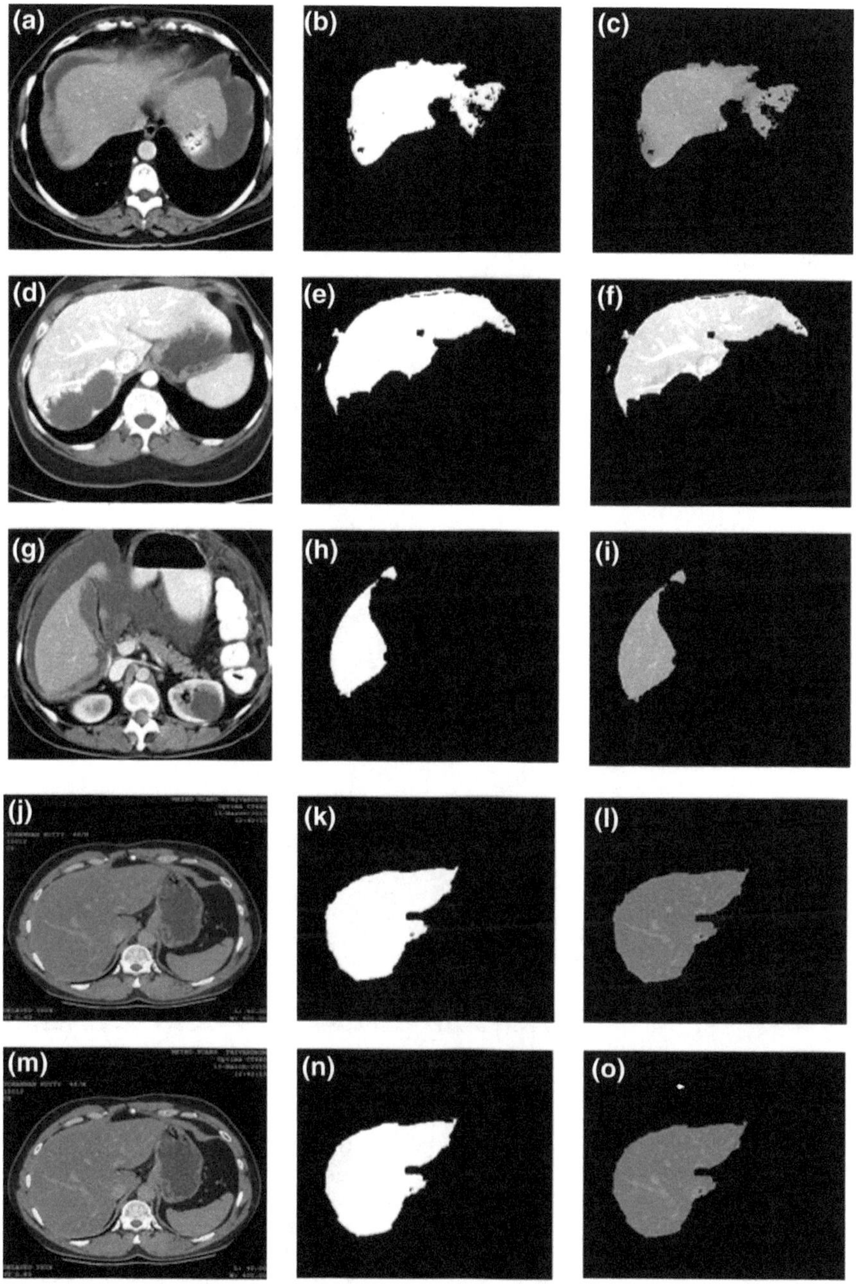

Fig. 8 ID3–ID8 liver segmentation results for CNN algorithm: The first column depicts the input images, and second and third columns depict the CNN segmentation results

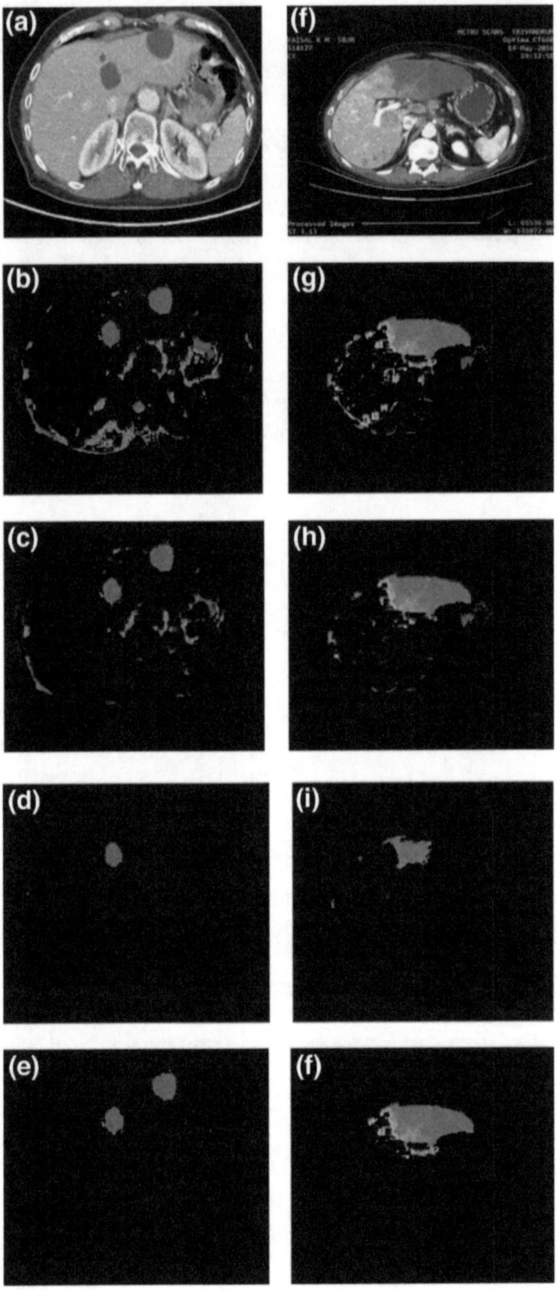

Fig. 9 ID1 and ID2 liver tumor segmentation results: The first row and column depict the input images, the second row depicts the BPNN results, the third row depicts the GMDH algorithm results, the fourth row depicts the decision tree algorithm results, and the fifth row depicts the CNN algorithm results

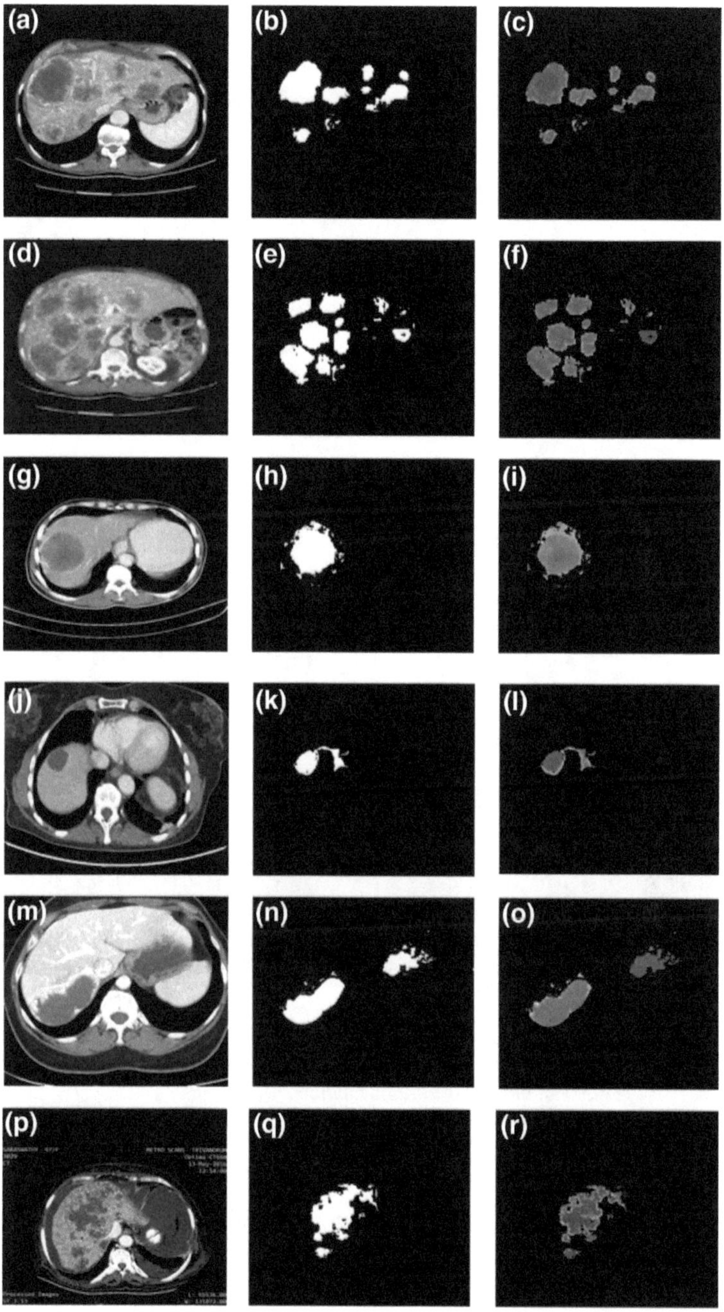

Fig. 10 ID3–ID8 liver tumor segmentation results for CNN algorithm: The first column depicts the input images, and second and third columns depict the CNN segmentation results

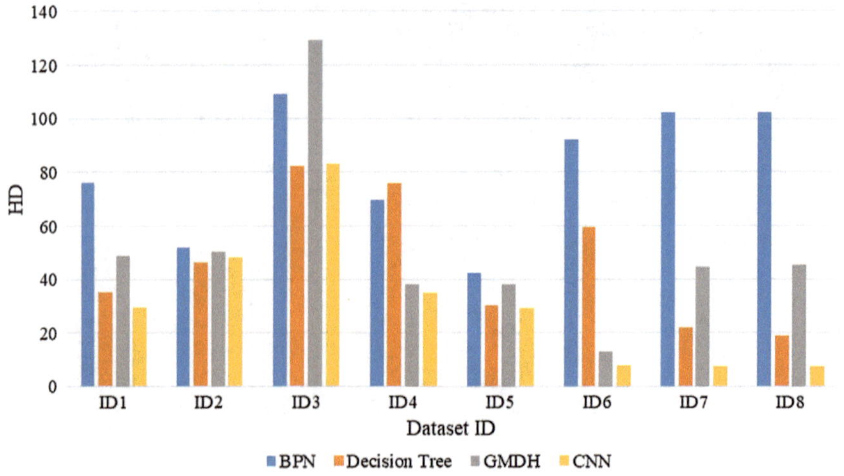

Fig. 11 Hausdorff distance plot for liver segmentation

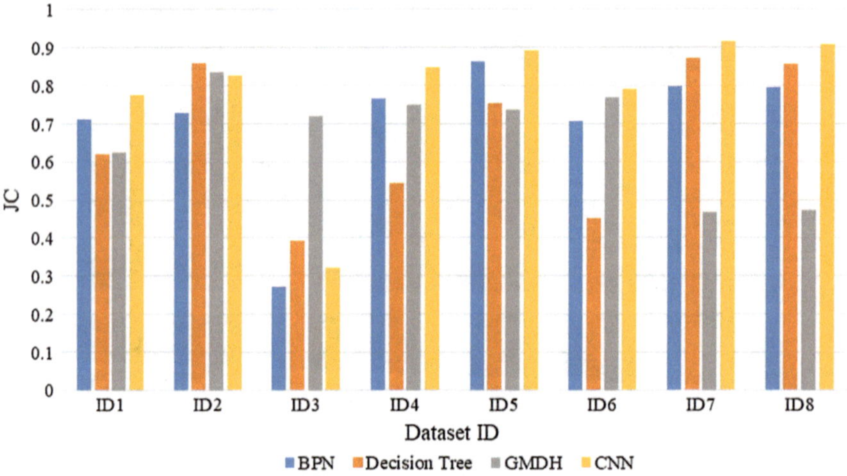

Fig. 12 Jaccard's coefficient plot for liver segmentation

sensitivity, specificity, and precision plots for liver tumor segmentation are depicted in Figs. 26, 27, and 28. The results reveal the superiority of CNN in liver tumor segmentation. The SVM and fuzzy SVM were employed for the tumor classification of abdomen CT datasets (Figs. 29 and 30).

The one-way ANOVA statistical test is used here for the analysis of results. The null hypothesis is as follows:

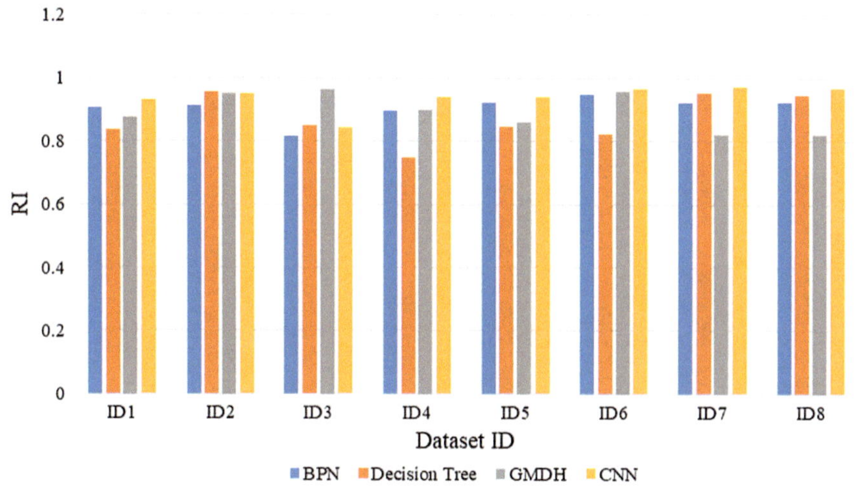

Fig. 13 Rand index plot for liver segmentation

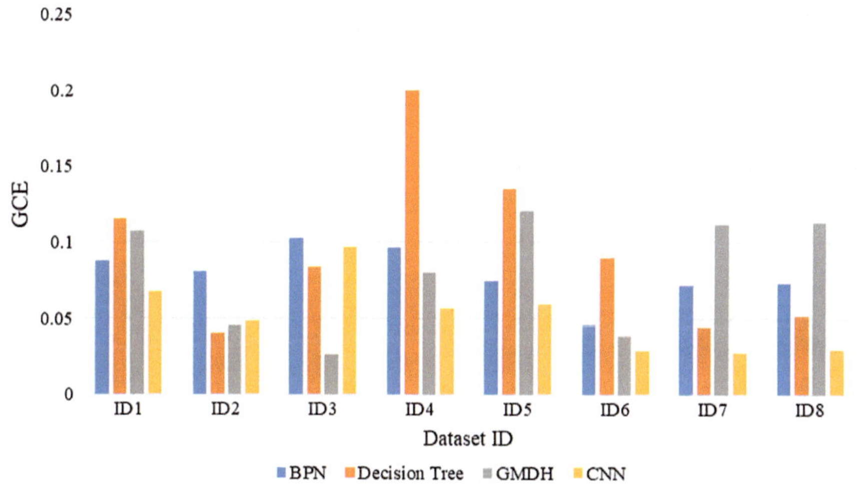

Fig. 14 GCE plot for liver segmentation

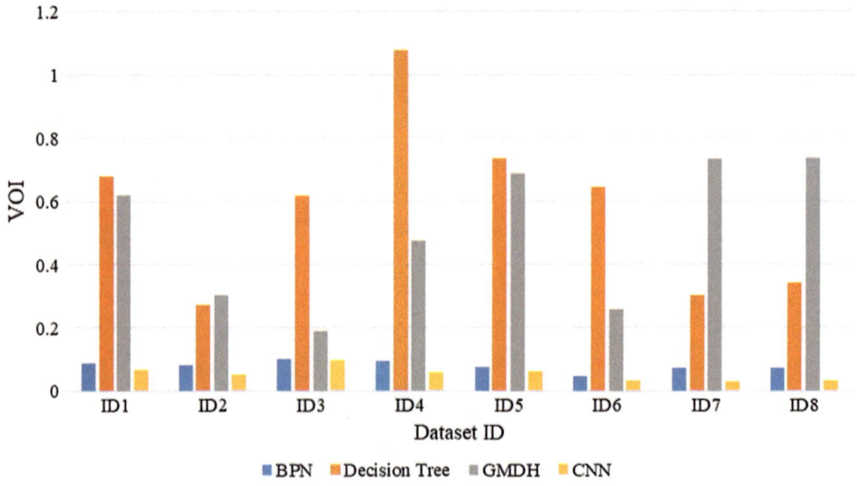

Fig. 15 VOI plot for liver segmentation

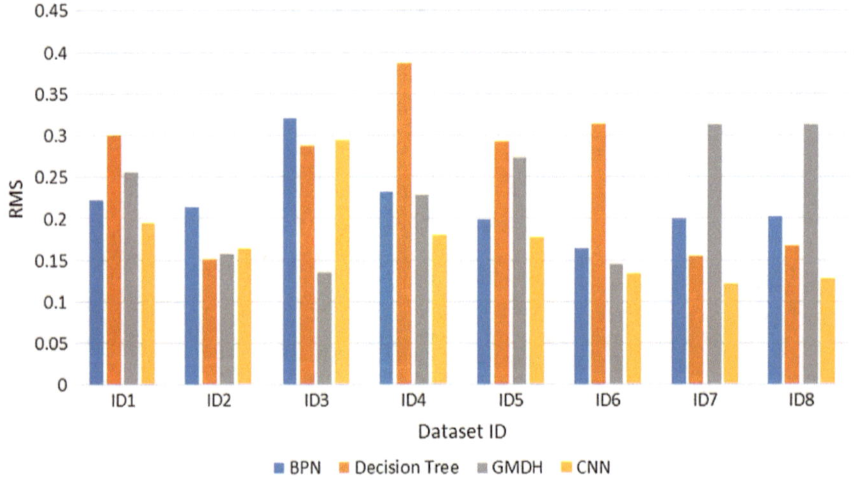

Fig. 16 RMSE plot for liver segmentation

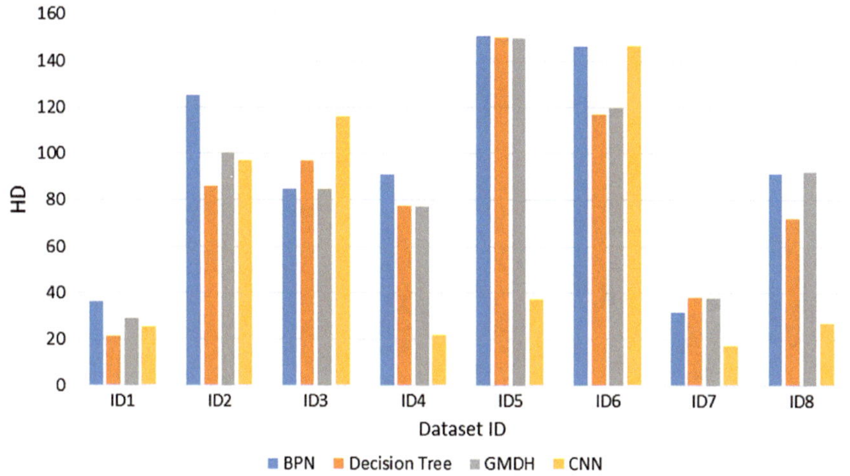

Fig. 17 Hausdorff distance plot for liver tumor segmentation

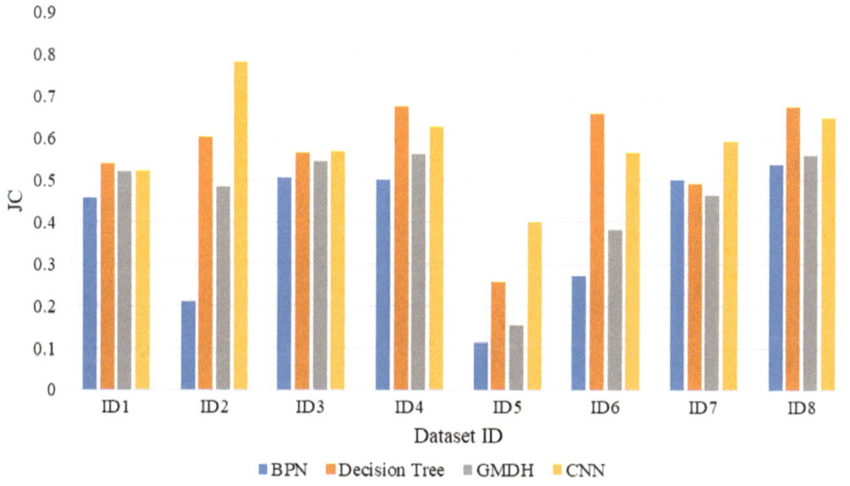

Fig. 18 Jaccard's coefficient plot for liver tumor segmentation

$$H_o: \mu_1 = \mu_2 = \mu_3 = \mu_4$$

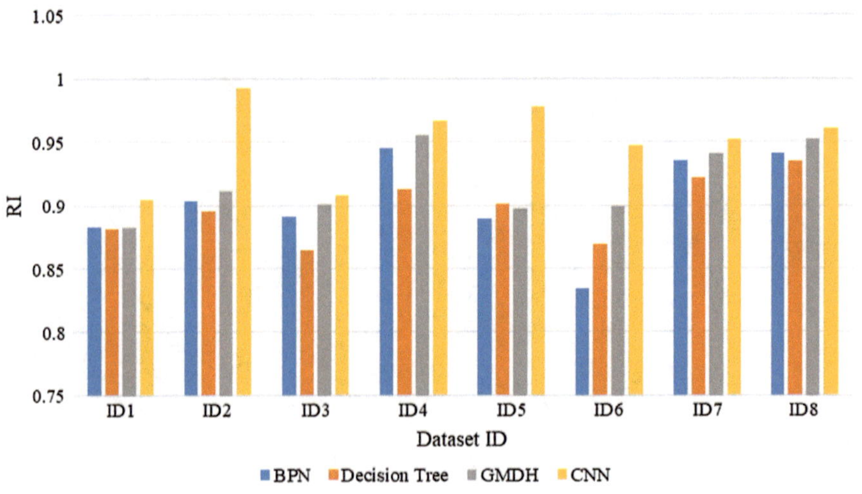

Fig. 19 Rand index plot for liver tumor segmentation

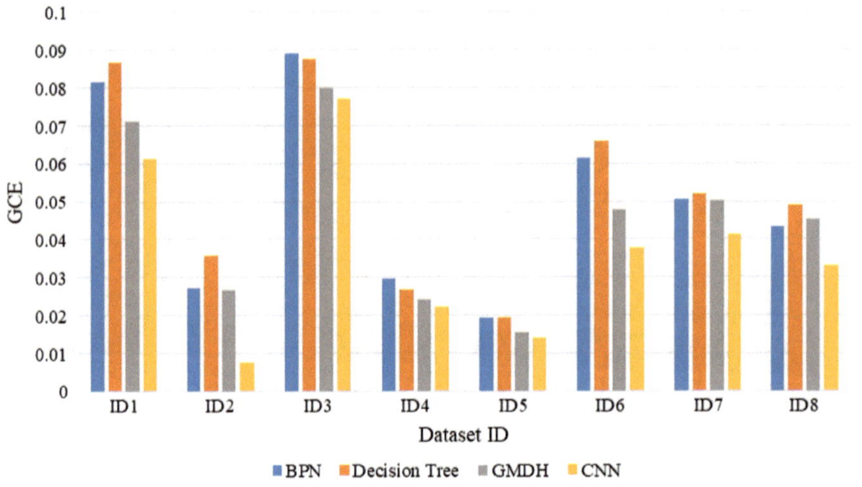

Fig. 20 GCE plot for liver tumor segmentation

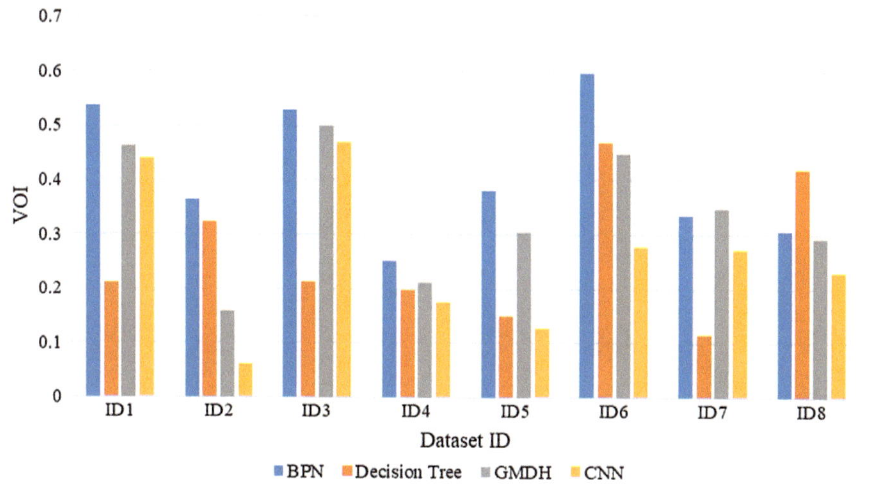

Fig. 21 VOI plot for liver tumor segmentation

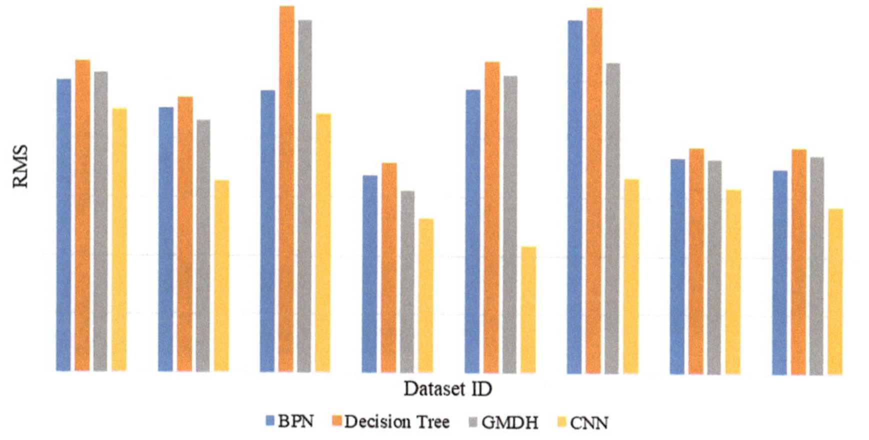

Fig. 22 RMSE plot for liver tumor segmentation

In general, the null hypothesis

$$H_o\colon \mu_1 = \mu_2 = \mu_3 = \cdots \mu_k \tag{41}$$

where k is the number of levels of independent variables.

The null hypothesis states that the population means from which samples are selected are equal. Here eight samples are taken for algorithm; hence, $\boldsymbol{\mu_k} = \boldsymbol{\mu_4}$.

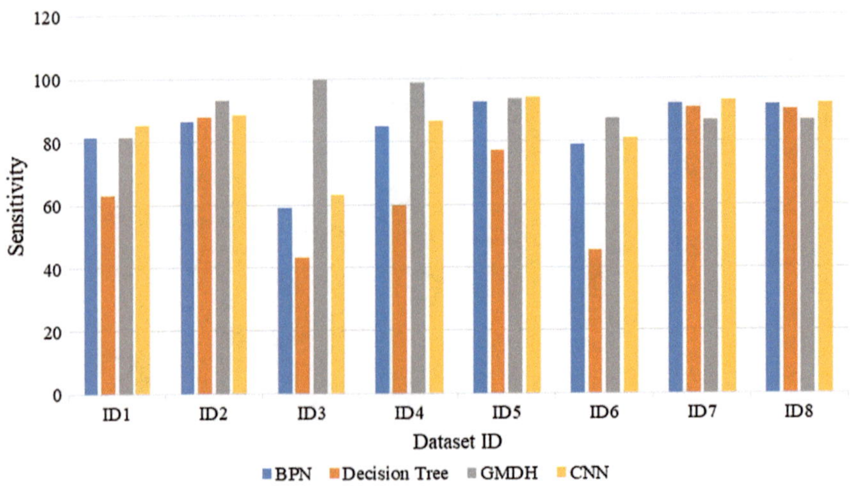

Fig. 23 Sensitivity plot for liver segmentation

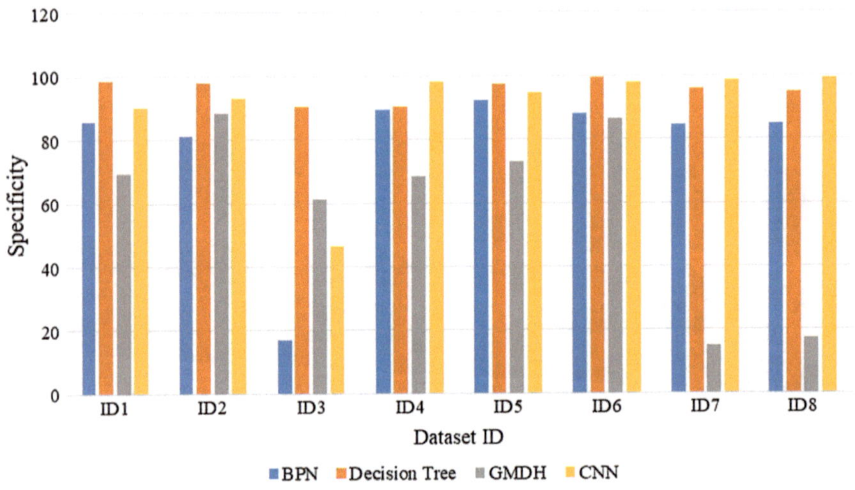

Fig. 24 Specificity plot for liver segmentation

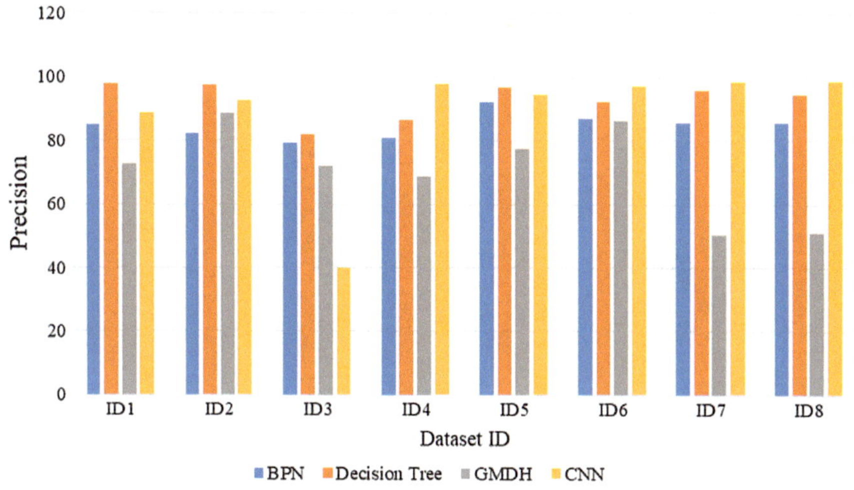

Fig. 25 Precision plot for liver segmentation

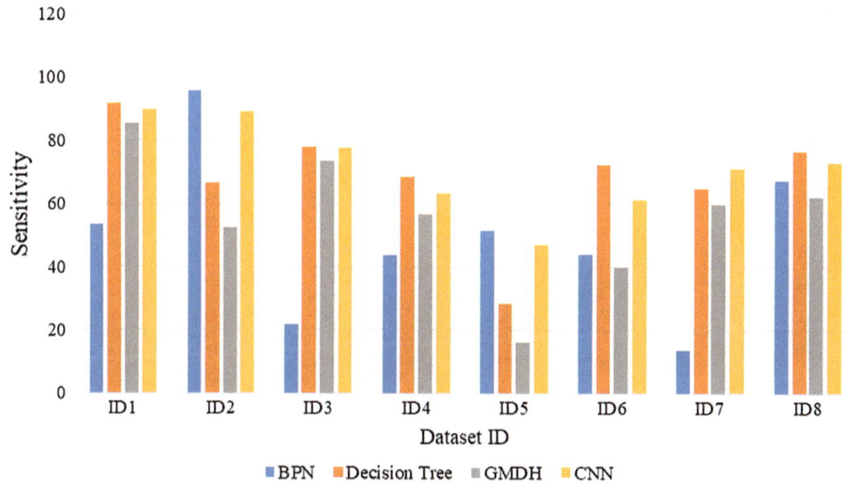

Fig. 26 Sensitivity plot for liver tumor segmentation

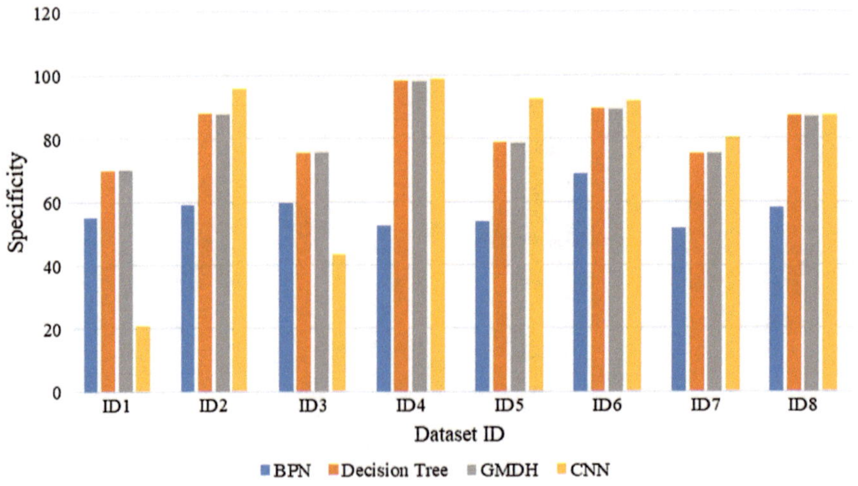

Fig. 27 Specificity plot for liver tumor segmentation

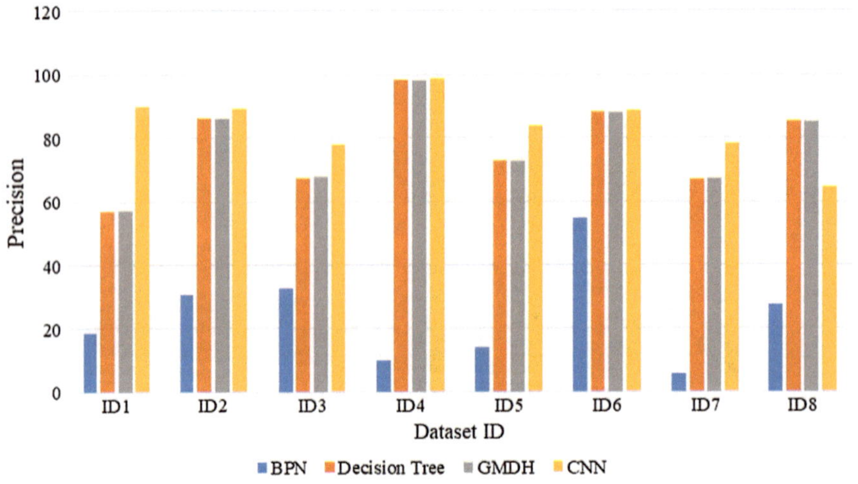

Fig. 28 Precision plot for liver tumor segmentation

	Normal	Benign	Malignant	Classification (Overall)	Accuracy (Precision)
Normal	119	2	6	127	93.70%
Benign	9	41	11	61	67.21%
Malignant	1	2	124	127	97.63%
Truth (Overall)	129	45	141	315	
Accuracy (Recall)	92.24%	91.11%	87.94%		
Overall Accuracy	90.15%				
Kappa	0.834				

Fig. 29 SVM classification results for abdomen DICOM CT datasets

	Normal	Benign	Malignant	Classification (Overall)	Accuracy (Precision)
Normal	74	1	3	78	94.87%
Benign	6	31	0	37	83.78%
Malignant	0	2	92	94	97.87%
Truth (Overall)	80	34	95	209	
Accuracy (Recall)	92.50%	91.17%	96.84%		
Overall Accuracy	94.25%				
Kappa	0.908				

Fig. 30 Fuzzy SVM classification results for abdomen DICOM CT datasets

The alternate hypothesis is that at least one group mean significant differs from other group means

$$H_a\colon \mu_i = \mu_k \quad \text{for some } i, k \tag{42}$$

where i and k indicate unique classes (Tables 3, 4, 5, and 6).

Table 3 Performance metrics analysis for liver segmentation by ANOVA

Performance metrics	Source variance	SS	DF	MS	*P*-critical	*F*-critical
Hausdorff distance	Between methods	10,502.50	3	3500.83	0.009	4.653
	Within methods	21,066.48	28	752.37		
	Total	31,568.98	31			
Jaccard's coefficient	Between methods	0.0696	3	0.0232	0.543	0.7296
	Within methods	0.8908	28	0.0318		
	Total	0.1066	31			
Rand index	Between methods	0.0195	3	0.0065	0.1244	2.0873
	Within methods	0.0871	28	0.0031		
	Total	0.1066	31			
Global consistency error	Between methods	0.0079	3	0.0026	0.1372	1.9972
	Within methods	0.0369	28	0.0013		
	Total	0.0449	31			
Variation of information	Between methods	1.8421	3	0.614	0.00001	19.6449
	Within methods	0.8752	28	0.0313		
	Total	2.7173	31			
Root mean square error	Between methods	0.0288	3	0.0096	0.1237	2.095
	Within methods	0.1283	28	0.0046		
	Total	0.1571	31			

Table 4 Performance metrics analysis for liver tumor segmentation by ANOVA

Performance metrics	Source variance	SS	DF	MS	*P*-critical	*F*-critical
Hausdorff distance	Between methods	5016.03	3	1672.01	0.4790	0.84863
	Within methods	5516.82	28	1970.24		
	Total	60,182.86	31			
Jaccard's coefficient	Between methods	0.2012	3	0.0671	0.028	3.49753
	Within methods	0.5369	28	0.192		
	Total	0.7381	31			
Rand index	Between methods	0.0138	3	0.0046	0.007143	4.9275
	Within methods	0.0262	28	0.0009		
	Total	0.04	31			
Global consistency error	Between methods	0.0012	3	0.0004	0.5655	0.69043
	Within methods	0.0166	28	0.0006		
	Total	0.0178	31			

(continued)

Table 4 (continued)

Performance metrics	Source variance	SS	DF	MS	*P*-critical	*F*-critical
Variation of information	Between methods	0.1328	3	0.0443	0.69746	2.629
	Within methods	0.4715	28	0.0168		
	Total	0.6044	31			
Root mean square error	Between methods	0.0294	3	0.0098	0.0124	4.334
	Within methods	0.0634	28	0.0023		
	Total	0.0928	31			

Table 5 Segmentation algorithms efficiency analysis for liver segmentation by ANOVA

Performance metrics	Source variance	SS	DF	MS	*P*-critical	*F*-critical
Sensitivity	Between methods	1973.068	3	657.689	0.16339	4.057
	Within methods	4539.11	28	162.111		
	Total	6512.18	31			
Specificity	Between methods	5950.75	3	1983.58	0.0104	4.5213
	Within methods	12,284.10	28	438.717		
	Total	18,284.84	31			
Precision	Between methods	2143.60	3	714.53	<0.0119	4.3815
	Within methods	4566.22	28	163.07		
	Total	6709.83	31			

Table 6 Segmentation algorithms efficiency analysis for liver tumor segmentation by ANOVA

Performance metrics	Source variance	SS	DF	MS	*P*-critical	*F*-critical
Sensitivity	Between methods	2682.77	3	894.25	0.111406	2.190
	Within methods	11,432.80	28	408.31		
	Total	14,115.58	31			
Specificity	Between methods	3439.70	3	1146.56	0.0106	4.5018
	Within methods	7131.35	28	254.69		
	Total	10,571.05	31			
Precision	Between methods	18,589.84	3	6196.61	<0.00001	338,273
	Within methods	5129.14	28	183.18		
	Total	23,718.98	31			

In the tables, *DF* means degree of freedom, *SS* means sum of the squares, and *MS* means mean sum of the squares due to the source. The *F* value is closer to '1', when the null hypothesis is true. A large value of *F* indicates that variation among the group means is high and the *P* value is determined from the *F* ratio. The significance level (alpha or α) is usually set to 0.05, and when the *P* value is higher the significance level, null hypothesis will be accepted

7 Conclusion

This research work proposes DLCNN for liver and anomalies segmentation. The DLCNN outperforms BPNN, GMDH, and decision tree algorithm, and the results are validated in terms of performance metrics. For the classification of tumor stages, SVM and fuzzy SVM were employed. The fuzzy SVM results were superior when compared with the SVM results. The outcome of this research work will be an aid for the researchers for the analysis of existing machine learning algorithms for ROI extraction and diagnosis of diseases. The DLCNN and fuzzy SVM give promising results, and hence can be applied for other medical imaging modalities for segmentation and classification of anomalies.

Acknowledgements This work was supported by Department of Science and Technology under Instrument Development Programme.

References

1. Schmidhuber, J.: Deep learning in neural networks: an overview. Neural Netw. **61**, 85–117 (2015). https://doi.org/10.1016/j.neunet.2014.09.003
2. Alpaydm, E.: Combined 5 × 2 cv F test for comparing supervised classification learning algorithms. Neural Comput. **11**(8), 1885–1892 (1999). https://doi.org/10.1016/j.neunet.2014.09.003
3. Sharma, N., Ray, A.K., Sharma, S., Shukla, K.K., Pradhan, S., Aggarwal, L.M.: Segmentation and classification of medical images using texture-primitive features: application of BAM-type artificial neural network. J. Med. Phys./Association of Medical Physicists of India **33**(3), 119 (2008). https://doi.org/10.4103/0971-6203.42763
4. Soltanian-Zadeh, H., Rafiee-Rad, F.: Comparison of multiwavelet, wavelet, Haralick, and shape features for microcalcification classification in mammograms. Pattern Recogn. **37**(10), 1973–1986 (2004). https://doi.org/10.1016/j.patcog.2003.03.001
5. Zhang, G.P.: Neural networks for data mining. In: Data Mining and Knowledge Discovery Handbook. Springer, Boston, MA (2009), pp. 419–444. https://doi.org/10.1007/978-0-387-09823-4_21
6. Kumar, S.N., Fred, A.L., & Varghese, P.S.: An overview of segmentation algorithms for the analysis of anomalies on medical images. J. Intell. Syst. (2018). https://doi.org/10.1515/jisys-2017-0629
7. Egmont-Petersen, M., de Ridder, D., Handels, H.: Image processing with neural networks—a review. Pattern Recogn. **35**(10), 2279–2301 (2002). https://doi.org/10.1016/S0031-3203(01)00178-9
8. Ozkan, M., Dawant, B.M., Maciunas, R.J.: Neural-network-based segmentation of multi-modal medical images: a comparative and prospective study. IEEE Trans. Med. Imaging **12**(3), 534–544 (1993). https://doi.org/10.1109/42.241881
9. Jain, A.K.: Data clustering: 50 years beyond K-means. Pattern Recogn. Lett. **31**(8), 651–666 (2010). https://doi.org/10.1016/j.patrec.2009.09.011
10. Havaei, M., Davy, A., Warde-Farley, D., Biard, A., Courville, A., Bengio, Y., Larochelle, H.: Brain tumor segmentation with deep neural networks. Med. Image Anal. **35**, 18–31 (2017). https://doi.org/10.1016/j.media.2016.05.004
11. Litjens, G., Kooi, T., Bejnordi, B.E., Setio, A.A.A., Ciompi, F., Ghafoorian, M., Sánchez, C.I.: A survey on deep learning in medical image analysis. Med. Image Anal. **42**, 60–88 (2017). https://doi.org/10.1016/j.media.2017.07.005

12. Jain, A.K., Mao, J., Mohiuddin, K.M.: Artificial neural networks: a tutorial. Computer **29**(3), 31–44 (1996). https://doi.org/10.1109/2.485891
13. Kohonen, T.: Learning vector quantization. In: Self-organizing Maps. Springer, Berlin, Heidelberg (2001), pp. 245–261. https://doi.org/10.1007/978-3-642-56927-2_6
14. Pack, C.: Optimized Multilayer Perceptron with Dynamic Learning Rate to Classify Breast Microwave Tomography Image. Theses and Dissertations, 1696 (2017) http://openprairie.sdstate.edu/etd/1696
15. Moreira, M.W., Rodrigues, J.J., Kumar, N., Al-Muhtadi, J., Korotaev, V.: Nature-inspired algorithm for training multilayer perceptron networks in e-health environments for high-risk pregnancy care. J. Med. Syst. **42**(3), 51 (2018). https://doi.org/10.1007/s10916-017-0887-0
16. Kwon, K., Kim, D., Park, H.: A parallel MR imaging method using multilayer perceptron. Med. Phys. **44**(12), 6209–6224 (2017). https://doi.org/10.1002/mp.12600
17. Devi, C.A., Rajagopalan, S.P.: Brain stroke classification based on multi-layer perceptron using watershed segmentation and Gabor filter. J. Theor. Appl. Inf. Technol. 56(3) (2013)
18. Middleton, I., Damper, R.I.: Segmentation of magnetic resonance images using a combination of neural networks and active contour models. Med. Eng. Phys. **26**(1), 71–86 (2004). https://doi.org/10.1016/S1350-4533(03)00137-1
19. Zhao, Y., Zan, Y., Wang, X., Li, G.: Fuzzy C-means clustering-based multilayer perceptron neural network for liver CT images automatic segmentation. In: 2010 Chinese Control and Decision Conference (CCDC). IEEE, May 2010, pp. 3423–3427. https://doi.org/10.1109/ccdc.2010.5498558
20. Selver, M.A., Guzelis, C.: Semiautomatic transfer function initialization for abdominal visualization using self-generating hierarchical radial basis function networks. IEEE Trans. Visual Comput. Graphics **15**(3), 395–409 (2009). https://doi.org/10.1109/TVCG.2008.198
21. Rostami, M.T., Ezoji, M., Ghaderi, R., Ghasemi, J.: Brain MRI segmentation using the mixture of FCM and RBF neural network. In: 2013 8th Iranian Conference on Machine Vision and Image Processing (MVIP). IEEE, Sept 2013, pp. 425–429. https://doi.org/10.1109/iranianmvip.2013.6780023
22. Zhang, D.Q., Chen, S.C.: A novel kernelized fuzzy c-means algorithm with application in medical image segmentation. Artif. Intell. Med. **32**(1), 37–50 (2004). https://doi.org/10.1016/j.artmed.2004.01.012
23. Maglogiannis, I., Sarimveis, H., Kiranoudis, C.T., Chatziioannou, A.A., Oikonomou, N., Aidinis, V.: Radial basis function neural networks classification for the recognition of idiopathic pulmonary fibrosis in microscopic images. IEEE Trans. Inf Technol. Biomed. **12**(1), 42–54 (2008). https://doi.org/10.1109/TITB.2006.888702
24. Nayak, D.R., Dash, R., Majhi, B., Prasad, V.: Automated pathological brain detection system: a fast discrete curvelet transform and probabilistic neural network based approach. Expert Syst. Appl. **88**, 152–164 (2017). https://doi.org/10.1016/j.eswa.2017.06.038
25. Mala, K., Sadasivam, V., Alagappan, S.: Neural network based texture analysis of CT images for fatty and cirrhosis liver classification. Appl. Soft Comput. **32**, 80–86 (2015). https://doi.org/10.1016/j.asoc.2015.02.034
26. Song, T., Jamshidi, M.M., Lee, R.R., Huang, M.: A modified probabilistic neural network for partial volume segmentation in brain MR image. IEEE Trans. Neural Netw. **18**(5), 1424–1432 (2007). https://doi.org/10.1109/TNN.2007.891635
27. Othman, M.F., Basri, M.A.M.: Probabilistic neural network for brain tumor classification. In: 2011 Second International Conference on Intelligent Systems, Modelling and Simulation (ISMS). IEEE, Jan 2011, pp. 136–138. https://doi.org/10.1109/isms.2011.32
28. Ozekes, S., Osman, O., Ucan, O.N.: Nodule detection in a lung region that's segmented with using genetic cellular neural networks and 3D template matching with fuzzy rule based thresholding. Korean J. Radiol. **9**(1), 1–9 (2008). https://doi.org/10.3348/kjr.2008.9.1.1
29. Döhler, F., Mormann, F., Weber, B., Elger, C.E., Lehnertz, K.: A cellular neural network based method for classification of magnetic resonance images: towards an automated detection of hippocampal sclerosis. J. Neurosci. Methods **170**(2), 324–331 (2008). https://doi.org/10.1016/j.jneumeth.2008.01.002

30. Wang, S., Fu, D., Xu, M., Hu, D.: Advanced fuzzy cellular neural network: application to CT liver images. Artif. Intell. Med. **39**(1), 65–77 (2007). https://doi.org/10.1016/j.artmed.2006.08.001
31. Mortazi, A., Bagci, U.: Automatically designing CNN architectures for medical image segmentation. In: International Workshop on Machine Learning in Medical Imaging. Springer, Cham, Sept 2018, pp. 98–106 https://doi.org/10.1007/978-3-030-00919-9_12
32. Kumar, S.N., Lalitha Kumari, S., Lenin Fred, A., Anchalo, B.S.M., Sebastian, V.P.: Feed forward neural network based automatic detection of liver in computer tomography images. Int. J. Pharm. Tech. Res. **9**, 231–239 (2016)
33. Takao, S., Kondo, S., Ueno, J., Kondo, T.: Deep feedback GMDH-type neural network and its application to medical image analysis of MRI brain images. Artif. Life Robot. **23**(2), 161–172 (2018). https://doi.org/10.1007/s10015-017-0410-1

An Optimized EMG and GSR Biofeedback Therapy for Chronic TTH on SF-36 Scores of Different MMBD Modes on Various Medical Symptoms

Rohit Rastogi, D. K. Chaturvedi, Santosh Satya, Navneet Arora, Mayank Gupta, Himanshu Verma and Himanshu Saini

Abstract The biofeedback therapies have been popular for long for the treatment of various mental challenges, stress and chronic headache like TTH Weakened cognitive and physiological control is due to TTH-originated stress and results in performance reduction and can lead to symptoms like headache, gastrointestinal disturbances, elevated blood pressure, chest pain, insomnia, peptic ulcers, sexual dysfunction, skin ailments, etc. When health-related applications are considered, MMBD is more efficient than IoT as development activities are focused on scalar sensor data. Complexity and unique nature of MMBD increment the application of IoT by using biofeedback sensors. The efforts of present chapter are made to analyze two biofeedback therapies GSR and EMG on various modes like audio, visual and audio-visual mode in the treatment of TTH over a period of 12 months. The physical, mental and total scores of subjects have been analysed on both the therapies of electromyograph and galvanic skin responses in auditory, visibility and combined process. In the present chapter,

R. Rastogi (✉) · H. Verma · H. Saini
Department of CSE, ABESEC, Ghaziabad, India
e-mail: rohit.rastogi@abes.ac.in

H. Verma
e-mail: himanshu.17bcs1053@abes.ac.in

H. Saini
e-mail: himanshu.17bcs1062@abes.ac.in

D. K. Chaturvedi
Department of Electrical Engineering, DEI, Agra, India
e-mail: dkc.foe@gmail.com

S. Satya
Department of Rural Development, IIT, Delhi, India
e-mail: ssatya@rdat.iitd.ernet.in

N. Arora
Department of ME, IIT, Roorkee, India
e-mail: navneetroorkee@gmail.com

M. Gupta
Tata Consultancy Services, Noida, India
e-mail: mayankgupta3090@gmail.com

S. Bhattacharyya et al. (eds.), *Hybrid Machine Intelligence for Medical Image Analysis*, Studies in Computational Intelligence 841,
https://doi.org/10.1007/978-981-13-8930-6_8

the author team has made efforts to check the hypothesis and check the efficacy of both biofeedback techniques on possible individual and combined modes.

Background and Purpose

Nearly 90% of all headaches are subset of tension-type headache (TTH) Rubin (Biofeedback and binocular vision. J Behav Optom 3(4):9598, 1992 [1]). Efficacy of electromyography (EMG) biofeedback (BF) has been extensively studied on tension-type headache and is proved with the help of Big Data. However, insufficient amount of study can be spot on accuracy of galvanic skin resistance (GSR) biofeedback. So far, there are no studies on the accuracy of secluded visual, audio and combined EMG or GSR biofeedback TTH Mullaly et al. (Efficacy of BF in the treatment of migraine and tension type headaches. Pain Physician 12:1005–1011 2009 [2]) and Fumal and Scohnen (Tension-type headache: current research and clinical management Lancet Neurol 7:70–83, 2008 [3]). Stress due to TTH is a very frequent occurrence in all our lives. A TTH symptom experienced for long by an individual with no relief or with increased frequency is termed as “distress”. Acute and chronic stress may affect physiological responses to TTH. Stress-related disorders are called as “psychosomatic” disorders in which body and mind of an individual are involved. Prime negative effect induced in one’s life is the result of events like deaths of close family members, surgery and divorce. In past years, such incidents have been passably related to attribute of headache. Biofeedback therapies include the use of non-pharmacological treatments that utilize scientific instruments to evaluate, amplify and feedback physiological information to the patients being monitored, thereby promoting control and manipulation of physiological parameters. No doubt, everyone will accept the key role or researchers in academic industrialization and “collegialization” of research. There is a vast increment in the research methods in enterprises and industries due to commercialization of academic process and its related organizations. Since, Internet is available in the easy rates to everybody, this facility has its own boons and curses because knowledge of different aspects are available to individual now a days in bigger amount than earlier time. Many tools for the research purposes are being created, cross-checked by validation smart techniques, and academy and industries are in race to get the patents for them so that they can create another mean for moneymaking and business in scientific creations. Many developers are working tirelessly to create the tools as per these company specifications. No doubt the SF-36 score and related questionnaire form are much popular and so the organizations with the trust for medical outcomes, quality metric designing, laboratory for assessing the health standards, incorporation of Optum have prepared the similar policies for usage of SF36 data globally. These corporations provide the license agreements for individual scholars for commercial and research purposes and do evaluation of data, responses consistence and consistency internally, and they also provide for ease the scoring and interpretations of data for the general users and researchers. The SF-36 score can also be published on Internet to enlarge its reach. The data collected can be analysed with the help of Big Data by any organization. Multimedia technology has shifted the approach we gaze at computers. The initial computers were considered as single-purpose machines that could only solve any complex mathematical problem. In 1960, huge corporate database was managed by mainframe computers. It has been observed that the use of devices related to multimedia increased at an exponential rate in all formats—audio, visual or audio-visual). Billion-dollar market is set on the bottom of electronic gadgets. For technocrats, the new challenge is to handle the huge generation of data, which is created per millisecond. This chapter deals with data related to MMBD, like reliability and scalability, accessibility, heterogeneity and quality of services for mental health.

Keywords Electromyograph · Galvanic skin response · Meditation · Auditory · Visual and audio-visual · Short Form Health Survey, i.e. SF 36 scores · Mental and physical scores · Tension-type headache · Biofeedback · Mental health

1 Introduction

1.1 Biofeedback

The idea of reducing stress with the help of self-control is known as biofeedback (Cassel 1985). It involves the use of defined techniques which are used to reduce tension. These techniques involve:

- Twilight learning
- Permissive concentration
- Competency
- Autogenic feedback training.

For a better chance for overall wellness, one needs to gain self-control by using the above-stated techniques. The practising of biofeedback signal can be helpful in reducing stress symptoms, in addition to chronic pain symptoms [4, 5].

Biofeedback works extraordinarily in case of young adults and children. A study was conducted in a college where students attended workshops, kept a diary and had worked in individual sessions. Changing their sleeping patterns along with a stress-control log, an increase in GPA and focus were observed in the students who participated. Several other case studies have also observed a reduction in stress, in its participants [6, 7].

1.2 Mental Health Introduction

The system of human is completely destroyed by lifestyle change and drug treatment. About 25% of population of the UK suffer from mental health.

The USA consists of highest percentage of people spot with mental health [8, 9].

1.3 Importance of Mental Health, Stress and Emotional Needs and Significance of Study

Behaviours, feelings and perceptions that regulate an individual's comprehensive level of excellence, success, effectiveness and happiness establish functioning of a person, said by Komhauser (1965).

It relies on improvement and confinement of objectives to consent. An individual who is minutely confused by stress and strains of day-to-day life and is firm in his intentions is basically, a mentally healthy person. Thus, "Gestalt" view of an individual is the primary concept of mental health. It consolidates the concept of behaviour and personality distinctive. It can also be related as the behavioural distinctive of an individual [10].

1.4 Meaning for Mental Health

According to WHO, mental health includes "personalized benefit, grasp self-control, self-government, capability, boomer subservience and self-realization of an individual's sentimental and psychological potential among others" [11, 12].

WHO further states that the happiness of a person is encircled in the recognition of their strength, handling ordinary stresses of life, advantageous effort and grant to their nation. However, the mental health is majorly characterized by the silent appraisal, challenging professional and cultural differences [13, 14].

1.5 Definitions

Mental illness is described as know-how of harsh and dreadful psychological syndrome in exchange of damaging normal working. Such syndrome is depressed mood, delusions, anxiety, hallucinations and obsession thinking (Fig. 1).

1.6 Factors Affecting Mental Health

According to Joel L. Young M.D, nine lifestyle components that can affect our mental health are as follows: performance and enterprise level, smoking, diet, substantial activity, injustice, communal and public activities, liaison, introspection and other rumination approach and vigorous sleep [15–18].

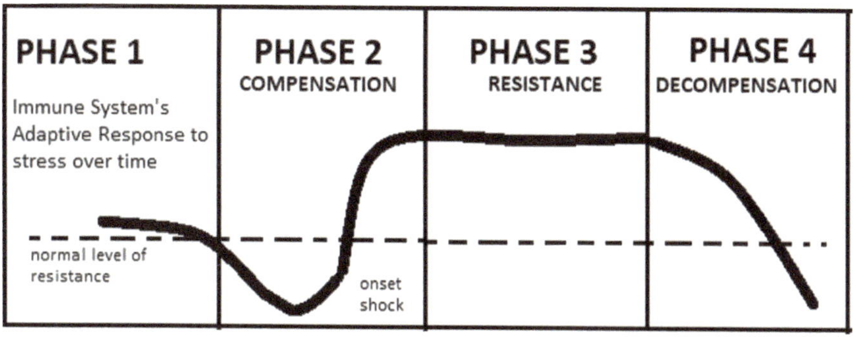

Fig. 1 Phases of stress [27]

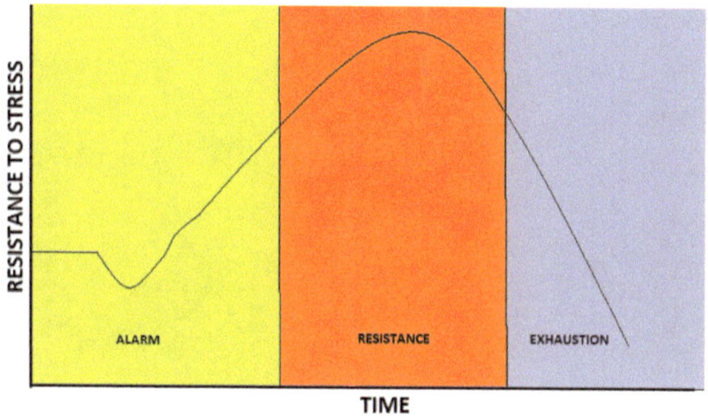

Fig. 2 General adaptation syndrome [33]

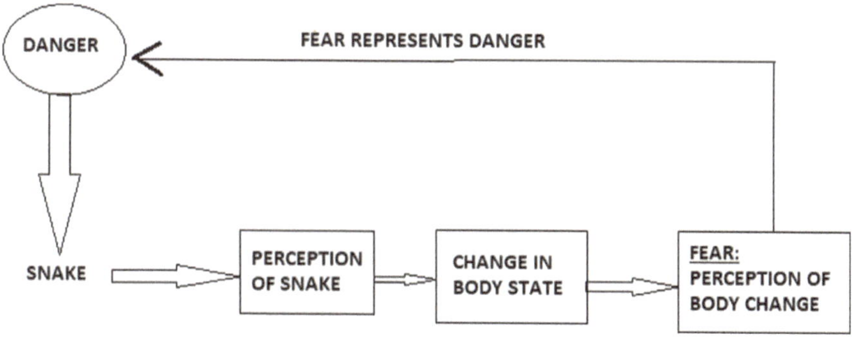

Fig. 3 Stress growth simulation

1.7 *Models of Stress: Three Models in Practice*

1—General adaptation syndrome (Figs. 2 and 3). Stages: alarm, resistance, exhaustion

2—Selye eustress and distress (Fig. 4)

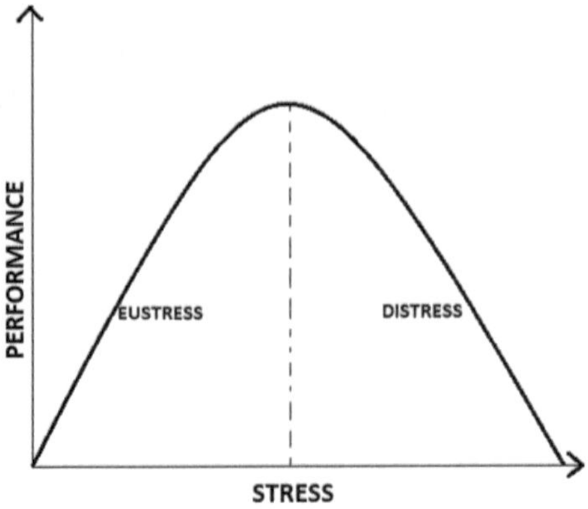

Fig. 4 Stress performances [30]

Fig. 5 Stress type mood states [32]

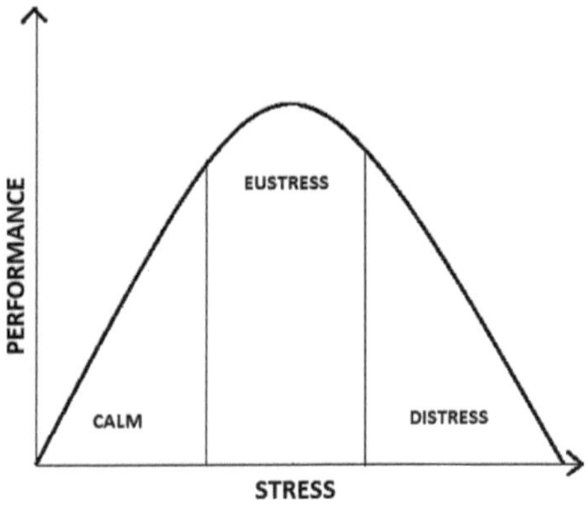

3—Lazarus: cognitive appraisal model.

1.7.1 Variety of Stress

Stress management can be difficult and complex, as there are several types of stress: (1) intense stress (emotional problem, muscular pressure), (2) anecdotal intense stress, and (3) incurable stress, each with its own temperament, indication, span and treatment access (Fig. 5).

1.7.2 Causes of Stress

Korchin (1965) have distinguished some classes of stressful situations as follows:

1—uncertainty and under stimulation, 2—informative overload, 3—danger, 4—ego control failure, 5—ego-mastery failure, 6—self-esteem danger and 7—other esteem danger [19, 20].

1.8 Symptoms of Stress

Stress warning signs and symptoms are as below

(1) Cognitive symptoms: incapability to concentrate, memory problem, poor wisdom, seeing only the negative, anxious or extending thoughts and steady concern
(2) Emotional symptoms: irascibility, moodiness, anxiety, inability to relax, feeling submerge and sense of unhappiness
(3) Physical symptoms: anguish and ache, diarrhoea, vomiting, vertigo, rapid heartbeat, low of sex drive, chest pain and frequent colds
(4) Behavioural symptoms: ignoring responsibilities, alienating from others, sleeping suspiciously or unsuspiciously, consuming more or less, nervous habits (e.g. nail biting, pacing) and using drugs, alcohol and cigarettes to destress [21–23].

1.9 SF-36 Introduction and Background

RAND was the factor which helped in the advancement of Short Form Health Survey; it is an ingredient part of Medical Outcomes Study which is known as MOS [1]. SF-36 is a series of collective, consistent, and efficiently conducts quality-of-life system. These systems depend on subject self-describing and are now extensively handled by educated care grouping and by Medicaid for conventional control and appraisal of responsible conclusion in grown-up patients. SF-36 is feasible in all leading world languages like English, Mandarin, etc [24].

For people groups other than Asian immigrants, combinations on different scales and perception of mental and substantial fitness have been vastly studied. Judgment of SF-36 Health Survey wishes to address for artistic divergence in those communities and finds many interesting conclusions [25, 26].

1.10 The Mechanism and Techniques of SF-36

To analyse the experiments of health sector, the Short Form Health Survey is very much important and helps a lot. “SF-36 health survey” occurred, and it discovered more than 9000 items by PubMed [27].

The SF-36 evaluates eight scales:

- Physical functioning (PF)
- Mental health (MH)
- Bodily pain (BP)
- Social functioning (SF)
- Vitality (VT)
- General health (GH)
- Role emotional (RE)
- Role physical (RP).

Fundamental analyses demonstrated that there are couple of different theory restrained by SF-36: a mental scope, characterized by Mental Component Summary (MCS), and a physical scope, characterized by Physical Component Summary (PCS). All ranges do grant in various dimensions to the count of the pair MCS and PCS allotment. The factual estimation of SF-36 arbitrary calculates MCS, and PCS demands the practice of certain conclusion [28, 29].

Eight scaled scores are the evaluated quantity of the query in their category. Each scale is honestly mutated into a 0–100 scale on the presumption that each query carries equal substance. Higher the score, lesser the disability. A total of zero is proportionate to superlative disability, and a total of 100 is proportionate to no disability. To measure the scores, it is mandatory to understand the concept of Big Data. Researcher needs to figure out number of scores, as pricing depends on these scores only.

“SF-36 Total/Global/Overall Score” does not support SF-36 scoring manual to forecast particular part of health-related satisfaction. According to its creators, it is useless to merge two SF-36 arbitrary measures to construct a total count of health-related satisfaction. In antipathy of this, some researchers seek to adopt and to hypothesize clumsily from such measures [30].

Method

For scoping scrutiny, Preferred Reporting Items for Systematic Reviews and Meta-Analyses (PRISMA) method along with Big Data is adapted. Scoping reviews grant the involvement of several study designs, while an efficient review may involve few study architectures. Benefit of SF-36 Total/Overall/Global Score is covered in this chapter [31].

Guidance for transforming a person score to z-scores and to contribute regulated mixed scores (standard deviation 10, mean 50) for certain populations is published on Australian Longitudinal Study on Women’s Health. A fascinating mark of the chapter is to calculate physical health scores and mental health scores using Big Data [32].

1.11 Uses and Limitations of SF-36

Uses

Evaluating individual patient's health status
Investigating the moneymaking of an analysis
Comparing and monitoring disease anxiety.

Limitations

Sleep variable is not considered.
Low feedback rate of >65 population.

1.12 Deep Meanings and Applications with Variations of SF-36 in Variety of Fitness Scores

Relationship between health state utility values and the minimally important difference is stated in case of the SF-6-D 2003 by Stephen J. Walters and John E. Brazier in Health and Quality of Life Outcomes, Vol. 1, No. 1, p. 4.

1.13 SF-36v2 Health Survey

The SF-36v2 Health Survey inquires 36 arguments to test practical health and health from the subject's perspective. It is a rational, dependable and authentic measure of physical and mental health that can be concluded in five to ten minutes. It can also be referred as a universal health survey as compared to a disease-specific health survey, as it can be used across age (18 and older) and for any disease, whereas disease-specific health survey targets on a peculiar disease [33].

The present chapter is relevant to administrators, psychoanalyst, patients and investigator, and all have different applications. Applications include: predicting medical expenses, calculating health enhancement or downturn, analysing disease anxiety across community and appraising treatment performance [34].

1.14 Types and Other Versions of SF-36

The SF-36v2® scores Mental Component Summary (MCS) and Physical Component Summary (PCS) and separately for the eight-well-being sphere. Resemblance can be made with other universal generic health surveys as SF-36 is based on norm-based scoring. A choice-based service index, known as SF-6D, is also feasible to recognize commercial interest. Numerous modes such as online and PDA are offered. The SF-36v2® is convenient in beyond 170 transcription in both intense one-week and ideal four-week duration [35].

1.15 Distinctions Among SF-36 and RAND-36

Calculation of pain scales and general health is the major variation. The variation in scoring is encapsulated by Mazel, Hays and Sherbourne (Health Economics, 2: 217–227, 1993).

The Short Form (36) Health Survey 12 is frequently manipulated in health finance as it is an estimate of health status. SF-36 is originated from Medical Outcome Society (MOS), done with the help of RAND Corporation. Commercial version of SF-36 has been released by group of researchers, while the original license-free (free from RAND) SF-36 is convenient in public territory. SF-12 is a shorter version [26, 36].

2 Previous Studies (Literature Review)

2.1 Tension-Type Headache and Stress

A US study has displayed that the ratio of absentees for TTH to migraine in population was three to one, which was further proved by a couple of Danish studies. In a study by Fuh et al. 2008 where a cohort study was supervising to study the outcome of elderly patients with long-term tension-type headache (CTTH) in a span of 13 years, the authors found 30% of patients with CTTH evolved to persistent migraine (CM) or patchy migraine (EM) [37, 38].

Tension-type headache is objectively and pathophysiologically miscellaneous. The multiplex interrelation of the assorted pathophysiological factors of TTH makes this chaos often tough to treat. Different therapeutic measures have been recommended to be used in sequence or in combination. Subdivided therapies for TTH are short term which include futile treatment of each attack (generally pharmacological) and long term, which include deterrent treatments (pharmacological also non-pharmacological) [39].

Several non-pharmacological treatments have been recommended for management of TTH, being physical therapy, craniocervical training, oromandibular treatment acupuncture, relaxation therapies, cognitive training biofeedback, etc. However, the results of alternative therapies for efficacy of most treatment modalities are scant as per scientific study.

Biofeedback is the leading method for handling behavioural headache. It is frequently used in handling tension-type headaches and migraine. Biofeedback is found to be fruitful in diminishing the severity and prevalence of headaches. Reduction on medical utilization is also observed, when biofeedback is used [40].

2.2 Internet of Things and MMBD for Health Applications

In the present time, we may have a weapon named science which may work as deity or devil. Though it is sure that science cannot cure us from diseases, definitely it may give us the cure mechanism in its easiest form, say it accessing the treatment resource, doctor or medications. The technology may also reduce our heavy medical bills and may help us to decide the routine for regular health check-up from our place (office or home) to doctor at hospitals. The technology may also help the doctors for right diagnosis so that it diminishes the helplessness of patients for being hospitalized [17].

The Internet of things is presenting a new paradigm shift and has vast no. of applications in various health domains, say physical or mental. The advance progress of medical industry and scientific gadgets have made life simpler, providing patients a quality treatment. In long term, it makes the social and national life easy and supports the higher happiness and health index of general human being. For a developed nation, the health services are integrated component and technology with IoT and MMBD has made it easier, accessible and available to all [17, 41].

2.3 Multimedia Big Data

Mobile is an exceptionally marvellous gadget which helps us to efficiently represent various gestures and actions. In gadget world, variety of data (millions of terabytes) is generated in a single day.

This data needs to be handled efficiently. The different sectors of Big Data are

- Storage
- Analytics
- Indexing
- Mining and retrieval.

A little progress is seen in the field of managing Big Data, and because of the critical issues like public handling and security of the data, less research is continuing towards Big Data analysis. The present chapter focuses on questions related to managing huge amount of data which is produced from the interactions of IoT devices.

Surveys conducted by team of experts behaves as a path between the issues related to Multimedia media and the proposed solutions given by Big Data Researchers,

The major issues to be addressed are

- Frameworks
- Applications
- Advantages and disadvantages of existing model
- Efficient guideline
- Directions for future.

Only those surveys which aim at recent skills, extensive data and multimedia management techniques will help in advancements of Big Data.

It has been observed that with the advancements in smart phones and tablets, younger generation is spending lot of time on social media and Internet, to communicate and to share information. Multimedia data is usually noisy and unstructured as compared to traditional data which consists of texts and numbers. Hence, conventional data analysis is not suitable to handle multimedia (complex) data. Big Data analysis requires (Adler 2016).

- Heavy and efficient storage
- High end processor
- Maintenance
- Manipulation.

Large and unstructured multimedia data needs to be managed in more sophisticated and comprehensive way (Shyu et al. 2007).

2.4 Big Data Techniques for Large-Scale MM Data

Large-scale MM data extraction, wherein data is extremely large, can cause a serious waste of time and storage. Advent of new technology and digital MM has enabled the users to access the data in cost-effective storage system efficiently. Oracle has enlisted the benefits of digital MM in broad as follows:

(1) Digitally stored MM data is less prone to loss.
(2) This kind of archival storage enables devices connected to IoT without human intervention.
(3) Storage of meta-data in digital databases makes searching MM file easier.
(4) Network of data servers and archives enables quick data exchange.
(5) Interoperability with various media devices made sharing of media content from different sources possible.
(6) Digital MM capturing devices are ubiquitous. Several technologies are proposed to capture MM data effectively. The lead tools capture MM data from wide variety of media foundations such as network and DV cameras, webcams, TV tuners and capture cards in real time. It is necessary to apply systematic approach to capture and represent different multimedia type, which are as follows:

Visual data: Visual data, including images and videos, is rich in information and the semantic contents form almost 80% of all unstructured Big Data (Venter and

Stein 2012). The huge size of visual data compared to textual data is its biggest challenge, which is the reason for the use of Big Data solutions being used.

Advancement of mobile technologies, high-performance cloud computing, low-cost storage and sharing website have been the reason for the exponential growth of visual data in recent years. Video surveillance systems, autonomous vehicles, video and image retrieval systems, health care and use of "Spyder grams" to interpret SF-36 are few applications of visual data analytics.

Audio data: Audio or speech data is another data type widely seen in MM applications. Social media, industrial machines and medical devices are a few examples of the Big Data sources that need real-time audio analytics.

The procedure of retrieving meaningful information from unstructured aural data is known as audio analytics. Call centres are an example where audio and speech analytics are used to recognize spoken words from thousands of calls and to improve the interaction with costumers.

Big Data techniques such as Hadoop and Spark are being utilized to efficiently mine unstructured data and get a better understanding of users' needs.

Text data: Structured or unstructured multimedia data such as meta-data, web pages, social network feeds and surveys are included in this. In context to MM, large percentage of data is unstructured, because of which text analytic techniques have been developed to extract relevant data from large-scale unstructured text sources and transform it to a structured format (Ye et al. 2016). Text analytic can analyse social network feeds on a specific entity to extract and predict users' opinions and emotions towards such an entity, which can be utilized for further decision-making, further which is used in biofeedback (Gandomi and Haider 2015).

Sensor data: Technologies such as the cloud, cheap storage, and fast processors have boosted sensor data tremendously, which causes huge analysing and processing challenges (Zaslavsky et al. 2013).

Examples of such applications in Big Data include monitoring environmental conditions, airplane engines, industrial equipment, meteorological patterns and patient vital statistics (Philbin et al. 2007). Similarly, biofeedback uses computation of brain frequencies at different levels of consciousness to determine the state of sense of a person. It is estimated that the number of connected things over IoT can reach 50 billion by 2020 (Reed et al. 2012).

With the development of multimedia data analytics, some benchmark and popular large-scale datasets are emerging. Image-related datasets include Oxford 5K and Flickr ImageNet (Image-net. Org 2016), NUS-WIDE (Chua et al. 2009), Tiny Images Dataset (Tiny 2016) and so on. The video datasets contain YouTube-8 M (Abu-El-Haija et al. 2016), MediaEval 2015-SAVA (MediaEval 2015), TRECVID 2016-AVS (Trecvid 2016) and so on.

2.5 *Benefits of IoT*

Real-time monitoring and reporting is life-saving in the event of emergency situation such as heart asthma attacks and failure. Real-time monitoring via smart medical devices through smart phone apps and data connection can help by the exchange of information with the physician.

A study conducted by Centre for Connected Health Policy has shown a reduction of 50% in re-admission rate of heart failure patients within 30 days due to remote monitoring system.

The data collected by IoT devices such as blood sugar levels, blood pressure and oxygen level, weights and ECGs are stored in cloud and are shared with an authorized person to monitor the data regardless of place, time or device.

IoT helps in automating the patient workflow by healthcare mobility and next-gen tech facilities. Features such as interoperability, information exchange, machine-to-machine communication and data movement make healthcare service distribution effective [18].

Connectivity Protocols
Technology-driven set-ups such as Bluetooth LE, Wi-fi, Z-wave and Zigbee reduce the cost, by limiting unnecessary visits, utilizing better-quality resources and improving the allocation and planning.

Data Assortment and Analysis
IoT devices can collect, report and analyse the data in real time over cloud where providers get access to final reports and graphs only and hence cut the need to store the raw data. Moreover, vital healthcare analytic and data-driven insights help the organizations to speed up decision-making and reducing errors.

Tracking and Alerts
Devices use IoT to collect important data and transfer it to doctor for real-time tracking, in case of life-threatening circumstances [18].

On-time reports help doctors make well-versed decisions about the health of patients.

That is, IoT allows hands-on treatments and provides better results of patients.

Remote Medical Assistance
Mobility solutions in health care will enable the medics to instantly check for patient's ailments. Also, numerous healthcare delivery organizations are working to build next-generation tech machine that deliver drugs based on patient need and data available related to ailments over cloud. IoT will help to improve the patient's treatment in hospitals and reduce medical expenses [40, 42].

2.6 Challenges

Data Security and Privacy

Data security and privacy are among the most significant threats to IoT. Since IoT devices collects and transfers data in real time, it lacks data protocols and standards and has a significant ambiguity regarding data ownership regulation.

Cybercriminals can take advantage of these loopholes and can hack the system to compromise with information of both doctor and patients.

Patient's data may be misused to create fake identities in order to procure medical equipment and drugs and sell those later.

Integration (Multiple Devices and Protocols)

Connecting multiple devices might cause interference in the implementation of IoT due to unavailability of uniprotocol for manufacturers all over the globe. These differences in communication standards complicate and deter the process of data aggregation. This slows down the whole process and reduces the scope of implementation of IoT in health care.

Data Overload and Accuracy

We know that different communication protocols and standards make data aggregation difficult. Yet the data collected by IoT devices is used to gain important results, but its huge amount makes it extremely difficult for authorities to analyse, which greatly affects the quality of decision-making.

Cost

The high healthcare cost has become a headache for everyone. Due to high cost facilities, patients from developed countries prefer healthcare facilities of developing countries to cut their costs to as less as one-tenth which gives rise to "Medical Tourism". IoT in health care is a fascinating concept and promising idea.

3 Independent Variable: Emotional Need Fulfilment

Anjana Sharma (2000) said that the ability of an individual to control his anger and to stable his mood or to achieve success in work varies as per one's capabilities. Majorly 80%, it depends on emotional intelligence and 20% on one's verbal intelligence. Basic human needs for emotional fulfillment are universal and include: love, acceptance, affection, feeling valued, appreciated, secure, companionship, admiration, trust, respect, understanding, conversation and communication [22, 42].

Maslow in 1954 has given a hierarchy of human needs in his "Theory of Self-Actualization" which are psychological, safety, belonging, love, esteem needs and need of self-actualization.

4 Effective Spiritual Tool Meditation with Approach of Biofeedback EEG

See Figs. 6, 7 and 8.

5 Sensor Modalities

5.1 Biofeedback-Based Sensor Modalities

Three professional biofeedback organizations, the Association for Applied Psychophysiology and Biofeedback (AAPB), Biofeedback Certification International Alliance (BCIA) and the International Society for Neurofeedback and Research (ISNR) (Fig. 8 and Table 1), arrived at a consensus definition of biofeedback in 2008: "Biofeedback is a process that make us to learn to change physiological activities around us and enable us to change performance and health (Fig. 7). It is measured by instrument such as breathing, muscle activity, heart function skin temperature and brainwaves (Fig. 6)."

The information presented in the paper is a depiction of change in emotion, behaviour and thinking along with physiological changes. These changes may be continuing without the use of instrument [7].

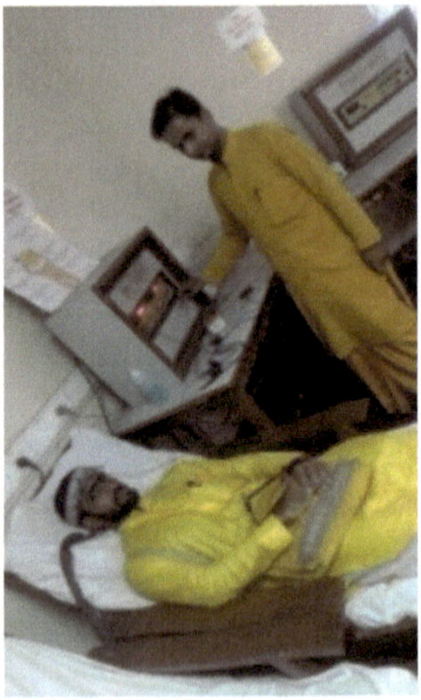

Fig. 6 Biofeedback measurements

DEPENDENT VARIABLE:-

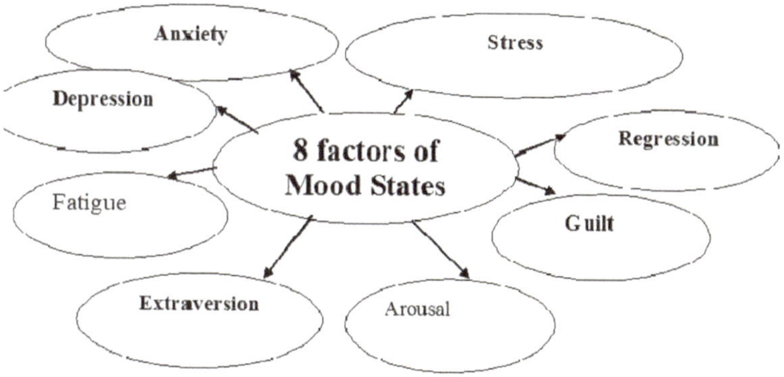

Fig. 7 Factors of mood states

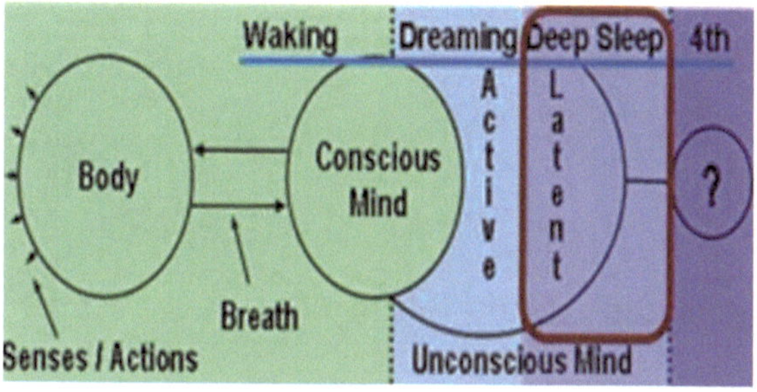

Fig. 8 Mind body and consciousness' [30]

5.2 *Electromyograph*

The "Muscle Whistler" shown here uses EMG electrodes is was earlier developed by DR. Harry Garland and DR. Roger Melen 1971 [8, 9] when the cells are electrically or neurologically activated, it is detected by electromyograph using one or more active electrodes placed over target muscle and reference electrode placed over active muscle. The SEMG is measured in microvolt [10, 11].

Table 1 Rays emitted by an individual in different mood states [30]

Name of level	Frequency in hertz/cycle per second	Description
Beta	14–30	Typical level of daily mental activity associated with tension or stress
Alpha	8–13	Relaxed, passive attention, often consider the goal of relaxation exercises. While this is a very relaxing state and useful to be practised, it is sometimes incorrectly thought to be the goal of Yoga Nidra
Theta	4–7	Normally considered to be unconscious, possibly drowsy or half asleep. This level is also sometimes incorrectly considered to be the level of Yoga Nidra, where there is still the experience of images and strems of thought
Delta	0.5–3.5	Considered to be unconscious dreamless. **In Yoga Nidra, the brain waves are at this level**, as practitioner is in conscious deep sleep, beyond the activities experienced at the other level

6 Experiments and Result—Study Plot (Figs. 6 and 7)

6.1 Study Design and Source of Data

The trial on clinical surrounding was performed in an idealistic atmosphere which was a single blinded but prospectively controlled environment, and it was random in nature.

Subjects were belonging to various clinics with specialty of neurology and subjects referred by neurologists to the outpatient of department of physiotherapy and department of psychology of Dev Sanskriti Vishwavidyalaya affiliated with Haridwar-based Shantikunj Ashram, and they were used in biofeedback techniques.

6.2 The Total Time for the Experiment and Agreement with All Patients

Subjects under this study were appointed, interviewed and given consent to be part of study from mid-July 2016 up to December 2018 and followed up till January 2019. Firstly, the written consent was given to them and all subjects were counselled to be participant in this study of National Importance, Ethical committee of ABESEC, Ghaziabad, was responsible to check the validity, process and ethics follow up in the recruitment process of subjects. The study was also registered and authorized from the Clinical Trials Registry, Department of Health of Government Of India. The ABESEC affiliated to AKTU, Lucknow, had granted the permissions and ethical clearance required for conduction of the experiment.

6.3 Sampling Design and Allocation Process

Randomized sampling technique and the method of lottery were adopted to avoid any clashes, and the decision and fixation of groups for the subjects were completed by this way. It established the fairness of the experiment. The study was primarily focused for the subjects who were suffering from chronic TTH/stress/anxiety. There were some eligibility limits for the subjects participating in the experiment, and those who could not follow it were asked to be out from the game. All eligible candidates were allotted their respective group with the lottery method to ensure the randomization process.

In a bowl, the experimenters kept the paper pieces with written as sequence number of groups all subjects were informed to take one piece of paper. The patients were allotted the same group, which number they obtained in their papers.

Four groups were defined like first was EMGa which showed the auditory group, second was denoted as EMGv as the visual group, third was EMG auditory-visual (EMGav) group and last was control group on which no experiment was conducted and their observation was carefully recorded, typically known as control group.

6.4 Estimated Measure of the Sample

By below method, sample estimation was calculated.

$$N = 2(Z\alpha + Z\beta)2 \times pq/(p1 - p2)2 \tag{1}$$

Probability of type I error was set at 0.05

Power of the study was set at 80% (0.8)

$p1 = 1.0$ and $p2 = 0.75$ were the mean differences of pre- and post- (baseline to one year) average frequency of headache per month in the EMG biofeedback training group and pain management group, respectively, from a study by Mullay et al.

$p = 0.875$ was calculated as $(p1 + p2)/2$, and $q = 0.125$ was calculated as 1-p. These are used in Eq. 1.

The sample size thus calculated was 26.6 per group. To entertain for dropouts, the sample size was chosen as 30 per group.

6.5 Study Population

6.5.1 Inclusion Criteria

Subjects included in the study were:

- Subjects with headache fulfilling the benchmark for TTH proposed by International Headache Society
- Both males and females between 18 and 65 years.

6.5.2 Exclusion Criteria

Subjects excluded from the study were:

- Subjects who underwent complementary alternative medicine interventions in the past 6 months
- Subjects with other headache types as described in International
- Classification of Headache Disorders (ICHD)—II classification
- Subjects with the presence of more than one type of headaches in addition to tension-type headache
- Subjects whose headache began after the age of 50 years
- Subjects with serious somatic or psychiatric disease

Subjects with history of drug abuse or use of analgesics and triptans >10 days per month.

6.6 Intervention

A meeting was conducted for all the subjects to inform them the procedure for individual treatment, brief of biofeedback machines.

The whole experiment was conducted in a very serious atmosphere of research laboratory of Department of Psychology. In addition to this, DSVV—Haridwar University arranged a minimum interruption-free zone in terms of a least problem of noise and lightening process.

There were at max seven sessions for each subject of electromyograph and galvanic resistance with 7 sessions and each of twenty minutes.

No subject was allowed to miss any session without prior approval of competent authority.

If any single person will quit the show, he/she will be informed the latest date and he/she will undergo similar training again and cover the agenda. If any of the subjects misses a session, the methodology of biofeedback will be conducted later and that subject will appear in the interview. And all are expected using EMG-IR Re-trainer.

EMG BF machine provided auditory and visual feedback. Auditory feedback was in the form of clicks which increased in frequency and became a continuous sound with increase in frontalis muscle tension and no sound with relaxation of frontalis muscle. Visual feedback on the display monitor was in the form of glowing bars along with a numerical display which displayed the relative EMG activity of frontalis muscle in figures. The number of glowing bars was directly proportional to tension in the frontalis muscles. The training was given at 2% sensitivity with actual GSR. No changes in sensitivity were made throughout the training sessions.

Cleaning the skin was done using cotton pad which was soaked into spirit. The skin ready surface EMG electrodes made by Ag-AgCl, triode electrodes were applied nearby and at the centre of each eyebrow. There were 118 subjects who were placed including the no medication/control group. They all were belong to our club spirituality and yoga club also.

We expect that skin preparation was done as per that of very old aged too. Due to blind review (single) in case of machine experiment too, subject will never know the groups of biofeedback machines of EMG and GSR therapy. The whole info was with material to the instructor, and whole info was given to instructor who will pass the instructions, respectively.

EMGa group received respective auditory feedbacks. The subjects were instructed to reduce the tone and frequency of the sound which would help them achieve relaxation. During the session, the monitor on which the visual display was present was moved away from the field of vision of the subject.

Similarly, EMGv BF visual groups were instructed to reduce the number of glowing bars. During the treatment session, the volume of the auditory tone was muted.

EMGav audio-visual groups were instructed to reduce the tone and frequency of sound and decrease the number of bars simultaneously.

Subjects in the control group were not asked to manipulate either the visual or auditory display. They were only informed that their stress levels were being recorded through the machines.

The subjects were requested and suggested in terms of discipline to practise relaxation at home, both during the course of therapy and at the end of 15 sessions, in a way similar to the relaxation during the biofeedback therapy sessions. However, compliance of the subjects in the home programme was not monitored. All subjects were allowed to take the medications prescribed by their treating physicians especially if they were preventive/prophylactic medications. The subjects were also made careful by principal investigator of experiment to take any decision related to consumption of analgesic and other meditations, e.g. abortive and palliative, and these will be only then used when the prescribed medication was not much useful and the headache was painful and unbearable.

6.7 Primary and Secondary Variables

The AHS—American Headache Society—Behavioural Clinical Trials Workgroup has issued certain recommendations in 2005, so as per this the selected variables (primary) for this particular experiment were:

Avg. prevalence of headache (counted as per week)
Avg. extent of headache (counted as per week)
Avg. prevalence of headache (counted as per week).

The endorsement for headache prevalence broadcasting is consistent with the IHS guidance for controlled trials of drug treatments [20, 21] and will aid meta-analyses and other contrast across studies of various mediations [24, 27].

In this experiment, the variables (secondary) were targeted as short form of survey variable popularly called as SF-36, scores for physical and mental and total capabilities and QoL—the quality of life.

6.8 Analgesic Consumption

Referred as non-headache measures by the secondary variables and standardized by the Behavioural Clinical Trials Workgroup of AHS—American Headache Society, 2005.37.

6.9 Measurement of Resultant Parameters

Data was collected in whole demographics which was being collected irrespective to any sex, TTH intensity and gender discrimination. All subjects were treated for clinical trial.

6.10 Record Keeping for the Pain Parameter

The AHS has issued the instruction through the Workgroup responsible for clinical trial the guidelines, all subjects were asked for maintaining the events of migraine or chronic TTH, and they were supposed to record the duration, intensity and frequency of these incidents in regular and honest and disciplined manner.

1. Mean of the occurrence of chronic stress on weekly basis.
2. Mean duration or time details for occurrence of headache weekly basis in hours per week.
3. Mean value of the VAS scores for chronic TTH on weekly basis.

4. The consumption of medications was also reported through pain diary.
5. Short Form Health Survey form with 36 questionnaire was used in English and local mother language (Hindi here in this case).

SF-36 is a versatile, 36-question Short Form Health Survey which produce a psychometrically based physical and mental health summary measures and a total score.

Its authenticity has been estimated to exceed 0.80 using both internal consistency and test–retest practices. The consumption of medications was also reported through pain diary with consult of each subject's doctor.

6.11 Acquisition of the Data

Below time limit was checked to collect the data required.

Starting of the Experiment (Baseline): The score of the first, third, sixth and twelfth month was collected at the end of month. Also, the score of primary and secondary variable was collected a week before the experiment conducted.

6.12 Mathematical Description

The percentage of the subjects in our intergroup analysis with $p < 0.05$ timely responses. The process was done that the baseline data was reduced with individual scores.

Capturing of Data and feel the graphical representation
All the data was collected in a pain diary which later on was maintained in a Excel sheet and the visualization was done with licensed Tableau software.

To present effectively and analytically, various visuals are used for representation; the different visuals are used for analysis and representation purposes. Re-trainer (Chattanooga group Inc, U.S.A.) and GSR-BF has contributed to get visuals and GSR machine, GPF-2000 from organizers was used for experiments with references available.

6.13 Estimation of Hypothesis

There is a big effect of medication on stress and mental health and also a great reduction using therapies like EMG and GSR on chronic TTH over a period of 12 months. However, there is no difference seen in mental health using EMGa, EMGv and EMGav BF therapy. So there is no relation between stress and EMGa, EMGv and EMGav BF therapy.

7 Process for Meditation and Data Collection Procedure

Subjects were asked to correct data given in the following instructions by the instructor

- Adopt.
- Try.
- Prepare.
- Keep the head neck and spine upright.
- Close the eyes and relax the whole body.
- Many experiments are based on the Alpha-One, Alpha-Waves. On Machine readings, reduced waves are indicator of calm, cool and steady and concentrated domain state. Readings exhibit that there are many persons who can perform well and sincerely.

There is 32-bit channel machine on which operation shall occur and students performing experiments should not share the activity of one's brain with others.

In the record keeping, one should give the memento in one's booklet. The experiment team expects the record keeping diary from each subject. The diary of the subjects was used to record the average intensity, duration and frequency of chronic TTH (Fig. 9).

8 Description of Resultant Value, Analysis in Complete Swing and Data in All Modes

8.1 The Trends on Different Scores of EMGa Versus GSRa

The complete experiment lasted for 1 year of duration on different patients suffering with acute stress problem, and their Short Form Health Survey scores with mean value of mental, physical and complete scores were considered. The below graph is

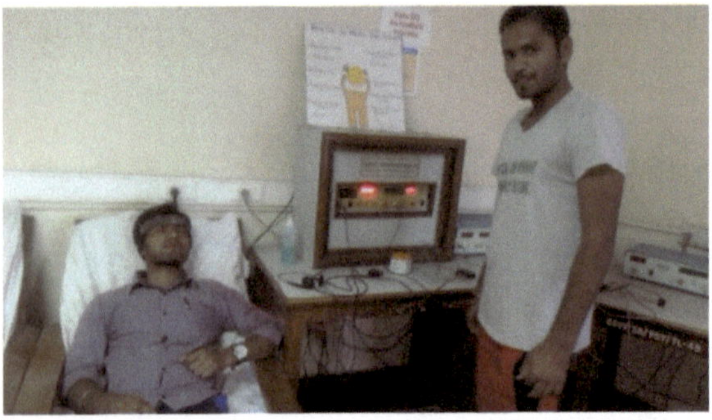

Fig. 9 Biofeedback experiment on different subjects

representing the behaviour progression of different modes of auditory techniques in electromyograph and galvanic skin resistance machine (Fig. 10).

GSRa: Initially for the first month, GSR is showing the increment in the physical scores. The above graph also shows that out of the 8 components of SF-36, the patients' fluctuation is very high in GSRa method. It is continuously varying in 3-month and 6-month time duration, and after 6 months, a rapid downfall shows the disinterest of patients from the concerned therapy, showing GSRa less effective.

It also shows the continuous dropout in the therapy throughout the experimental period.

The physical score value (calculated with the help of Big Data) of the subjects under study also increases from 53.38 to 74.57.

EMGa: On the other hand, the EMGa is showing gradual but consistent increment in the physical score of the subjects. From the day 1 up to total experiment time period, the EMGa is showing nice performance in the achievement of physical scores by the subjects under study.

With the help of Big Data, we concluded that the dropout's ratio and tendency are less compared to GSRa and the physical score difference of the subjects is also very high from 58.41 to 82.83. Also, the Tableau graphical analysis shows a higher subject dropout for treatment of chronic TTH in GSRa compared to EMGa.

So, it can be concluded that EMGa is better than GSRa in parameters like fitness and muscle strength. The EMGa gives higher and faster increment to the subjects in the same time period, so it is more efficient (Fig. 11).

GSRa: If we see the graphical comparison for mental fitness, the GSRa is showing very fast increment straightaway in initial 6 months of experiment but rapid downfall in the post 6 month of time duration.

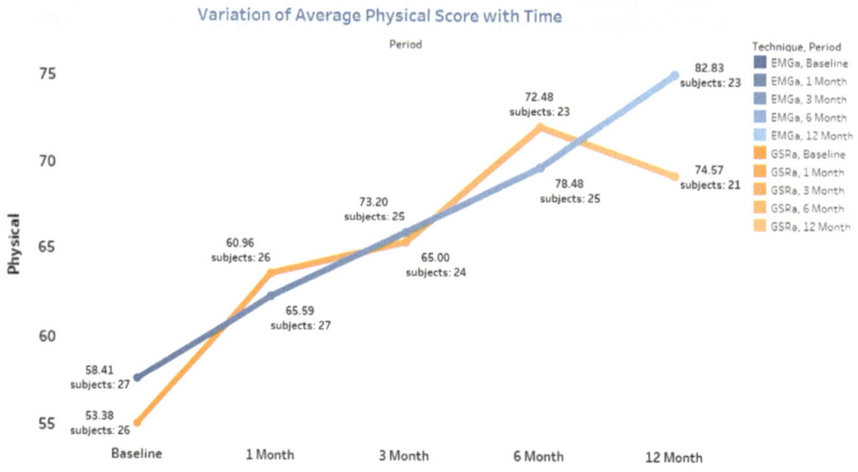

Fig. 10 Trend of average of physical scores

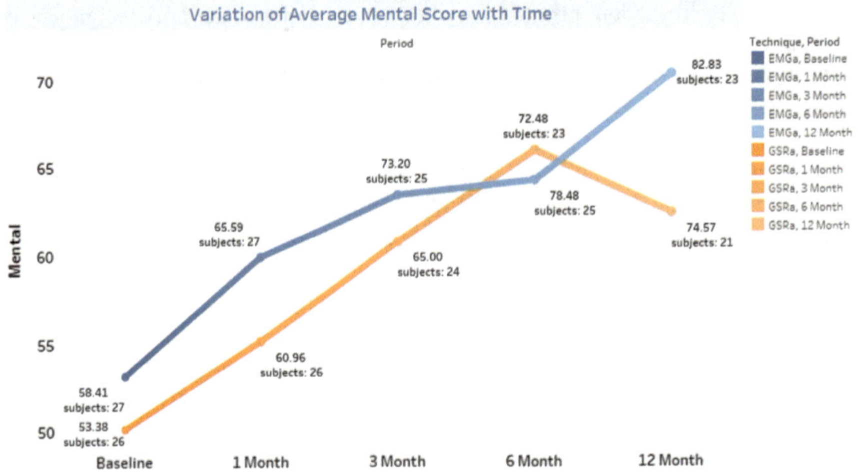

Fig. 11 Trend of average of mental scores

The complete data was analysed with the help of Big Data, and the results show us that, he subject dropout is from 26 to 25 means just 1 in initial 6 months but rapid dropout of 4 in next 6 months which shows the less efficiency of the therapy. Also, the GSRa and EMGa start near about similar physical scores of the subjects 53.38 and 58.41.

EMGa: EMGa for the increment of mental scores of the subjects under audio mode shows the drastic increment and less dropouts as per the graph. The mental fitness, vitality and personal satisfaction of the subjects have been increased up to 58.41 to 82.83. So, the chronic TTH has been improved. From the above facts, one can easily conclude that EMGa is showing here too better performance than GSRa in mental score achievements (Fig. 12).

The total scores are subjective and simply calculated by sum of physical and mental scores.

The above graph of total scores clears depicts that the rate of increment of total score is less in GSRa as compared to EMGa and in terms of subject dropouts in given time period and the value of the total scores achieved by the subjects, the EMGa therapy is showing the better performance than GSRa.

EMGa is showing the continuous improvement in the scores on its subject group in straightforward manner; rather, the GSRa is fluctuating a bit for its lot. So overall both are effective, but EMGa is better than GSRa.

The GSRa therapy has lost the confidence by the fluctuation and indefinite behavioural pattern as in almost all the cases, it has shown good performance in first 3 months and then uncertain attitude in remaining experimental life.

EMGa has maintained the confidence value, and it is relied on the subjects continuously for treatment of chronic tension-type headache. So, by overall analysis,

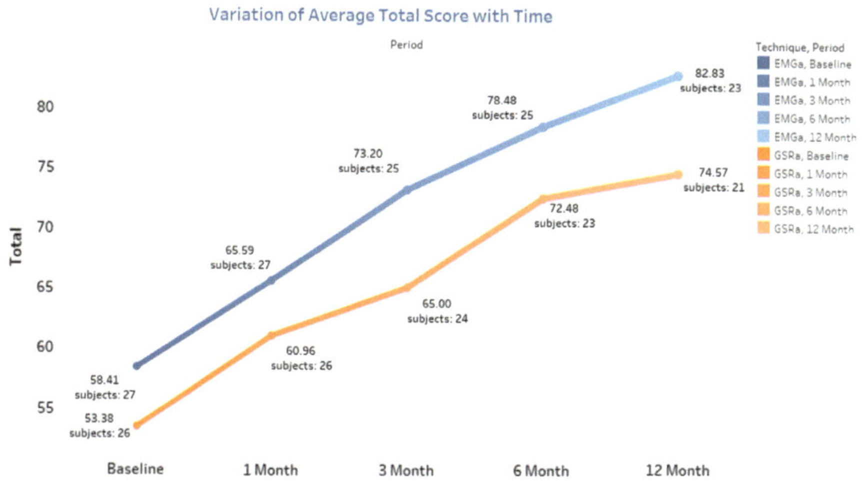

Fig. 12 Trend of average of total scores

one can easily conclude that EMGa is performing better in reduction of TTH than GSRa.

8.2 *The Trends on Different Scores of EMGv Versus GSRv*

The complete experiment lasted for 1 year of duration on different patients suffering with acute stress problem and their Short Form Health Survey scores with mean value of mental, physical and complete scores were considered. The below graph is representing the behaviour progression of visual techniques in electromyograph and galvanic skin resistance machine (Fig. 13).

There is clear depiction of betterment of EMGv technique over the GSRv technique.

In the same time period of 1-year duration, huge amount of data is collected which went under analysis and the results reveal that the dropouts are 1 less in GSRv but the physical score value is much higher in EMGv therapy showing the clear improvement in the batch of EMGv.

Also, the rate of increment of physical score, vitality and fitness are lesser in GSRv compared to EMGv There is continuous gradual increment, but the rate of improvement is always in positive side (Fig. 14).

In above-said graph of mental scores, we can observe a massive amount of data which is investigated with the help of Big Data; the EMGv batch is continuously winner in respect of the subject dropouts, rate of increment of mental fitness, rate of variation in subject dropouts and rate of variation of mental scores of the subjects as compared to those of GSRv group subjects in treatment for tension-type headache.

Both therapies are starting from near about similar mental scores of their allotted batches, but after the designated stipulated time of experiment and looking over the annals by Big Data, we concluded that, the EMGv batch is having a good amount of gap of mental scores that is 82.7 over the 76.33 of GSRv.

So it automatically proves the betterment of EMGv method over the GSRv therapy (Fig. 15).

The total score is designated by the sum of special scores of physical and mental scores.

In above graphical representation of set of near about subjects (27 in each) for both the therapies, GSRv is showing the slow and continuous increment in first half time (6 months) but very drastic improvement in next half.

In the similar pattern, the EMGv group is showing the effectiveness of total score gradually in first 3 months but drastic in 3–12 months period (observed using Big Data). So EMGv is showing better efficiency and fast response on the subjects in lesser time period.

By chance, total score in the initial as well as final phase are same for both the therapies and both here show neck-to-neck competition in total scores of the subjects. In terms of rate of improvement, EMGv is better performing than GSRv in the area of TTH reduction.

The GSRv therapy has lost the confidence by the fluctuation and indefinite behavioural pattern as in almost all the cases, it has shown good performance in first 3 months and then uncertain attitude in remaining experimental life.

EMGv has maintained the confidence value, and it is relied on the subjects continuously. So the overall result can be concluded that EMGv is a better performer in reduction of TTH than GSRv.

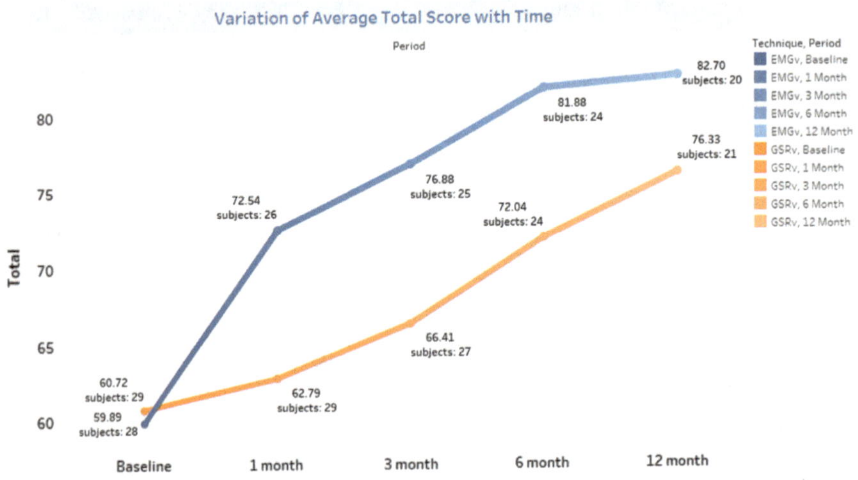

Fig. 13 Trend of average of physical scores

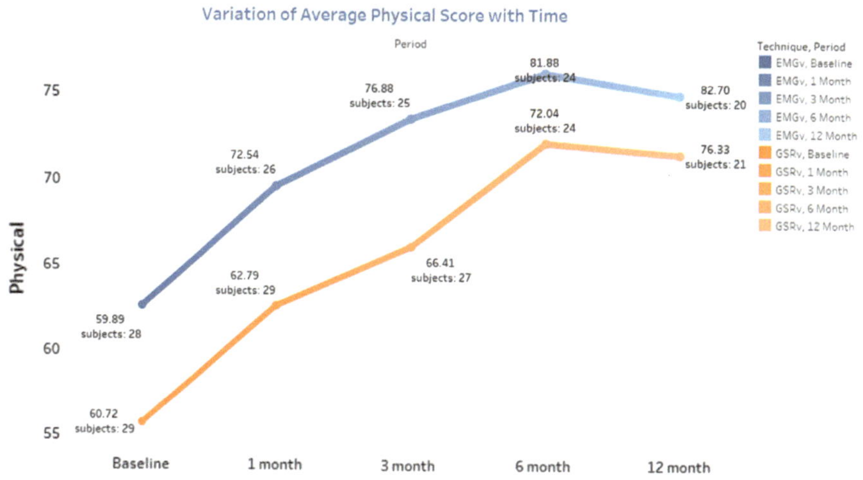

Fig. 14 Trend of average of mental scores

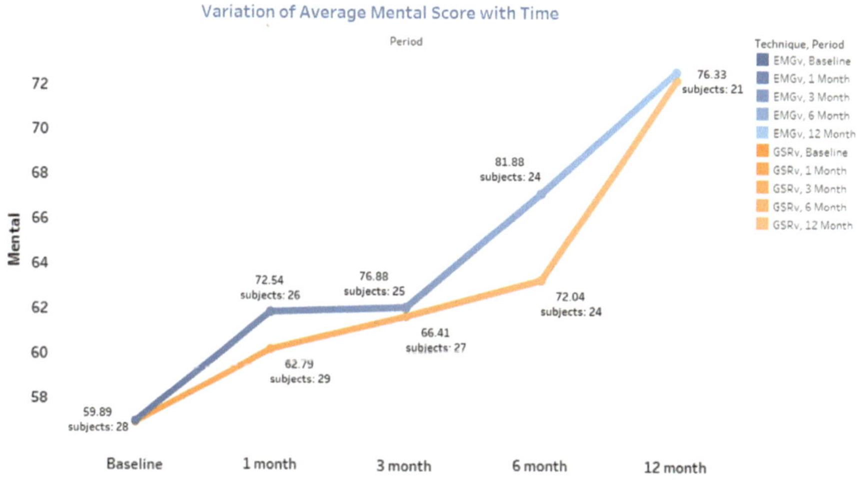

Fig. 15 Trend of average of total scores

8.3 *The Trends on Different Scores of EMGav Versus GSRav*

The complete experiment lasted for 1 year of duration on different patients suffering with acute stress problem, and their Short Form Health Survey scores with mean value of mental, physical and complete scores were considered. The below graph is representing the behaviour progression of audio-visual techniques in electromyograph and galvanic skin resistance machine (Fig. 16).

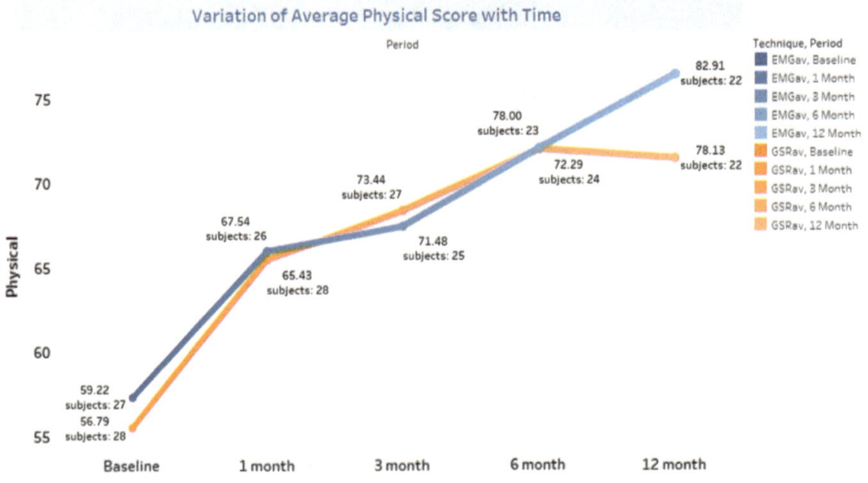

Fig. 16 Trend of average of physical scores

The audio-visual method therapy is more effective and powerful than individual audio or visual methods.

Analysing the graph for physical scores with the support of Big Data, one can easily find cut-throat competition in performance of EMGav and GSRav groups' subjects in the same stipulated time duration. GSRav group defined the constant gradual increment in first half of the experiment (6 months), and even EMGav was in lesser performance in this time.

But over the second half period of experiment, EMGav showed clear-cut winning over GSRav as GSRav became stagnant and stable and did not show any improvement due to possibility of subject disinterest. Also, the subject dropout ratio was higher in GSRav than EMGav. EMGav showed the continuous remarkable performance in area of TTH by improving physical fitness and vital score of the subjects, and its speed was gradual and constant. At the end, the final score achieved with the help of Big Data shows that there was a significant gap of scores between both from 82.91 to 78.13 so it clearly indicates that EMGav is better and efficient than GSRav therapy (Fig. 17).

The graphical analysis done with the help of Big Data is self-explanatory that the EMGav is always higher from starting to end of experiment period over GSRav. Also, the rate of fluctuations is also recorded in GSRav for mental scores means in between time period, GSRav subjects depicted less mental fitness than those of subjects treated under EMGav.

EMGav has shown continuous and straight increment in the mental scores and health. Though both the therapies are showing similar subject dropouts, the score difference at the end of experiment (calculated with the help of Big Data) clearly reveals that EMGav is better performing therapy than GSRav therapy in acute TTH problem (Fig. 18).

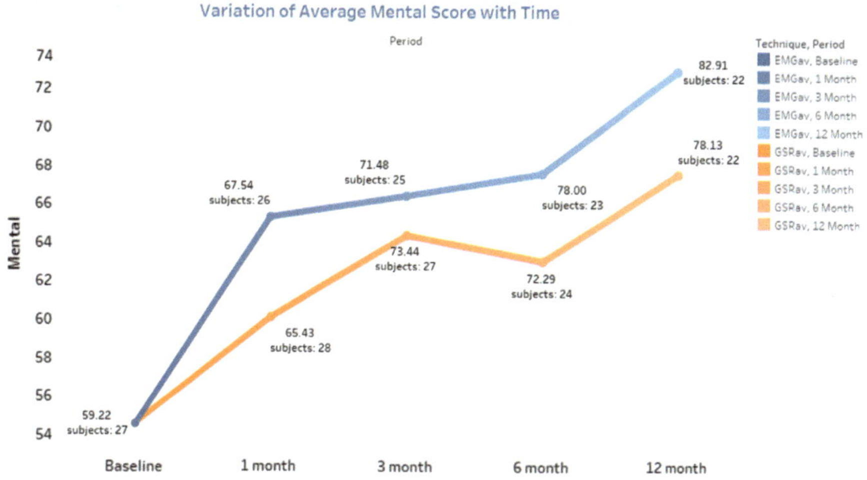

Fig. 17 Trend of average of mental scores

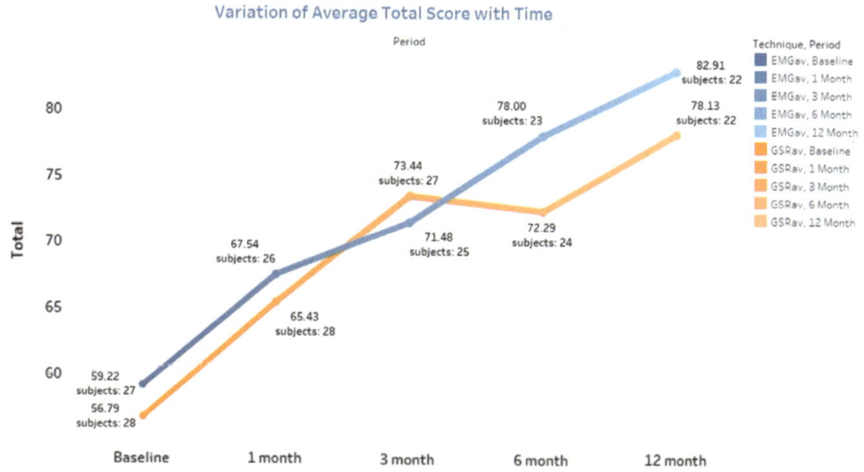

Fig. 18 Trend of average of total scores

Total score is almost the reflection of physical and mental scores.

As per the attached graph (made with the help of Big Data), GSRav has performed well up to 3 months but fluctuated and degraded up to next 3 months, showing the disinterest and less confidence of the subjects over the therapy. After 6 months of experiments, it again showed some improvement. The total difference in Total score value in GSRav is recorded as $78.13 - 56.79 = 22.74$.

On other side, the EMGav groups have shown continuous and drastic but gradual improvement on the total scores of SF-36 for the subject batch. The subject's dropouts are also less in this, showing the clear interest and benefits from the therapy. The difference in initial and final stage of experiment was recorded as 82.91−59.22 = 23.69

Also, there is no fluctuation in EMGav therapy results like to GSRav. The GSRav therapy has lost the confidence by the fluctuation and indefinite behavioural pattern as in almost all the cases, it has shown good performance in first 3 months and then uncertain attitude in remaining experimental life.

EMGav has maintained the confidence value, and it is relied on the subjects continuously.

So, by overall analysis done by applying Big Data, one can easily conclude that EMGav is performing better in reduction of TTH than GSRav.

8.4 The Trends on Different Scores of EMGa Versus EMGv Versus EMGav

The complete experiment lasted for 1 year of duration on different patients suffering with acute stress problem, and their Short Form Health Survey scores with mean value of mental, physical and complete scores were considered. The below graph is representing the behaviour progression of different modes of auditory, visual and combined techniques in electromyograph machine (Fig. 19).

The analysis on three modes on the same therapy of EMG has interesting results.

On deeper analysis done by implementation of Big Data, one can easily deduce the betterment of EMGav group (audio-visual therapy) as the all three therapies on the peer groups start with nearly same physical scores of 59.22, 59.89 and 59. 22 but

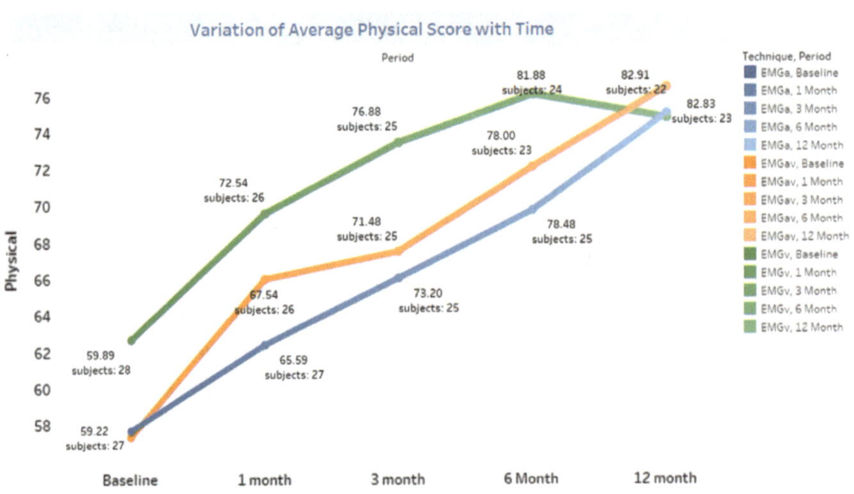

Fig. 19 Trend of average of physical scores

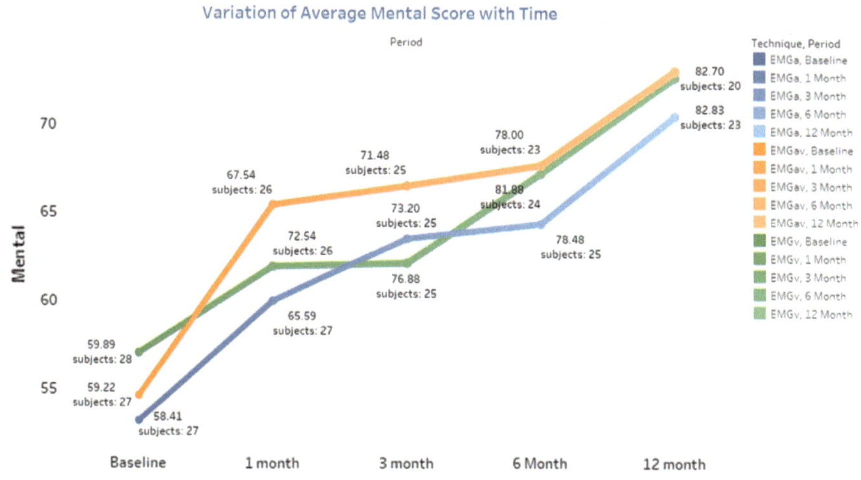

Fig. 20 Trend of average of mental scores

after 1 month, there is drastic increment in EMGv 72.54 but slower to EMGa and EMGav of values 65.59 and 67.54, showing very fast improvement rate in EMGv groups. EMGv has continuously shown good results in first half but has fluctuating tendency. After 6 months of duration, EMGv has a poor performance 82.83 and shows the decrement as compared to EMGa of 82.83 which has slowed but continuous increasing. EMGav has highest performance 82.91 than EMGa, and also the subject dropouts are nearly same in all.

So, EMGav has maintained the confidence value and it is relied on the subjects continuously.

So, by overall analysis done by practising Big Data, one can easily conclude that EMGav is performing better in reduction of TTH than EMGa and EMGv (Fig. 20).

The graph obtained by the analysis of Big Data is very clearly depicting that subjects are experiencing better mental health with EMGav group therapy. From day 1 to entire experimental time period, the EMGav has shown fantastic performance of 82.7 score than EMGa and EMGv of 82.7 and 82.83. The rate of increment is always towards upper side, and after 3 months, the fast increment is recorded in EMGav as compared to EMGa and EMGv. Though the dropouts are higher in EMGv than others due to different reasons, one can believe that EMGav may provide better solution for TTH reduction (Fig. 21).

Total score is almost the reflection of physical and mental scores.

As per the attached graph obtained by evaluation of Big Data, EMGv has performed well up to 3 months (72.54) but comparatively slowly increased up to 81.88 and degraded up to next 6 months up to 83.7 even the highest total score among all 3 therapies, showing the disinterest and less confidence of the subjects over the EMGa and EMGav therapy.

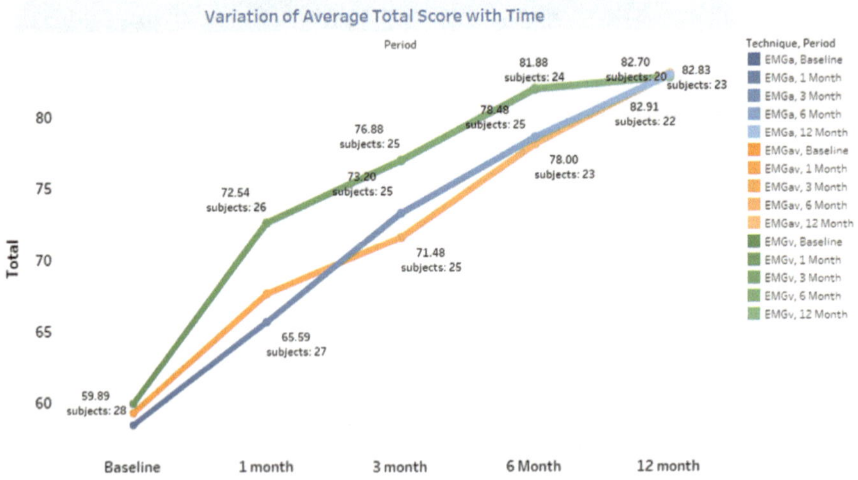

Fig. 21 Trend of average of total scores

The difference in initial and final stages of experiment was recorded as for EMGv is $82.7-59.89=22.81$.

On other side, the EMGav groups have shown continuous and drastic but gradual improvement on the total scores of SF-36 for the subject batch. The subjects' dropouts are also less in this, showing the clear interest and benefits from the therapy. The difference in initial and final stages of experiment was recorded as $82.83-59.89=22.94$.

Also, there is no fluctuation in EMGav therapy results like to EMGv.

The EMGa and EMGv therapies have lost the confidence by the fluctuation and indefinite behavioural pattern as in almost all the cases, it has shown good performance in first 3 months and then uncertain attitude in remaining experimental life.

EMGav has maintained the confidence value, and it is relied on the subjects continuously.

So, by overall analysis done using Big Data, one can easily conclude that EMGav is performing better in reduction of TTH than EMGa and EMGv.

9 Novelty in the Proposed Work

In the coeval study, the fallout of EMGav, EMGa and EMGv are designed on contrasting patients considering the incurable description of TTH stress. Stress extremely is an ordinary term which we observe in our daily life. The word "Stress" is a method "to draw tight" and is manipulated to describe complications, burden, force, calamity and a lot more. Nearly every people encounters stress as a consequence of their hustling life and demanding to salvage time for their friends and family. All of us encounter stress in our life.

There are a couple of bigger mental processes. The study is crucial as there is an essential need to study the chronic-type TTH stress with the comfort of backup therapies such as EMG on distinctive modes on divergent aged group, gender, caste in the grounds of scientific studies. The study is done precisely and is one of a kind to show the influence of BF therapies in distinctive modes and rate of improvement in stress/TTH reduction was recorded. First time the chronic TTH is being quantized with the help of physical, mental and total scores, and standard, reliable and validated instruments like SF-36 are being used to measure it [43–45].

The result of instrument like SF-36 has introduced many methodologies for the advancement and development of Big Data analyst. It tells us about the large number of Big Data challenges and techniques in about 200 Web article and publications. Multimedia analyst is provided with framework and tools, yet the problems are faced by analyst in multimedia data. Many applications have been enriched by the great contribution of multimedia analysis. So, multimedia Big Data analysis has great opportunities for advancement.

10 Future Scope, Limitations and Possible Applications

Here, there is no need to roam as the outsider students and there is an important opportunity venture to ask for the BF therapy (asked in the TTH)-based stress issue (recharging may be arranged outside as we have limited capacity). AKTU University is affiliating this event.

The study evaluation was done at the eleventh hour to study the physiological, psychological and therapeutic effect of positive thinking and meditation done on focused attitude [22, 45, 46].

- Control group and long-duration study suggested us to define the exclusion and inclusion criteria.
- Study was meant for larger sample and experimental conduction.
- Due to bigger large sample, we are able to see the advantages.
- These experiments have their own beauty as it is comparing two human beings in different states, socio-economic status and varying gender.
- One can facilitate the education on results explored and through meditation, one will attain the relaxation.
- Study was slightly done and detailed results were obtained and got checked.

Study its effect not only on mental and physical but on consciousness level too.

However, the process of biofeedback is not affordable as of now. To successfully take advantage and apply IoT stakeholders must make it economic otherwise it will remain in the hands of high class.

There is variety of applications of MMBD in Big Data and IoT in health care which can be implemented. The rise of IoT has various scopes which have made it exciting. In health care, it has several applications like tracking of patients, staffs availability of hardware, reducing emergency room wait time.

In other fields, it can be used for:

- Maintaining record of biofeedback machine
- Predicting the future possibilities of TTH level in the patient
- Using the data of EMGa, EMGav and EMGv, the treatment of patient can be improved.

IoT has introduced many portable wearing devices that have made life easier for patient. Some of these are as follows.

Hearables are new devices which have changed the world for the people who suffered hearing loss. Nowadays, they are compatible with your smart phone and send the data directly to your phone.

Ingestible is really a hard thing to believe. They are small-sized sensors which go inside body and warn for any irregularities.

These sensors are boon for diabetes patients as they provide early warning. Proteus Digital Health is one example.

There are few devices like Moodables that change our mood. It may look like science fiction but it is real.

Thync and Halo Neuroscience have made significant progress in it. Moodables are attached to our brain that sends low-intensity voltage and enhance our mood.

The advancement in AI has grown the drone technology which takes decision by making visual perception. Drones like Skidoo uses computer vision to navigate by finding obstacles. Same technology can be used by visually impaired person for navigation. Healthcare charting in IoT devices such as Aude mix has reduced the work of doctor while sketching. It uses voice commands to capture the patient's data. Make is accessible and saves doctors time.

11 Conclusions

It is clear from our study that the effect of audio mode of EMG is more efficient than the visual mode and audio-visual mode in treatment of TTH due to stress based on various parameters. It was also seen that the person with EMG therapies has seen great success in performing their work with more calmness and concentration, while the person in control group has seen no change.

A person behaviour is dependent on the fact that an event raised is raised as a challenge or as an opportunity. After these therapies, person now has better coping capacity. Person with low intelligence level has poor decision-making skill and poor skill to perform their job. Hence, with overall scenario of the results, it can be concluded EMGa has better performance than EMGv and EMGav. Seeing Tableau visualizations, it can be concluded that stress and mental health are directly related.

IoT had transformed the way of transferring facilities to the healthcare industry. These technologies improve the product by focusing on minor changes. Despite all the challenges, ever-increasing multimedia Big Data provides great insights into human behaviours and sentiments, which in turn leads to great opportunities to making great progress in many fields (Michael and Miller 2013; Chenand Zhang 2014). The following are some examples:

A high number of mobile users could provide data for characterizing shopper behaviour and aid in the decision-making process. It will also help product improvement and target marketing (Michael and Clarke 2013). Thanks to the advanced technologies in data storage and wireless communication, mobile visual search is one of the most popular applications nowadays (Tan et al. 2016). With the rapid growth of multimedia data, the demand for more efficient search and retrieval techniques is stronger than ever, especially for mobile applications.

Analytics of patient data could lead to the detection of devastating diseases at an early stage and provide timely cures or even the prevention of other diseases (Strickland 2013). In addition, the technology of mobile-based health interventions that can help people is proliferating.

Corporations could utilize the Big Data technology, increase the productivity and make best use of their man power. In addition, at ACM Computing Surveys, Vol. 51, No. 1, Article 10. Publication date: January 2018, page 24 10:24, S. Pouyanfar et al., business intelligence involving rich geospatial and spatial–temporal information has been improved using multimedia Big Data techniques (Vidal-García et al. 2017).

Climate science also utilizes big multimedia spatial–temporal data to analyse and monitor the temperature and climate system of the Earth (Yang et al. 2017). Moreover, astronomy Big Data can study the universe more efficiently. This not only helps the researcher record and understand the universe but also helps to protect the Earth from impact from other space bodies.

Other opportunities include the formation of strategic direction, the development of better customer service, the identification of new products and services, and so on (Manyika et al. 2011; Yang et al. 2017).

To summarize, the proper utilization of multimedia Big Data will improve productivity, transform economics and bring new insights to the world (Chen and Zhang 2014). As mentioned earlier, remarkable effort will be needed to face the future challenges with multimedia Big Data. Several future directions and solutions are listed below.

Comprehensive Conclusion

In the study presented in this chapter, the authors have tried best to design two hypothetical statements. Null and alternate to prove the problem statement that whether on different patient chronic TTH parameters, there is any correlation of auditory, visual and combined modes under electromyograph machine.

Analysing the null hypothetical statement proved correct, the authors found that the biofeedback techniques are highly correlated in positive direction for chronic and acute headache.

The auditory mode was found more effective than both rest modes in this discussion, and chronic distress issue was decreasing in auditory mode comparatively faster than both rest modes [25, 47–49].

Acknowledgements The authors would like to thank the management of Dev Sanskriti Vishwavidyalaya, Haridwar, and Shantikunj for their support and also the faculties and management of ABESEC, Ghaziabad, for their support in the research work.
The Spirituality, Yoga and Ethics Committee of ABESEC, Ghaziabad, has guided us in every step and directed for consent forms of willingness for experiment from the participants. We are also thankful to its members.

References

1. Rubin, A.: Biofeedback and binocular vision. J. Behav. Optom. **3**(4), 9598 (1992)
2. Mullaly, J.W., Hall, K., Goldstein, R.: Efficacy of BF in the treatment of migraine and tension type headaches. Pain Physician **12**, 1005–1011 (2009)
3. Fumal, A., Scohnen, J.: Tension-type headache: current research and clinical management. Lancet Neurol. **7**, 70–83 (2008)
4. Millea, J.P., Brodie, J.J.: Tension type headache: Am. Fam. Physician. **66**(5), 797–803 (2002)
5. Boureau, F., Luu, M., Doubrere, J.F.: Study of experimental pain measures and nociceptive reflex in Chronic pain patients and normal subjects. Pain **44**, 131–138 (1991)
6. Scott, D.S., Lundeen, T.F.: Myofascial pain involving the masticatory muscles: an experimental model. Pain **8**, 207–215 (1980)
7. Haynes, S.N., Griffin, P., Mooney, D., Parise, M.: Electromyographic BF and relaxation instructions in the treatment of muscle contraction headaches. Behav. Ther. **6**, 672–678 (1975)
8. Kropotov, J.D.: Quantitative EMG, Event-Related Potentials and Neurotherapy. Academic Press, San Diego, CA (2009)
9. Cassel, R.N.: Biofeedback for developing self-control of tension and stress in one's hierarchy of psychological states. Psychol.: J. Hum. Behav. **22**(2), 50–57 (1985)
10. Turk, D.C., Swanson, K.S., Tunks, E.R.: Psychological approaches in the treatment of chronic pain patients—When pills, scalpels, and needlesare not enough. Can. J. Psychiatry **53**(4), 213–223 (2008)
11. Valdes, M.R.: A program of stress management in a college setting. Psychother. Priv. Pract. **6**(2), 43–54 (1988)
12. Wenk-Sormaz, H.: Meditation can reduce habitual responding. Adv. Mind-Body (2005)
13. Carlson, Neil: Physiology of Behavior. Pearson Education, Inc., New Jersey (2013). ISBN 978-0-205-23939-9
14. Haddock, C.K., Rowan, A.B., Andrasik, F., Wilson, P.G., Talcott, G.W., Stein, R.J.: Home-based behavioural treatments for chronic benign headache: a meta-analysis of controlled trials. Cephalalgia **17**, 113–118 (1997)
15. McCrory, D., Penzien, D.B., Hasselblad, V., Gray, R.: Behavioural and Physical Treatments for Tension Type and Cervocogenic Headaches. Foundation for Chiropractic Education and Research, Des Moines, IA (2001)
16. Holroyd, J.C., Penzien, D.B.: Client variables and the behavioural treatment of recurrent tension Headache: a meta-analytic review. J. Behav. Med. **9**, 515–535 (1986)
17. Rastogi, R., Chaturvedi, D.K., Arora, N., Trivedi, P., Mishra, V.: Swarm intelligent optimized method of development of noble life in the perspective of Indian Scientific Philosophy and Psychology. In: Proceedings of NSC-2017(National system conference), DEI Agra, Dec. 1–3 (2017)
18. Rastogi, R., Chaturvedi, D.K., Satya, S., Arora, N., Singh, P., Vyas, P.: Statistical analysis for effect of positive thinking on stress management and creative problem solving for adolescents, In: Proceedings of the 12th INDIACom; 2018 ISSN 0973–7529 and ISBN 978-93-80544-14-4, pp. 245–251
19. Collet, L., Cottraux, J., Juenet, C.: GSR feedback and Schultz relaxation in tension headaches: a Comparative study. Pain **25**, 205–213 (1986)

20. Tfelt-Hansen, P., Block, G., Dahlof, C., et al. (2000) Guidelines for controlled trials of drugs in migraine, 2nd ed. Cephalalgia **20**, 765–778
21. Schoenen, J., Boureau, F., Kunkel, R., et al.: Guidelines for trials of drug treatments in tension-type headache, 1st ed. Cephalalgia **15**, 165–179 (1995)
22. Rastogi, R., Chaturvedi, D.K., Arora, N., Trivedi, P., Singh, P., Vyas, P.: Study on Efficacy of electromyography and electroencephalography biofeedback with mindful meditation on mental health of youths. In: Proceedings of the 12th INDIACom; 2018 ISSN 0973–7529 and ISBN 978-93-80544-14-4, pp. 84–89
23. Rastogi, R., Chaturvedi, D.K., Arora, N., Trivedi, P., Chauhan, S.: Framework for use of machine intelligence on clinical psychology to study the effects of spiritual tools on human behavior and psychic challenges. In: Proceedings of NSC-2017(National system conference), DEI, Agra, Dec. 1–3, 2017
24. Andrasik, F., Lipchik, G.L., McCrory, D.C., Wittrock, D.A.: Outcome measurement in headache research: headache parameters and psychosocial outcomes. Headache **45**, 429–437 (2005)
25. Rastogi, R., Chaturvedi, D.K., Sharma, S., Bansal, A., Agrawal, A.: Audio visual EMG & GSR biofeedback analysis for effect of spiritual techniques on human behavior and psychic challenges. In: Proceedings of the 12th INDIACom; 2018 ISSN 0973–7529 and ISBN 978-93-80544-14-4, pp. 252–258
26. Rastogi, R., Chaturvedi, D.K., Satya, S., Arora, N., Yadav, V., Yadav, V., Sharma, P., Chauhan, S.: Statistical analysis of EMG & GSR biofeedback efficacy on different modes for Chronic TTH on various indicators. International Journal of Advanced Intelligence Paradigms, Inderscience Publishers Ltd., ISSN online 1755-0394, ISSN print 1755-0386
27. McCrory, D.C., Rebecca, N., Gray, R.N., Tfelt-Hansen, P., Steiner, T.J., Taylor, F.R.: Methodological issues in systematic reviews of headache trials: adapting historical diagnostic classifications and outcome measures to present-day standards. Headache **45**, 459–465 (2005)
28. Garratt, A.M., Schmidt, L., Mackintosh, A., Fitzpatrick, R.: Quality of life measurement: bibliographic study of patient assessed health outcomemeasures. Br. Med. J. **324**, 1417–1419 (2002)
29. Monzón, M.J., Láinez, M.J.: Quality of life in migraine and chronic daily headache patients. Cephalalgia **18**(9), 638–643 (1998)
30. Guitera, V., Munoz, P., Castillo, J., Pascual, J.: Quality of life in chronic daily headache: a study in a general population. Neurology **58**, 1062–1065 (2002)
31. Ware, J.E., Kosinski, M., Bayliss, M.S., McHorney, C.A., Rogers, W.H., Raczek, A.: Compar ison of methods for the scoring and statistical analysis of SF 36, health profile and summary measures: summary of results from the Medical Outcome Study. Med. Care **33**(4), AS264–79 (1995)
32. Kleiman, D.L., Vallas, S.P.: Science, capitalism, and the rise of the "knowledge worker": the changing structure of knowledge production in the United States. Theor. Soc. **30**, 451–492 (2001)
33. Saris-Baglama, R.N., Dewey, C.J., Chisholm, G.B., et al.: Quality metric health outcomes™ scoring software 4.0, p. 138. Quality Metric Incorporated, Lincoln, RI, (2010)
34. National Center for Biotechnology Information, U.S. National Library of Medicine, http://www.ncbi.nlm.nih.gov/pubmed (accessed 30 October 2015)
35. Optum. SF-36healt survey, http://www.webcitation.org/6cfdiZOJI (accessed 30 October 2015)
36. SF 12 "Archived copy". Archived from the original on 2015-03-18. Retrieved 2015-11-28
37. Jenkinson C.: A shorter form health survey: can the SF-12 replicate results from the SF-36 in longitudinal studies? J. Public Health Med. **19**, 179–186 (1996-11-03). PMID 9243433. Retrieved 2015-11-28
38. Rastogi, R., Chaturvedi, D.K., Satya, S., Arora, N., Gupta, M., Yadav, V., Chauhan, S., Sharma, P.: Book chapter titled as Chronic TTH Analysis by EMG & GSR biofeedback on various modes and various medical symptoms using IoT, Paperback ISBN: 9780128181461, Advances in ubiquitous sensing applications for healthcare, Book-Big Data Analytics for Intelligent Healthcare Management

39. ACM Computing Surveys, Vol. 51, No. 1, Article 10. Publication date: January 2018, p. 9
40. Abramova, V., Bernardino, J.: NoSQL databases: MongoDB versus cassandra. In: Proceedings of the International C* Conference on Computer Science and Software Engineering. ACM, pp. 14–22 (2013)
41. Rastogi, R., Chaturvedi, D.K., Satya, S., Arora, N., Singhal, P., Gulati, M.: statistical resultant analysis of spiritual & psychosomatic stress survey on various human personality indicators. In: The International Conference proceedings of ICCI 2018
42. Rastogi, R., Chaturvedi, D.K., Satya, S., Arora, N., Sirohi, H., Singh, M., Verma, P., Singh V.: Which one is best: electromyography biofeedback efficacy analysis on audio, visual and audio-visual modes for chronic TTH on different characteristics. In: The Proceedings of ICCIIoT-2018, 14–15 December 2018 at NIT Agartala, Tripura, ELSEVIER- SSRN Digital Library (ISSN 1556-5068)
43. Rastogi, R., Chaturvedi, D.K., Satya, S., Arora, N., Yadav, V., Chauhan, S., Sharma, P.: Analytical comparison of efficacy for electromyography and galvanic skin resistance biofeedback on audio-visual mode for chronic TTH on various attributes. In: The Proceedings of the ICCIDA-2018 on 27 and 28th October 2018, CCIS Series, Springer at Gandhi Institute for Technology, Khordha, Bhubaneswar, Odisha, India
44. Rastogi, R., Chaturvedi, D.K., Satya, S., Arora, N., Saini, H., Verma, H., Mehlyan, K.: Comparative efficacy analysis of electromyography and galvanic skin resistance biofeedback on audio mode for chronic TTH on various indicators, In: The proceedings of ICCIIoT- 2018, 14–15 December 2018 at NIT Agartala, Tripura, ELSEVIER- SSRN Digital Library (ISSN 1556-5068)
45. Rastogi, R., Chaturvedi, D.K., Satya, S., Arora, N., Trivedi P., Singh, A., Sharma, A., Singh, A.: Intelligent analysis for personality detection on various indicators by clinical reliable psychological TTH and stress surveys. In: The Proceedings of CIPR 2019 at Indian Institute of Engineering Science and Technology, Shibpur on 19th-20th January 2019, Springer-AISC Series
46. Rastogi, R., Chaturvedi, D.K., Satya, S., Arora, N., Bansal, I., Yadav, V.: Intelligent analysis for detection of complex human personality by clinical reliable psychological surveys on various indicators, in the national Conference on 3rd MDNCPDR-2018 at DEI, Agra On 06–07 September, 2018
47. Rastogi, R., Chaturvedi, D.K., Satya, S., Arora, N., Yadav, V., Chauhan, S., Sharma, P.: SF-36 Scores Analysis for EMG and GSR Therapy on Audio, Visual and Audio Visual Modes for Chronic TTH, in the proceedings of the ICCIDA-2018 on 27 and 28th: CCIS Series. Springer at Gandhi Institute for Technology, Khordha, Bhubaneswar, Odisha, India (2018)
48. Rastogi, R., Chaturvedi, D.K., Satya, S., Arora, N., Saini, H., Verma, H., Mehlyan, K., Varshney, Y,: Statistical analysis of EMG and GSR therapy on visual mode and SF-36 scores for chronic TTH, in the proceedings of UPCON-2018 on 2–4 Nov 2018 MMMUT Gorakhpur, UP
49. Rastogi, R., Chaturvedi, D.K., Satya, S., Arora, N., Yadav, V., Chauhan, S.: An optimized biofeedback therapy for chronic TTH between electromyography and galvanic skin resistance biofeedback on audio, visual and audiovisual modes on various medical symptoms. In The National Conference on 3rd MDNCPDR-2018 at DEI, Agra On 06–07 September, 2018

Proficient Reconstruction Algorithms for Low-Dose X-Ray Tomography

G. Nagarajan and B. S. Sathish Kumar

Abstract Computer tomography (CT) has been progressively increasing demand in the field of medicine and industry due to its accuracy and fast image acquisition. CT scanner emanates X-Ray to the human being during the image acquisition, and it is evident that patient may be at huge risk in inducing cancer on human beings. Even though the radiation risk in CT, the increased attention must be needed during the investigation and problem of low radiation dose is fetching an area of focusing. In a wide ranging, it is consisting of two methods to reduce CT. Initially, the current density of the X-Ray tube and angle of projection are reduced without modification in revelation dose of each direction. Subsequently, the radiation dose is reduced, the noise increases gradually, in turn, it will affect the quality of image. In such case, reconstruction of image happens to be an ill-posed problem. Landweber iterative algorithm is proposed in this chapter and simulated, and its benefits are compared with existing iterative algorithms. The quality of the image reconstructed is calculated in terms of mean absolute error and correlation coefficient.

Keywords Computer tomography · Fan-beam tomography · Image reconstruction · Iterative algorithm · Parallel-beam tomography · Radiation dose

1 Introduction

Computational imaging is one of the most developing domains in the medical field. It has gained researchers from academic and industrial sectors. Tomography is a medical imaging method which allows the reconstruction of internal information about an object, using a sequence of X-Ray calculation obtained from lots of angles in the region of the object in noninvasive order. The combination of computer and

G. Nagarajan · B. S. Sathish Kumar (✉)
Department of ECE, Pondicherry Engineering College, Puducherry 605014, India
e-mail: bssathishkumar.79@gmail.com

G. Nagarajan
e-mail: nagarajanpec@pec.edu

S. Bhattacharyya et al. (eds.), *Hybrid Machine Intelligence for Medical Image Analysis*, Studies in Computational Intelligence 841,
https://doi.org/10.1007/978-981-13-8930-6_9

tomography is called computer tomography. The computational imaging are mixture of optical elements and signal processing techniques, which inturn increases their capabilities. There we focus on computational part and explore reconstruction algorithms which can estimate an underlying scene as accurately as possible.

1.1 Advantages of Computer Tomography

- CT scanning has the ability to distinguish bone, soft tissue, and blood vessels.
- It is used to measure as noninvasive and painless.
- Inner and outer geometry of the object is found.
- Easily measure the volume of data of high density.
- It is fast and accurate with a short scanning time.
- Less sensitive to patient movements.

1.2 Limitations in Computer Tomography

- The examination of CT is not suitable in soft tissue details like the brain, internal pelvic organs, and joints in knees and shoulders.
- CT examination is not suited for pregnant women because of potential risk to the fetus and also the patient whose weight exceeds the limit of the moving table of CT scanner.
- The size of CT scanning requires large space for their installation and also requiring the X-Ray shielding, climate-controlled housing of high-speed data and computer has a facility of graphics processing.
- It consists of complex and numerous influences in affecting measurements.

2 Generation of Tomography

In traditional X-Ray, it produces only the object's projection image. There is an overlap in projection path which leads to blur an image and the objects are not visually understandable. To crack that problem, computer tomography is used to reconstruct image. To accomplish this, the narrow X-Ray beam is sent through the object from source and place the detector on the other side, in order to detect the intensity [1]. Then, the location of a source and detector is changed around the object. The measurement process is repeated. In CT, it is seen that the measurement of attenuation distribution over an object is recorded [2]. Once the attenuation distribution in each and every direction around the object is found by the use of reconstruction algorithm, CT presents the cross-sectional information of object [3].

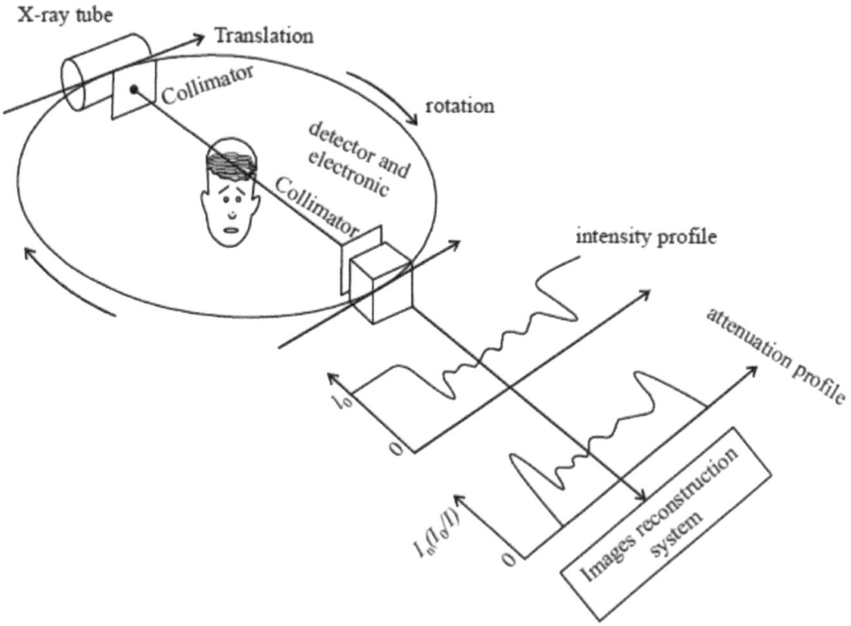

Fig. 1 Schematic of CT scan [7]

The system requirement for the sensor arrangements and imaging modality is different in industrial tomography than medical tomography [4]. The intensity of the X-Ray beam is detected in one direction called projection. The procedure to create the cross-sectional information of an object is called image reconstruction from projections [5]. In 1963, image reconstruction was introduced, and its benefits lead this technique widely used in the diagnostic medical field, i.e., provide the inner information of object without harmful. Simultaneously, acquired the sonogram data in a different direction and each has unique X-Ray energy distribution in CT [6]. Figure 1 shows the basic CT image.

The basic idea of image reconstruction is shown in Fig. 1. In X-Ray CT, "I_0" is X-Ray transmitted from source and "I_1" attenuated intensity and measured at detectors [8]. Using Lambert–Beer's law, the connection among the intensities and peculiar of object radiated by X-Ray is

$$I_1 = I_0 \exp\left(-\int_L \mu(x)\mathrm{d}x\right) \tag{1}$$

where $\mu(x)$—intrinsic attenuation coefficient of the material at location "x" and "L"—path of the X-Ray.

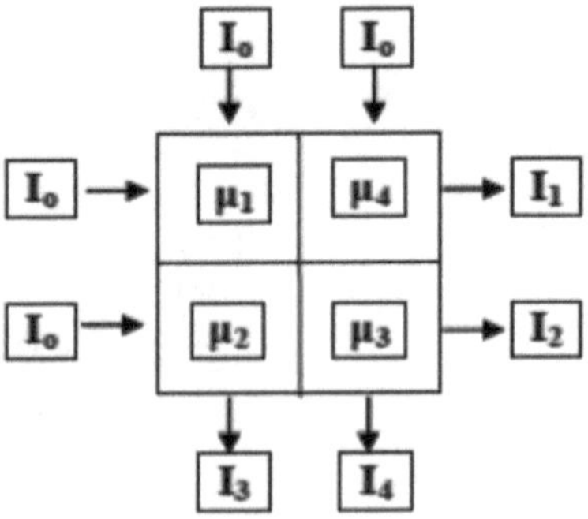

Fig. 2 Mapping of projections

As shown in Fig. 2, it consists of four unknown attenuation coefficient μ_1, μ_2, μ_3, and μ_4 in the middle region. Also, X-Ray source transmits four X-Ray beams in the middle region both in horizontal and vertical directions and four intensities I_1, I_2, I_3, and I_4 are measured as

$$I_1 = I_0 \exp\left(-\int_L \mu(x)\mathrm{d}x\right) \tag{2}$$

$$I_2 = I_0 \exp\left(-\int_L \mu(x)\mathrm{d}x\right) \tag{3}$$

$$I_3 = I_0 \exp\left(-\int_L \mu(x)\mathrm{d}x\right) \tag{4}$$

$$I_4 = I_0 \exp\left(-\int_L \mu(x)\mathrm{d}x\right) \tag{5}$$

For an assumption in above, they take four variables and expressed in four equations. These equations are easily solved and calculated the attenuation coefficients. Then, the image is reconstructed from the calculation. This is the basic principle of operation image reconstruction from the projections. In real-time applications, the reconstruction region is not limited by 2 × 2 images. It may be thousands of pixels which lead to more complication in solving all equations and finding the attenuation coefficient. "The attenuation coefficient features that the value at X-Ray is diluted by dispersing or absorption as they travel through the object. The coefficients recovering in the intensities I_0 and I_1 are equivalent to the image of the density division of the object." Rearranging Eq. (1),

$$-\frac{I_1}{I_0} = \int_L \mu(x)\mathrm{d}x \tag{6}$$

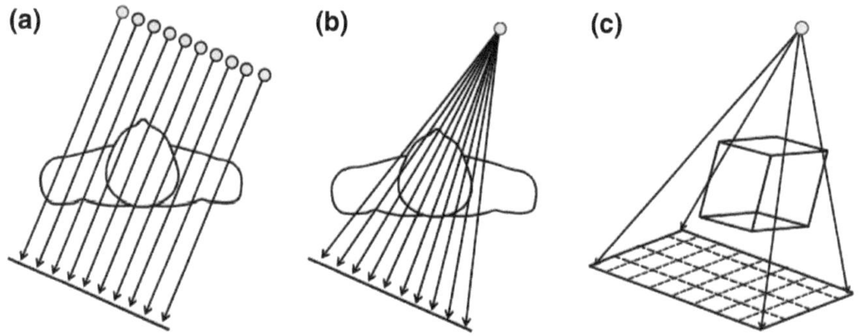

Fig. 3 Three X-Ray source arrangements: **a** parallel arrangements, **b** fan-beam arrangements, and **c** cone-beam arrangements

The customized gives a clear vision to the reconstruction difficulty. "For a given input and output X-Ray intensities, the line integral of I_0 and I_1 and the object attenuation coefficient beside the ray path may be determined. The CT is to find the attenuation coefficient function $m(x)$ at different positions depends on gathered data and received by the sensor at I_0 and I_1 in separate views. It will treat as a reverse task which guises the backward work on the line integration for significant data." The scanning direction of an X-Ray and the path is an important role in CT. It consists of three types of geometry used in CT: parallel beam, fan, and cone beam as shown in Fig. 3.

Parallel beam is one of the oldest methods used in CT, source, and detectors which are arranged on a one-to-one basis. Source produces a narrow-angle beam. It is simple and easy to implement. No switching control over sensors is needed. This is a time-consuming method. The conversion of the parallel beam to the fan beam is done by the process called rebinning [9]. Fan-beam geometry has a wide-angle beam, i.e., fan-shaped X-Ray from source and detectors is arranged as a ring around the object in another side. One source covers more projection range. Detection coverage is 100% of all part in a circle. It is much faster than the parallel beam [10]. In cone-beam configuration, source emits an X-Ray in views of all, and instrument is applied to trace the projections. The acquired times are decreased significantly.

3 Types of Image Reconstruction Algorithms

The reconstruction algorithms are used to establish the image from projections at a different angle. The advancement of hardware in X-Ray tomography enables the development of techniques in image reconstruction algorithm also. There are several properties to be an ideal algorithm to have: fast convergence rate, global convergence, simplicity, parallelizability, exibility to accommodate any type of system model and capability to enforce object constraints like nonnegativity. Consequently, it is highly

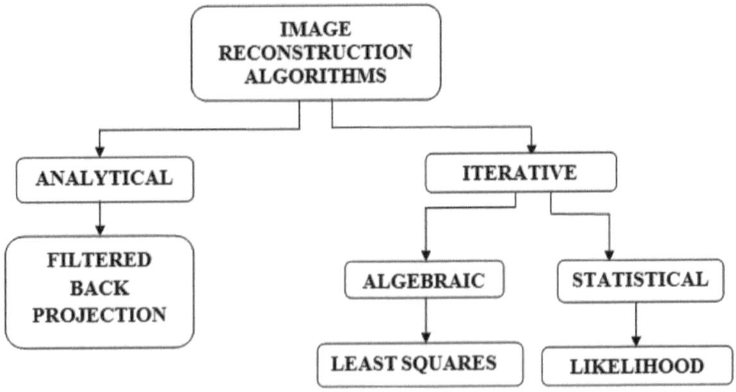

Fig. 4 Classification of reconstruction algorithms

desirable to improve the performance of the reconstruction algorithm to obtain the quality of image. There are two familiar methods which have been used in computer tomography shown in Fig. 4.

3.1 Analytical Reconstruction Algorithm

Filtered back projection (FBP) is a basic approach and most preferred in medical CT systems. It is a single-step reconstruction algorithm and also attempts to find a direct mathematical solution for an image from unknown projections. It tends to be more computational efficient with numerical stability. There are many ways in implementing the analytical model which are restricted to the choice of acquisition geometrics. It cannot tolerate limited angle or sparse acquisition, and it is sensitive to the effects of noise.

3.2 Iterative Reconstruction Algorithm

Iterative reconstruction (IR) methods integrate physical models to reduce noise and artifacts with respect to the degree of modeling. Iterative algorithm uses N iterations to achieve better solutions. As a result, computational complicity is much higher. Due to the tremendous development of computer technology and computational capacities available in the processor, the use of IR methods provides lesser reconstruction time. Among the various image reconstruction techniques, the iterative reconstruction method is one which is striking as well as valuable. But the reconstruction algorithm needs to be efficiently handling large, sparse, inconsistent data sets. Such a computational complexity in iterative reconstruction is a major problem in imple-

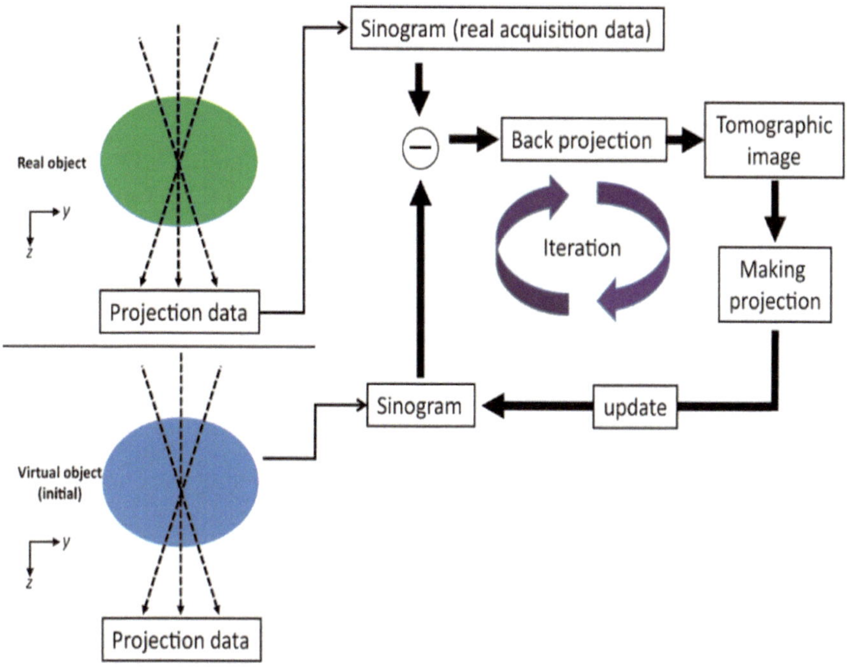

Fig. 5 Schematic representations of the steps of iterative reconstruction

mentation for simulated and real-time data. To reduce the complexity, optimization techniques and selection of projection may be employed to improve performance of reconstruction algorithm and to reduce the computational load.

The basic implementation of the iterative algorithm begins with an initial guess may be empty space or image acquired through the FBP method and iteratively apply corrections to the current estimation of image. The correction is a distinction between the known projection information and the projection calculated from the current estimate. This process will be repeated until this comparison is zero or within the limits. Several IR algorithms are developed for clinical applications [11, 12], and some of the techniques in IR are "simultaneous iterative reconstruction algorithm (SIRT), iterative least square technique (ILST)," etc. The variations among these techniques are the way in which corrections are carried in every iteration. As shown in Fig. 5, the basic operation of iterative reconstruction is as follows:

1. Creation of artificial raw data from the projection in forward.
2. Compare the raw data artificial and measured raw data real.
3. Apply projection in back to the estimation of the volumetric object.

Generally, the selection of the construction method compromises image noise and the computing power required to perform the reconstruction. The developed reconstruction algorithms are based on the assumptions of simulated data not for the real-time data. Reconstruction is not performed on real data due to the followings:

- The reconstruction of original or simulated images is not suitable to reproduce into optimum quality.
- The information received from the detector is not up to the mark to perform the line integral exactly.
- Measurements data are affected by statistical noise.

The issues of CT are desired to diminish the X-Ray radiation during the examination. Also, the radiation dose is an important aspect in deciding image quality with respect to investigative accuracy. Radiation dose is inversely proportional to noise and directly proportional to spatial resolution. But it raises patient risk. While a decrease in the radiation dose may lead to high noise and lack of accuracy. By knowing the relationship between image quality and radiation dose, the amount of radiation dose is determined.

4 Proposed Methods

The major drawback in conventional CT is the inherent connection of radiation dose with noise. For the lower-dose level, noise is more in the image space. The regeneration, rediscovery, and refinement of IR overcome these borders by achieving the constant image quality at a lower radiation dose in Landweber algorithm in comparison with SART, Cimmino, DROP, and CAV algorithms for image reconstruction [13]. The performance of computer tomography (CT) in reduced dose level is measured with the help of Landweber algorithm, and its benefits are compared with existing algorithms. The improvement in the quality of image is represented in terms of mean squared error and correlation coefficient.

4.1 SART Algorithm

The simultaneous algebraic reconstruction technique (SART) is joined the advantages of ART and SIRT. It has many advantages in spite of its computational cost. It requires less number of projections than FBP. Also, it is easily implemented in the ray of different geometry. Formula is expressed as

$$x_j^{(k+1)} = x_j^{(k)} + \frac{1}{A+j} \sum_{i=1}^{M} \frac{A_{i,j}}{A_{i,+}} \left(b_i - \tilde{b}_i \left(x^{(k)} \right) \right) \tag{7}$$

for $k = 0,1, \ldots.$

Step 1: Each ray i_θ is calculated the corrections $\alpha_{i\theta}$ which is carried by

$$\Delta P_{i\theta} = P_{i\theta} - \hat{P}_{i\theta}$$

Step 2: Find the correlation coefficient by

$$\alpha_{i\theta} = \sum_{j=1}^{N} w_{i\theta,j}^{2}$$

Step 3: Make the correction to each cell j of test field as

$$f_j^{\text{new}} = f_j^{\text{old}} + \frac{\lambda w_{i\theta,j} \Delta P_{i\theta}}{\alpha_{i\theta}}$$

where λ is relaxation factor.
Step 4: Calculate the new value of projection using

$$\hat{P}_{i\theta} = \sum_{j=1}^{N} w_{i\theta,jf}$$

Different iterative approaches are used to overcome the computation problems in weighting factors. The above procedure is completed until the optimum value of quality in an image.

4.2 DROP Algorithm

Diagonally relaxed orthogonal projections (DROP) permit the "diagonal component-wise relaxation in conjunction with orthogonal projections on to the individual of the system. This relaxation can be used to exploit the sparsity of the system." It differs from Cimmino's method in using diagonal component-wise relaxation. It is expressed in mathematical form as

$$x^{k+1} = x^k + \lambda_k U_{t(k)} \sum_{i \in B_t(k)} \frac{b_i - \langle a^i, x^k \rangle}{(a_l^i)^2} a_j^i, \tag{8}$$

The implementation procedure of DROP algorithm is as follows:

Step 1: Initialization of arbitrary $x_0 \in R_n$.
Step 2: Update the coefficient represented in [7].

4.3 Component Averaging (CAV) Algorithm

Component averaging (CAV) is an iterative algorithm for unraveling huge and sparse structures of linear equations. It produces a superior rate of convergence than ART. The parallel strategies based on domain decomposition and SPMD. It is decomposed as in the following stages:

Stage 1: Compute the forward projection.
Stage 2: Find an error between the experimental and calculated projections.
Stage 3: Refinement of the model by back-projection error.

The mathematical representation of CAV Algorithm as presented below

$$x_j^{k+1} = x_j^k + \frac{\lambda_k}{s_j} \sum_{i=1}^{m} \frac{b_i - \langle a^i, x^k \rangle}{\left\| \|a^i\| \right\|^2} a^i, \quad j = 1 \ldots n \tag{9}$$

4.4 Cimmino Algorithm

Cimmino's for inconsistent case, the converge is done by least square solution. "In Cimmino's algorithm, projecting the current vector x^k orthogonally onto the next hyper plane to obtain, we form next iteration by computing the reflection points and computing the center of gravity." Since the initial point and all the reflections whose center is solution of the linear system, the center of gravity of the system is a better approximation to the solution.

$$x^{(k+1)} = x^{(k)} + A^T M^{-1}(b - Ax^{(k)}) \tag{10}$$

where M is diagonal matrix. ART does not convert to a single vector, but to a limit cycle. Such as Cimmino's, they do converge to the least square solution in the inconsistent case.

4.5 Landweber Algorithm

To solve ill-posed problems and it is extended to solve nonlinear problems that involve constraints. It is an attempt to regularize the problem. Calculate the initial value in image space. It attempts to recover a signal x from measurements space. To calculate the minimum value of error by reduce the noisy measurements. The process is repeated to obtain the optimum value.

$$g_{k+1} = g_k + \alpha S^T (V - S_{gk}) \tag{11}$$

In higher-frequency component, an image convergence is slower, while at low frequency, image convergence is too faster.

5 Dose Reduction Strategies

In DRS, the CT dosage can be reduced in two ways. Initially for lower exposure dose, the current intensity of the X-Ray tube is reduced under every angle. For improved CT images quality, X-Ray tubes' current densities set to be huge. Reduction in the density of current may minimize radiation dose at every angle, which increases noise intensity. Another way, the number of projection angles is reduced with the same level of dose at every angle with a reduction in projection angles which reduce the overall dose level during CT examination, that is, projection data quantity is less for the image to be constructed. The task for image reconstruction will be an ill-posed problem [14]. Use of baseline scanning in repeat CT, the radiation dose is greatly reduced without adjustment of repeat scan [15]. Two ways to reduce X-Ray dose are done by using tube current and X-Ray beam interruption [16]. The advantage of iterative reconstruction is providing compensation of noise due to reducing the photon flux [17]. For low dose, the noise in the projections has affected the quality of reconstructed image [18]. Model-based reconstruction is used to reduce exposure to object [19]. Iterative reconstruction influences on image quality in case of faster acquisitions at a minimum number of projections or exposure times [20].

The quality of image is highly suffered by noise and inversely related to radiation energy. This reflects the dose reduction is dominant in obtain image quality. Hence, the selection of optimal parameters in the dose reduction is required [21]. Classification of parameters in dose reduction is shown in Fig. 6.

The benefits of iterative reconstruction are reducing the noise and artifacts [22]. The phantoms are used in the low-dose simulation [23]. The metallic artifact reduction is used in the phantom study under low radiation dose and iteration level [24].

6 Simulation Results

IR is suited in dose reduction of Chest CT examination [25]. Minimization of dose to patient requires less number of projections [26]. The tomographic image is reconstructed by algorithm like SART, DART, CAV, Cimmino, and Landweber. Adjust parameters at the current tube and X-Ray source in detector produce less radiation exposure. The number of projections is taken as small. It provides an image with fewer amounts of details. The parameters such as mean absolute error and correlation coefficient are measured with different iteration in test image, head image, and thorax image simulated in different iterative algorithms. Dose reduction is likely to reflect an optimization of algorithms [27].

This section provides the performance of the proposed system. More than 50% of reduction in radiation dose level, the iterative algorithm provides better result [28]. Different value of tube current and voltage influences is verified in the IR algorithm [29]. The simulation is carried out to find the correlation coefficient and mean square error for different iteration for the low dose in parallel-beam and fan-beam arrangements and compared with the exciting iterative algorithms.

Figures 7, 8, 9, and 10 show the metric curves of proposed and existing methods. The mean square error and correlation coefficient are measured in test phantom in different iterative algorithms. The proposed Landweber iteration algorithm produces much better image quality. Figures 11 and 12 present the performance comparison of head image configured in parallel-beam and fan-beam tomography simulated with iterative algorithms and plot the curves of the proposed and other iterative algorithms. It can be observed that the reconstructed image of the proposed method is better than other algorithms. As a lower number of iteration, less error is obtained between the original and reconstructed image with a fixed number of projections.

Figures 12, 13, 14, and 15 show the mean absolute error and correlation coefficient which are measured in thorax image performed by the proposed and existing iterative algorithms.

Table 1 shows the measurement of quality of image on simulating different iterative algorithms in the test phantom, head image, and thorax image, and parameters like structural content (SC), absolute difference (AD), maximum difference (MD), normalized absolute error (NAE), normalized cross-correlation (NCC), mean absolute error (MAE) and peak signal-to-noise ratio (PSNR) are found. Variety of efforts is taken to exhibit the better quality of image on less in projections and iterations. Also, Landweber iteration algorithm is proven that it has good in visual and error perception compared to other algorithms.

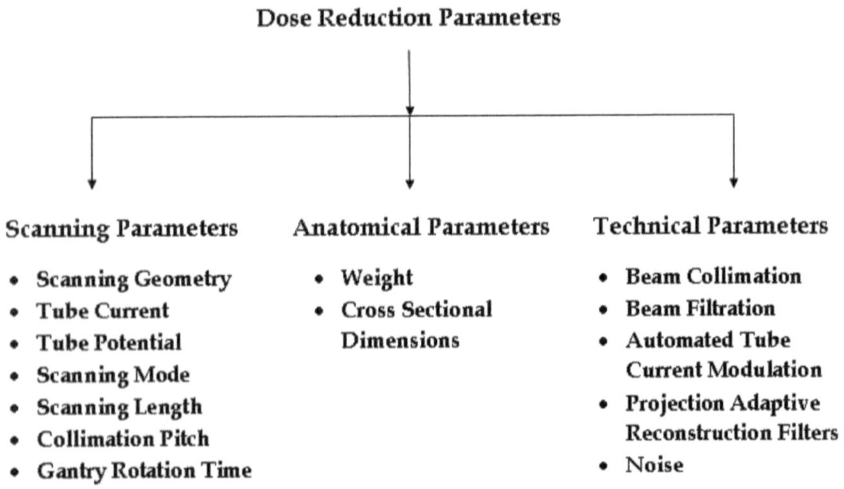

Fig. 6 Classification of dose reduction

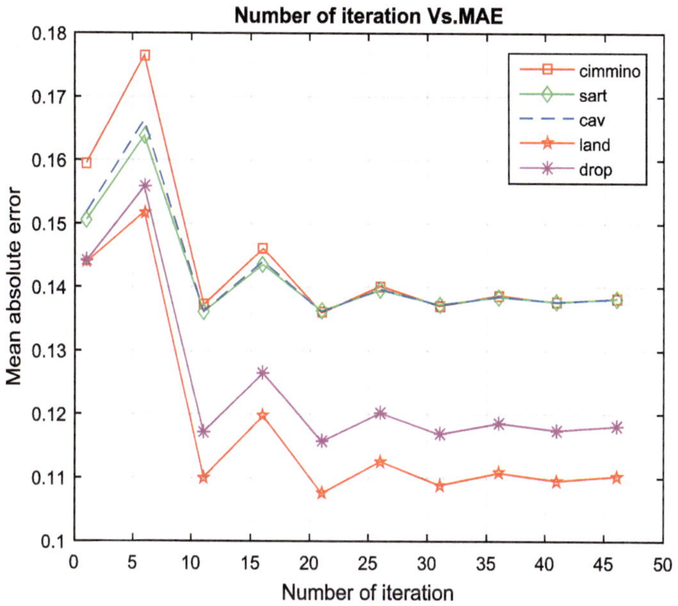

Fig. 7 Mean absolute error versus iteration (parallel beam—test image)

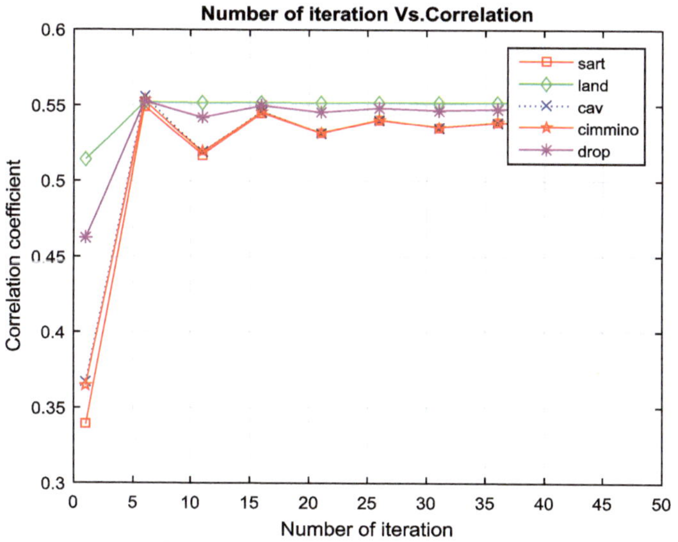

Fig. 8 Correlation coefficient versus iteration (parallel beam—test image)

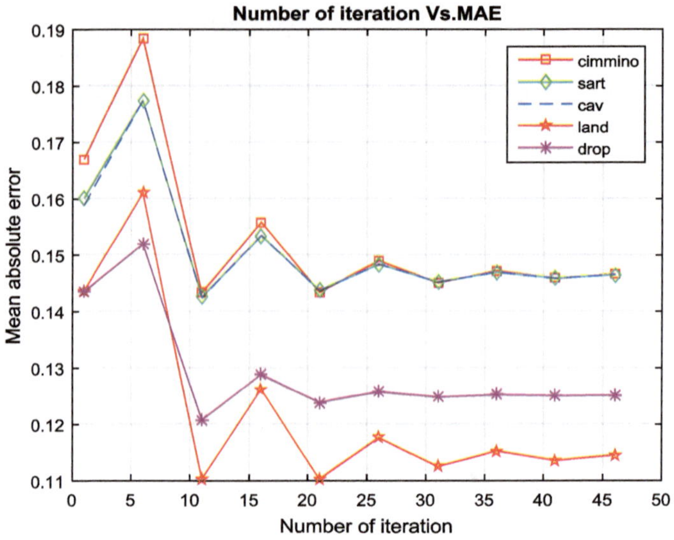

Fig. 9 Mean absolute error versus iteration (fan beam—test image)

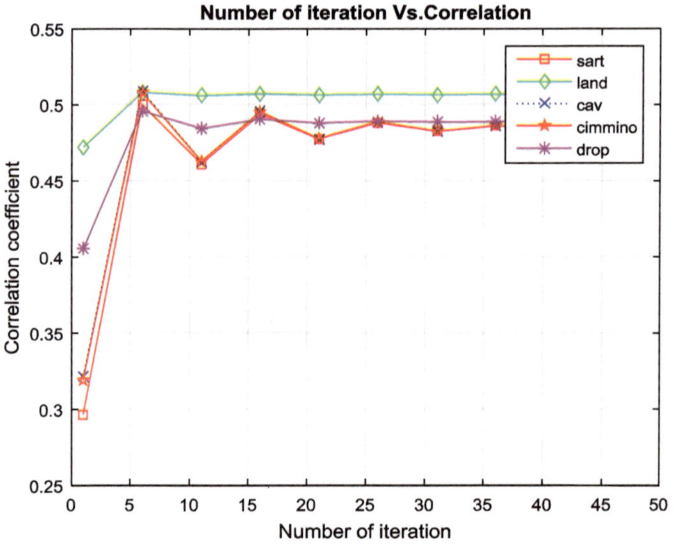

Fig. 10 Correlation coefficient versus iteration (fan beam—test image)

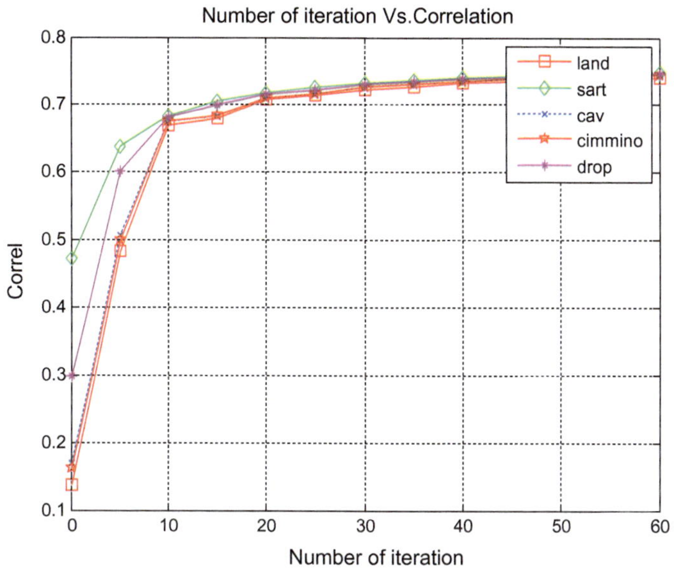

Fig. 11 Correlation coefficient versus iterations (fan beam—head image)

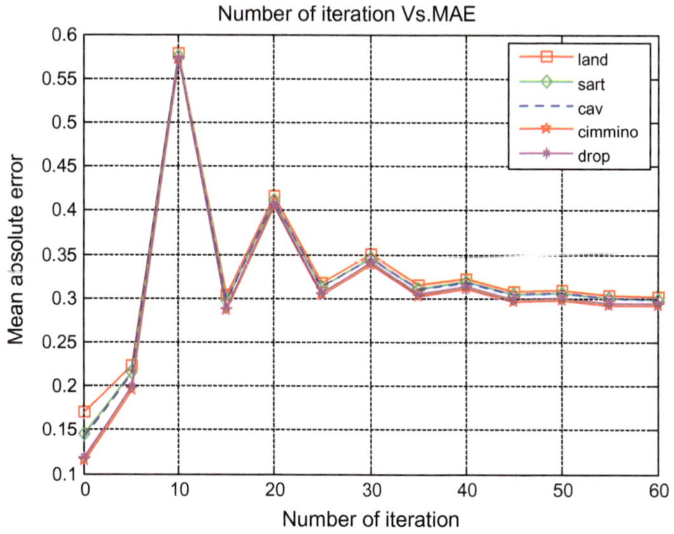

Fig. 12 Mean absolute error versus iterations (fan beam—head image)

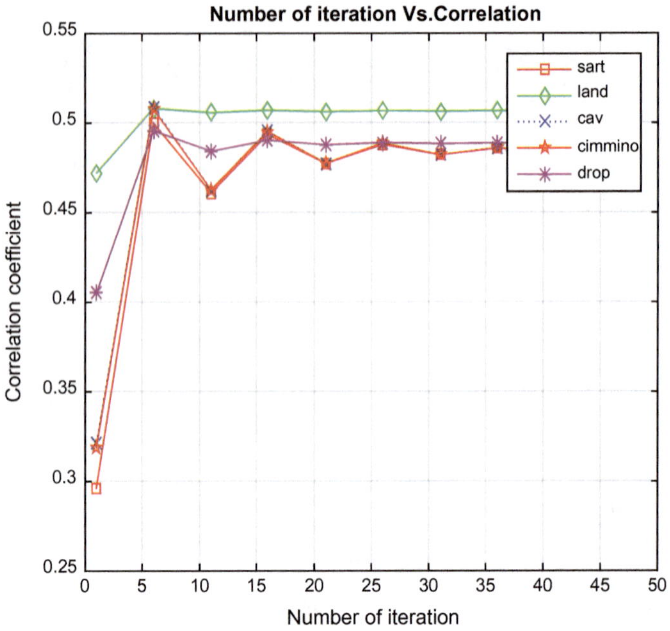

Fig. 13 Correlation coefficient versus iterations (fan beam—thorax image)

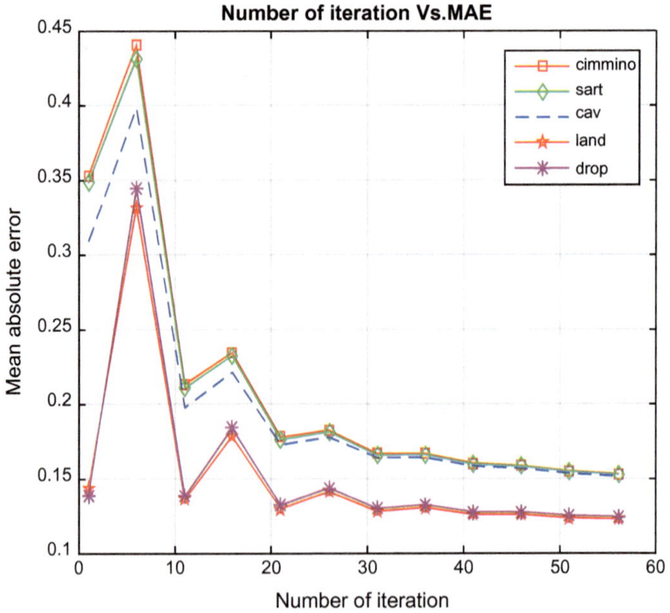

Fig. 14 Mean absolute error versus iterations (fan beam—thorax image)

Table 1 Performance of proposed scheme

Image/parameters	Reconstruction algorithm	SC	AD	NAE	MD	NCC	MAE	PSNR
Test image—fan beam	Landweber	2.1803	0.0141	0.8847	0.9959	0.4887	0.98	9.6262
	SART	1.8997	0.0370	0.8575	0.9629	0.5001	1.0	9.6262
	CAV	2.1802	0.0141	0.8847	0.9959	0.5001	0.98	9.6262
	Cimmino	2.1801	0.0141	0.8847	0.995	0.4587	0.96	9.6262
	DROP	1.9630	0.0032	0.8847	0.9911	0.4897	0.966	9.6262
Parallel-beam image	Landweber	2.4128	0.0177	0.9161	0.97	0.4145	0.933	9.6262
	SART	2.0165	0.0787	0.8909	0.93	0.4665	0.62	9.6262
	CAV	2.4125	0.0177	0.9162	0.94	0.4665	0.07	9.6262
	Cimmino	2.4125	0.0177	0.9162	0.95	0.4145	0.68	9.6262
	DROP	2.1011	0.087	0.9162	0.96	0.4488	0.065	9.6262
Thorax image	Landweber	2.1009	0.0358	1.0988	238.4187	.99	0.1219	78.9364
	SART	2.1010	0.0322	1.0988	238.4316	0.3994	0.1219	78.9364
	CAV	2.1010	0.0323	1.0988	238.4313	0.3994	0.1212	78.9243
	Cimmino	2.0987	0.0119	1.0990	238.4300	0.3996	0.1357	78.936
	DROP	2.1054	0.0806	1.0990	238.3946	0.3990	0.1405	78.996

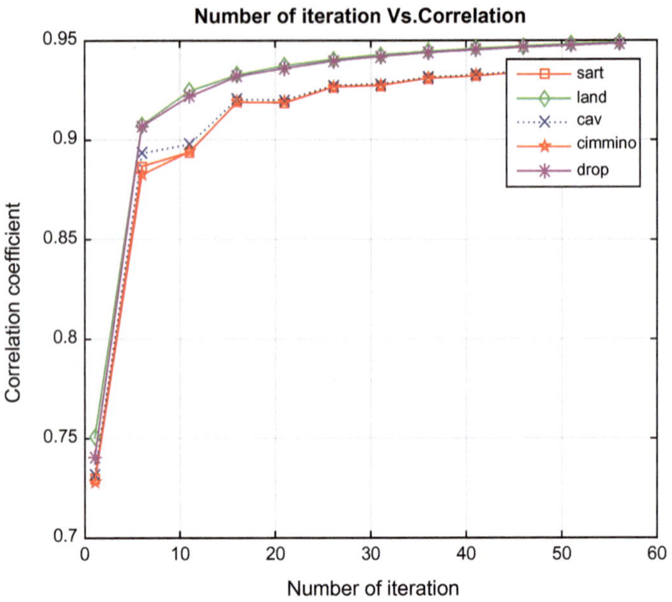

Fig. 15 Correlation coefficient versus iterations (parallel beam—thorax image)

The Landweber reconstruction algorithm is implemented in parallel and fan beam to attain a better quality of image in a reduced dose. In this chapter, different algorithms are investigated in reduced dose level. In case of reduce the dose level in case of higher image quality by setting higher iteration level [30, 31]. The value of X-Ray tube current and direction is varied to perform the reduction in dose. The improvement in quality is represented in terms of mean squared error and correlation coefficient. Results show that the proposed image reconstruction methods for computer tomography are performing well.

7 Conclusion

The advantages of iterative algorithm are excited in the application of medical imaging. The proposed algorithm is performed on a low dose for parallel beam and fan beam to obtain the quality of images, and the results are compared with existing algorithm. Dealing with the task in imaging is a deal by the trade-off among data size, image quality, and computation complexity to create the image in a low dose. In this chapter, SART, CAV, Cimmino, DART, and Landweber iterative algorithms are used in less dose level. With the prior setting of tube current, detector, and gantry direction, reduction in exposure is possible. Further, Landweber iteration reconstruction technique is providing better quality in dissimilar images.

References

1. Jackson, D.F., Hawkes, D.J.: X-ray attenuation coefficients of elements and mixtures. Phys. Rep. **70**(3), 169–233 (1981)
2. Xu, Q., Mou, X., et al.: Statistical interior tomography. IEEE Trans. Med. Imaging **30**(5), 1116–1128 (2011)
3. Eggemiont, P.P.B., Herman, G.T., et al.: Iterative algorithms for large partitioned linear systems with applications to image reconstruction. Linear Algebra Appl. **40**, 37–67 (1981)
4. Constantino, E.P.A., Ozanyan, K.B.: Tomographic imaging of surface deformation from scarce measurements via sinogram recovery. IEEE Sens. J. **9**(4), 399–410 (2009)
5. Kak, A.C., Slaney, M.: Principles of Computerized Tomographic Imaging, Chapter 2. IEEE Press (1988)
6. Gleason, S.S., Sari-Sarraf, H., et al.: Reconstruction of multienergy X-ray computer tomography images of laboratory mice. IEEE Trans. Nucl. Sci. **46**, 1081–1086 (1999)
7. Jiang, M., Wang, G.: Convergence studies on iterative algorithms for image reconstruction. IEEE Trans. Med. Imaging **22**(5), 569–579 (2003)
8. Prince, J.L., Links, J.M.: Medical Imaging Signals and Systems. Pearson Prentice Hall, Upper Saddle River, New Jersey (2006)
9. Peng, H., Stark, H.: One-step image reconstruction from incomplete data in computer tomography. IEEE Trans. Med. Imaging **8**(1), 16–31 (1989)
10. Al-Masni, M.A., et al.: A rapid algebraic 3D volume image reconstruction technique for cone beam computed tomography. Biocybern. Biomed. Eng. **37**(4), 619–629 (2017)
11. Liu, H.L.: Model-based iterative reconstruction: a promising algorithm for today's computed tomography imaging. J. Med. Imaging Radiat. Sci. **45**, 131–136 (2014)
12. Byrne, C.L.: Iterative image reconstruction algorithms based on cross-entropy minimization. IEEE Trans. Image Process. **2**, 96–103 (1993)
13. Jiang, M., et al.: Convergence studies on Iterative Algorithms for Image Reconstruction. IEEE Trans. Med. Imaging **22**(5), 569–579 (2003)
14. Zhang, H., Zhang, L., et al.: Low dose CT image statistical iterative reconstruction algorithms based on off-line dictionary sparse representation. Optik **131**, 785–797 (2017)
15. Shamul, N., Joskoicz, L., et al.: Radon space dose optimization in repeat CT scanning. IEEE Trans. Med. Imaging **36**(12), 2436–2448 (2017)
16. Kobler, E., et al.: Variational deep learning for low dose computed tomography. In: IEEE International Conference on Acoustics, Speech and Signal Processing, Sept 2018
17. Wu, D., et al.: Iterative low-dose ct reconstruction with priors trained by artificial neural network. IEEE Trans. Med. Imaging **36**(12), 2479–2486 (2017)
18. Liu, J., et al.: Data inconsistency for low dose CT imaging. IEEE Trans. Med. Imaging **36**(12), 2499–2509 (2017)
19. Aeed, S., et al.: Low dose CT perfusion of the liver using reconstruction of difference. IEEE Trans. Radiat. Plasma Med. Sci. **2**(3), 205–214 (2018)
20. Lin, Q., et al.: Optimization of image quality and acquisition time for lab-based X-ray microtomography using an iterative reconstruction algorithm. Adv. Water Resour. **115**, 112–124 (2018)
21. Aweda, M.A., Arogundade, R.A.: Patient dose reduction methods in computerized tomography procedures: a review. Int. J. Phys. Sci. **2**(1), 001–009 (2007)
22. Stiller, W., et al.: Basics of iterative reconstruction methods in computed tomography: a vendor-independent overview. Eur. J. Radiol. **109**, 147–154 (2018)
23. Moloney, F., et al.: A phantom study of the performance of model based iterative reconstruction in low dose chest and abdominal CT: when are benefits maximized. Radiography **24**(4), 345–351 (2018)
24. Zhang, H., et al.: Low dose CT image statistical iterative reconstruction algorithms based on off-line dictionary sparse representation. Optik **131**, 785–797 (2017)
25. Macri, F., et al.: Ultra-low-dose chest CT with iterative reconstruction does not alter anatomical image quality. Diagn. Intervent. Imaging **97**(11), 1131–1140 (2016)

26. Azencott, R., Bodmann, B.G., et al.: 3D-image reconstruction in highly collimated 3D tomography. IEEE Trans. Biomed. Eng. **2**(3), 1–9 (2007)
27. Hora, L.O., et al.: Iterative reconstruction and automatic tube voltage selection reduce clinical CT radiation doses and image noise. Radiography **24**(1), 28–32 (2018)
28. Pontana, F., et al.: CT Pulmonary angiogram with 60% dose reduction: influence of iterative reconstructions on image quality. Diagn. Intervent. Imaging **96**(5), 487–493 (2015)
29. Klink, T., et al.: Reducing CT radiation dose with iterative reconstruction algorithms: the influence of scan and reconstruction parameter on image quality and $CTDI_{vol}$. Eur. J. Radiol. **83**(9), 1645–1654 (2014)
30. Urikura, A., et al.: Objective assessment of low-contrast computed tomography images with iterative reconstruction. PhysicaMedica **32**(8), 992–998 (2016)
31. Kidoh, M., et al.: Low-dose abdominal CT: comparison of low tube voltage with moderate-level iterative reconstruction and standard tube voltage, low tube current with high-level iterative reconstruction. Clin. Radiol. **68**(10), 1008–1015 (2013)

DVAE: Deep Variational Auto-Encoders for Denoising Retinal Fundus Image

Biswajit Biswas, Swarup Kr Ghosh and Anupam Ghosh

Abstract Deep learning performs as a computational tool with various potential utilities in ophthalmology. Retinal infections of the eye need to analyze small retinal vessels, microaneurysms, and exudates in the diagnosis of retinal diseases. Due to the appearance of various noises in the fundus images, the retinal vasculature is too complicated to be analyzed for retinal conditions. In this work, we have focused on the field of advanced deep learning in which plethora of architecture is available with the increase in dimension and flexibility of the retinal fundus images. Removal of noise is an essential part to better visibility of noisy fundus and thus a deep learning method for degraded retinal fundus image restoration scheme has been suggested in this investigation. A deep convolutional denoising auto-encoder method based on total variational multi-norm loss function minimization with batch normalization approach has been introduced for restoration of the fundus. The proposed scheme is utilized to restore the perceptible structural details of fundus as well as to decrease the noise level. Moreover, the speed of the network for target noisy images is faster compared to that of other models after fine-tuning of the network with the dropout mechanism. The retinal image databases such as DRIVE, STARE, and DIARETDB1 have been adopted to assess image denoising effects. Our approach to increase the visibility of fundus images by noise reduction through a deep training method has significantly delivered better performance without losing image details along with having fast convergence rate.

B. Biswas
Department of Computer Science & Engineering,
University of Calcutta, Kolkata 700098, India
e-mail: biswajit.cu.08@gmail.com

S. K. Ghosh (✉)
Department of Computer Science & Engineering,
Maulana Abul Kalam Azad University of Technology, Kolkata 700064, India
e-mail: swarupg1@gmail.com

A. Ghosh
Department of Computer Science & Engineering,
Netaji Subhash Engineering College, Kolkata 700152, India
e-mail: anupam.ghosh@rediffmail.com

S. Bhattacharyya et al. (eds.), *Hybrid Machine Intelligence for Medical Image Analysis*, Studies in Computational Intelligence 841,
https://doi.org/10.1007/978-981-13-8930-6_10

Keywords Deep learning · Image denoising · Convolutional auto-encoder · Variational method · Retinal image

1 Introduction

Nowadays, deep learning has become a remarkably powerful tool for image processing and application [1]. It has considerable application in classification and prediction problem since the year 2012. Medical imaging is a challenging task due to illumination property of medical image such as brain MRI, chest X-ray, retinal image, lung MRI, and liver MRI. Medical imaging has two parts in general: first one is disease classification or anomaly detection from different medical images, and secondly, image pre-processing step. Image pre-processing is a crucial part in medical imaging since initial prediction is more important to the respective doctor for an initial assessment before going for automated diagnosis [1, 2]. There are several image pre-processing techniques such as image restoration, image enhancement, and denoising of image which increase the clarity of the image as well as give the details of the part of image (edge, nodes, vessels, etc.) [3, 4]. Before post-processing of images to detect some anomaly, pre-processing is equally important for further processing. Lots of image restoration, image enhancement, and image denoising methods have been developed by several researchers [5–7]. The concept of deep learning in image processing increases the accuracy as well as the robustness and hence we proposed a deep-learning-based image denoising method which is applied on retinal fundus image [8].

In this chapter, we have suggested a denoising model based on deep variational auto-encoders for denoising the retinal image. Lots of automatic anomaly detection algorithm X-ray images or MR images have been proposed since the last few decades. Diagnosis of retinal image is equally important to detect different diseases such as diabetic retinopathy, cataract, and glaucoma earlier to prevent blindness [9]. According to several ophthalmologist and eye experts, diabetes is predictable and considered to be one of the primary reasons of blindness, which is known as diabetic retinopathy (DR) [10]. Early diagnosis is very important for proper treatment because it prevents the visual impairment caused by diabetes. For diagnosis, DR causes damage to small blood vessels occurring in the retina [11]. Images of a patient with DR exhibit red and yellow spots indicative of hemorrhages and exudates. There are three types of DR: background diabetic retinopathy (BDR), proliferate diabetic retinopathy (PDR) and severe diabetic retinopathy (SDR). In BDR stage, arteries in the retina become dilated and show leakage, forming small, dot-like hemorrhages. In PDR phase, circulation problems cause areas of the retina to become oxygen-deprived or ischemic and new fragile vessels to develop to sustain the oxygen levels within the retina. It can cause blood leakage in the retina and vitreous, spots or floaters, along with decreased vision. In the SDR stage, more blood vessels are blocked, blocking blood supply to several areas of the retina along with continued abnormal vessel growth and scar tissue. Hard exudates is one of the primary characteristics of DR which belongs to non-proliferate diabetic retinopathy (NPDR) and the size can vary from tiny specks to large patches with sharp edges [9, 10].

Recently, deep learning has became a very popular tool for feature extraction and classification on large datasets since deep learning in contrast urges to solve the problem end-to-end and creation of abstract features is a nice property of it. Deep learning takes a crucial part in medical imaging such as image segmentation, image registration, image restoration, or different disease diagnosis for the large volume of data. The architecture of ANNs consists of neurons with activation functions and adjusted weight, and it is trained as supervised approach [12]. To adjust the weight in ANNs, back propagation neural network based on gradient decent algorithm takes a significant role in computational intelligence since last few decades. The main drawback of gradient decent is that it has an over fitting problem with the increase in the size of data since the algorithm is often trapped by its local minima. The deep neural network is popular in recent research since it overcomes the overfitting problem for large data as well as it can assign non-random initial value in pre-procedure training phase. In 2009, ImageNet's competition has drawn attention to many computer vision researchers all over the world and hence journey of deep learning has been started. There are four important deep learning architecture such as restricted Boltzmann machines (RBMs), deep belief networks (DBNs), convolutional neural networks (CNNs), and auto-encoder (AE) [13].

There is certain loss of information within the image during the capture of the image because of some limitation of X-ray hardware and therefore the image contains noise. Many vital structures do not seem to be perceived properly since medical images are poorly lit up. Lots of uncertainties are present within the image in imprecise form that leads to inaccurate diagnosis. Thus, it is essential to remove noise from the image or restore the image. The objective of denoising of an image is similar to restoring an image to another form that is appropriate for further processing [14]. Image denoising plays a crucial role in the image pre-processing stage in digital imaging applications as it restores the images from corrupted images [15]. The objective of image denoising is to restore the maximum feature details of the original image through the process of elimination of undesired noise. Various denoising methods have been suggested by several researchers, including neighborhood filtering, median filtering, total variation minimization (TV), and non-local mean (NLM) denoising algorithms [16, 17]. Conventional image denoising algorithms, such as Gaussian filter and median filters are satisfactory for pulling out image noise, still some blurred edge and texture regions are present in the image [15, 17]. Recently, deep-learning-based image denoising scheme has presented outstanding performance compared to other traditional image denoising methods. Zhang et al. (2014) proposed an adaptive bilateral-filter-based image denoising framework in which entropy-based feature extraction is used to restore medical image [5]. It is the most fashionable image domain-based denoising strategy which are used bilateral filter recently. A novel artifact suppressed dictionary learning scheme was suggested by Chen et al. (2015) for low-dose CT image processing [14]. A medical image denoising method using convolutional denoising auto-encoders presented by Gondara et al.[18]. All the aforementioned scheme have suffered from bias and variance issues as well as have long training time. In this investigation, we represent an image denoising strategy in the light of a deep variational denoising auto-encoder for retinal images called as DVAE and assess in applications in ophthalmology by looking at existing image denoising algorithms [19–21].

An auto-encoder (AE) is an alternative category of feed-forward neural network where the number of input layer and output layer are indistinguishable, called an autoassociator, that is designed for unsupervised learning. This can be one form of multilayer perceptron (MLP) in auto-association fashion that can be used for dimension reduction and data compression in an information system. Basically, AE is prepared to encode the input data and turn it into an abstract illustration by encoder function so that the input vectors are reorganized in that representation. Fundamentally, it tries to approximate the identity mapping in this procedure. Vincent et al. (2011) described the denoising auto-encoder (DAE) which deliberately adds noise into the training data and trains the AE with these adulterated information [22]. DAE can recover the original data and make it noise-free through the training process leading to a superior robustness. The main drawback of DAE is that we can be injected noise at input level and training criteria is intractable. In auto-encoder, the latent space where the encoded vectors lie may be discrete or may allow easy interpolation [20]. In this investigation, the design deep variational auto-encoder (DVAE) which is tractable with lower bound has been applied on the retinal image for removal of noise [19–21].

In this work, we have considered a deep denoising auto-encoder (DA) that supported variational technique to restore noise-free retinal image. First, the auto-encoder model maps the features vector into latent plane through its variational distribution function that reduces the dimension of features vector. For more noisy data, we have measured multiple-loss function to realize quick optimization in convolutional model throughout train periods. We get an inclination to verify overall performance of the proposed methodology and compare with different three state-of-the-art methods using a public standard benchmark database in an enormous satisfaction in training session. The experiential results determine that the planned methodology not only restores retinal images but additionally preserves image details a lot accurately and minimize number of epoch than most privileged technique. The primary contributions of this article are as follows:

- We analyze the deep denoising auto-encoder algorithms and propose a deep variational auto-encoder method (DVAE) for the removal of noise from retinal image and restoration to the exploitation of the benefits of pre-processing steps.
- Extracted features from convolutional model map into latent space which reduce the dimension of the data and restore the noise-free image with less number of features with robustness.
- Deep variational auto-encoder (DVAE) exploits the multi-level loss function using Kullback–Leibler divergence (KLD) in proposed scheme to discourse restoration for corrupted retinal images.
- We have designed DA based on total variation model to develop clarity of the retinal image for efficient elimination of noise with a very less number of epoch

and the accomplishment of the proposed scheme surpasses from state-of-the-art algorithms while achieving the best restoration results.

The rest of this chapter is structured as follows: In Sect. 2, we have discussed the preliminaries and mathematical model. The proposed methodology and designed retinal image restoration scheme in detail have been discussed in Sect. 3. The experimental results and discussion have been explored in Sect. 4, and finally conclusions of this work have been given in Sect. 5.

2 Preliminaries

This section concisely discusses the theories of deep variational auto-encoder (D-VAE) [19, 21] and the basic framework of retinal image denoisification method [20]. The proposed algorithm deep variational auto-encoder method (DVAE) is established on the basis of following perception.

2.1 *Denoise Model for Retinal Images*

According to mathematical definition [20, 22], we consider that a given $\mathbf{X} \in \mathbb{R}^{m \times n}$ noisy retinal image and $\mathbf{y} \in \mathbb{R}^{m \times n}$ is its corresponding noise-free image. Then, the interrelation between $\mathbf{X}$ and $\mathbf{y}$ can be written as follows [18]:

$$\mathbf{X} = \eta\left(\mathbf{y}\right)$$

where $\eta : \mathbb{R}^{m \times n} \rightarrow \mathbb{R}^{m \times n}$ denotes the extortion of the additive noise that distorts the normal retina. To recover the corrupted image, the noise reduction algorithm (viz. a function) is mainly converted into the original once while finding a function ϕ that holds the following criterion [14]:

$$\phi = \arg\min \|\phi\left(\mathbf{x}\right) - \mathbf{y}\| \tag{1}$$

where ϕ is deployed as the best approximation of η^{-1} and $\| \cdot \|$ is l_1 norm.

2.2 *Variational Auto-Encoder (VAE) for Noisy Retina*

For retinal image restoration, we have chosen unsupervised-learning-based deep variational auto-encoder (DVAE) which regenerates a noise-free image from its noisy version with self learning [21, 23]. Let $\mathbf{X}$ be a noisy retinal image and we have a vector of latent variables z taken from high dimensional space $\mathscr{Z}$ that can be easily sampled to some probability density function (pdf) $P(z)$ is defined over $\mathscr{Z}$. We can define a family of deterministic function $f(z;\theta)$ parameterized by θ as $f:\mathscr{Z}\times\Theta\rightarrow\mathscr{X}$. Our objective is to optimize θ such that we can sample z from $P(z)$ with high probability $f(z;\theta)$ and it is defined by [21]:

$$P(X)=\int P(X|z;\theta)P(z)dz \tag{2}$$

By the definition of total probability, $f(z;\theta)$ is substituted by the distribution $P(X|z;\theta)$. In variational auto-encoder (VAE), the choice of output distribution is often Gaussian distribution since it generates similar sample and hence we may write $P(X|z;\theta=\mathscr{N}(z|\mu,\sigma*\mathbf{I})$, where mean and standard deviation of sample are μ and σ respectively [20, 21] and $\mathbf{I}$ denote identity matrix. The network $P(X|z;\theta)$ is known as "generative network," and it is intractable due to complex nonlineraity of neural network.

The parameterized distribution from Eq. (2) has been introduced into the proposed neural network architecture for input z over the parameterized distribution X, where θ is the weight vector of the neural network. The result of VAE is also the approximate distribution $\bar{P}$ on the observation where X is defined as $\bar{P}(X|z;\phi)$ (called inference network) for variational parameter ϕ, also the weight vector of network and z and X are the output and input, respectively. The primary objective of VAE is to maximize the variational lower bound which is defined as [20, 21]:

$$\begin{aligned}\log\ P(X;\theta)\geq\mathbb{E}_{\bar{P}(z|X;\phi)}\left[\log\frac{P(X|z;\theta)}{\bar{P}(z|X;\phi)}\right]\\ =\mathbb{E}_{\bar{P}(z|X;\phi)}[P(X|z;\theta)-\mathbb{KLD}(\bar{P}(z|X;\phi)||P(z))]\end{aligned} \tag{3}$$

where $\mathbb{KLD}$ is the Kullback-Leibler Divergence is derived as: [6, 23]:

$$\mathbb{KLD}(\bar{P}(z|X;\phi)||P(z))=\mathbb{E}_{z\sim\bar{P}}[\log\ (\bar{P}(z)-\log\ P(z|X)]$$

The image restoration accuracy of an auto-encoder is calculated by the Eq (3) whose noise is injected in the hidden layer, i.e., output inference layer.

3 Retinal Image Restoration Scheme by Using Deep Variational Auto-Encoder

The proposed image reconstruction scheme is based on a deep learning approach that determines the sequences of noisy image as concurrent input and provides the sequences of noise-free output image (normal image) on a huge numbers of train-test data pairs, which consists of a train input sequence and also the desired target sequence. In this investigation, we have used deep variational auto-encoder for retinal

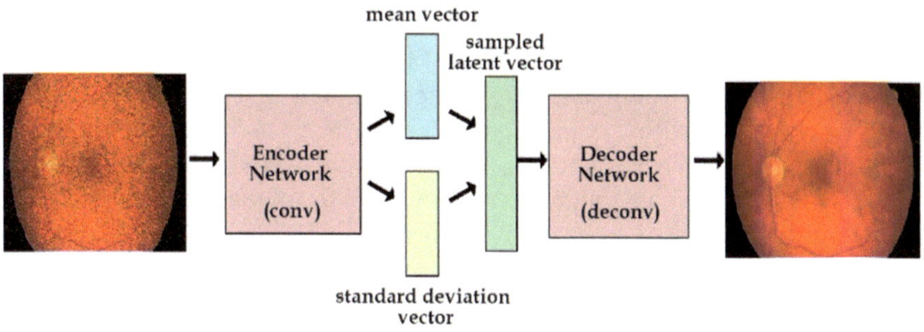

Fig. 1 Architecture of proposed auto-encoder

image restoration since mean (μ) vector and standard deviation (σ) vector of Gaussian distribution control the area of encoding of an input and the variation of area from mean. In encoding phase, image patches are selected at random and the decoder learns from all point in latent space, as well as nearby points of the reference latent space and hence the decoder is exposed to a range of variations of the encoding of the same input. Now, the model is designed with a certain degree of local variation by varying the encoding of one sample in latent space. Extracted features are mapped into the latent space which reduce the dimension of the features vectors using latent variables with probability distribution parameters in distribution function.

There are several values of vectors μ and σ for different samples, and hence the encoder can learn to produce different μ for different classes. So, a number of cluster generation depend on μ, and then minimize σ and μ accordingly. The decoder can efficiently reconstruct the training data based on the values of μ and σ. In order to optimize the model, we have suggested the Kullback–Leibler divergence ($\mathbb{KLD}$) into the loss function defined on Eq. (3). The KL divergence basically determines the divergence between two probability distributions. We have utilized KL divergence into loss function to closely resemble the target distribution, and optimize the probability distribution parameters μ and σ. The convergence rate of the proposed model is lesser than that of the other deep learning model due to dimension reduction property which reduces the computational cost.

We address the most deficiency by updating the DVAE architecture to incorporate "generative network" and "inference network" stages by batch normalization theme. Later, we continue this procedure by deploying the preparation of training data and also the perfect loss function with optimizer throughout training. Figure 1 shows the schematic block diagram of the proposed deep variational auto-encoder model (DVAE).

3.1 Training

In this section, we have illustrated data augmentation in order to construct the training data. Beginning with 300 slices of retinal image for each train sets during training period, for each image, we initiate a noisy image at one frame per data sets, adding the Gaussian additive noise in training phase [5]. We make 512×512 images during explanation, but the training is tested with resizing as 128×128 patch which are randomly chosen for each training sequence. A sequence of 50 consecutive frames has been used to give abundant temporal substances in the training phase.

The variational EM algorithm has been applied to train the model defined on Eq. (3) which is discussed in the following steps:

- **E-step**: We update ϕ such that the approximate distributation $\bar{P}(z|X;\phi)$ will be close to the posterior distribution $P(z|X;\theta)$ by keeping θ fixed.
- **M-step**: We update θ to increase the marginal log-likelihood by keeping ϕ fixed.

In order to train the whole network of DVAE, parameter ϕ is applied in back-propagation considering z as a function of Gaussian noise of the output of encoder. The parameter θ of generative network also rapidly updates in backpropagation step, and hence we can update both parameters θ and ϕ together by stochastic gradient descent algorithm.

In data normalization, we have randomly chosen small rotation of the training sequence in 2–3 degrees in train increasing capture more temporal details of changing directions. We additionally choose a random gamma correction within the range [0, 1] one by one to every color channel, and put on the network to the complete sequence [1].

There is a simple way of training the proposed DVAE with denoising principle in three steps as: First, a corrupted input is sampled as $\bar{X}^{(i)} \sim P(\bar{X}|X)$, where $\bar{X}$ is the corrupted retinal image of the original version X. Second, sample $z^{(j)} \sim \bar{P}(z|\bar{X}^{(i)})$ and third, the generative network $P(X|z^{(j)};\theta)$ reconstructs the sample and parameters are optimized by the Eq. (5).

3.2 Loss Function

The computed error between training targets and outputs from network during the training phase is described as a loss function. The most ordinary MSE (mean squared error) from the train image S and the target image T is defined by the $\mathscr{L}_2$ loss functions. However, another loss function, $\mathscr{L}_1$ loss instead of $\mathscr{L}_2$ can be reduced the blemish artifacts from reconstructed images [7]. As mentioned above, the first loss term is a spatial $\mathscr{L}_1$ loss in the basis of $\mathbb{KLD}$, denoted as $\mathscr{L}_z$ applicable for a distinct image in the training phase [14]. The second loss term is a spatial $\mathscr{L}_1$ loss, stand for $\mathscr{L}_s$ for a particular image in the training stage [18]:

$$\mathscr{L}_s = \frac{1}{N}\sum_{j=1}^{N} \|\mathscr{S}_j - \mathscr{T}_j\| \tag{4}$$

where $\mathscr{S}_j$ and $\mathscr{T}_j$ are jth pixel of the resultant and goal image correspondingly.

The loss function of DVAE is decided by lower bound of the variational model which is approximated by Monte Carlo sampling as:

$$\begin{aligned}\mathscr{L}_v &= \mathbb{E}_{P(\bar{X}|X)}\mathbb{E}_{\bar{P}(z|\bar{X})}\left[\log \frac{P(X,z;\theta)}{\bar{P}(z|\bar{X};\phi)}\right] \\ &\simeq \frac{1}{MN}\sum_{i=1}^{M}\sum_{j=1}^{N}\log \frac{P(X,z^{(j|i)};\theta)}{\bar{P}(z^{(j|i)}|\bar{X}^{(i)};\phi)}\end{aligned} \tag{5}$$

where $\bar{X}^{(i)} \sim P(\bar{X}|X)$ and $z^{(j|i)} \sim \bar{P}(z|\bar{X}^{(i)})$. Equation (5) represents the estimator of DVAE for all reconstruction parameters.

The $\mathscr{L}_2$ spatial loss gives a well image metric which is used to measure the overall outliers. To highlight the variances in adequate information (edge, corner, etc.), we use a Laplacian operator with square root of absolute $\mathscr{L}_1$ in the domain of gradient loss [7]

$$\mathscr{L}_l = \frac{1}{N}\sum_{j=1}^{N}\sqrt{\|\nabla \mathscr{S}_j - \nabla \mathscr{T}_j\|} \tag{6}$$

where individual gradient $\nabla(\cdot)$ is calculated based on high frequency error norm (HFEN), which is a metric used for medical image comparison [6]. The Laplacian of Gaussian kernel (3×3) has been used for edge detection, but it is sensitive to noise and hence to make image pre-smoothed, first Gaussian filter is used for edge detection [14]. To authenticate that the combined loss ends up in an enhancement above the spatial loss $\mathscr{L}_s$, it has been calibrated by the structural similarity metric (SSIM) [24] on tuning sequence of fundus after 50 epochs with 128 batch size of training. SSIM shown an enhancement for $\mathscr{L}_s$ from 0.9308 to 0.9736 for the fused loss.

4 Experiments

The effectiveness of our algorithm, along with comparisons is demonstrated on three standard publicly available databases, viz., DRIVE [25, 26] and STARE [27, 28], and DIARETDB1 [29] in this section.

4.1 Description of Datasets

In this work, we have used two standard publicly available databases DRIVE [25, 26] and STARE [27, 28]. The DRIVE database [25] consists of 40 color fundus images obtained from a diabetic retinopathy screening program in Netherlands. The images are compressed in JPEG format and are captured in digital form from a Canon CR5 non-mydriatic 3CCD camera at 45 degree field of view (FOV). The size of each image is 565 × 584 pixels, 8 bit per color channel and has a FOV of approximately 540 pixels in diameter. The 40 images have been divided into training set and test set, each containing 20 images.

The STARE [27, 28] database contains a total of 400 raw images of which small version of images can be viewed 100 at a time. The images are captured in digitized slides by TopCon TRV-50 fundus camera at 35 degree FOV. The size of each image is 700 × 605 pixels, 8 bit per color channel and has a FOV of approximately 550 pixels in diameter.

The DIARETDB1 [29] database consists of 89 fundus images. The images were captured with the same 50 degree FOV digital camera in Finland. The size of each image is 1500 × 1152 pixels, 8 bit per color channel. This database is mainly used to detect the hard exudates, hemorrhages, and microaneurysms.

4.2 Experimental Setup of DVAE

A total of 529 ordinary retinal images of size 565 × 584 obtained from aforementioned databases archive as normal-dose dataset in training phase. As stated above, the corresponding dataset of noise fundus images have been achieved by introducing Gaussian noise into the routine retinal images for the simulation of the proposed method DVAE. The salt-and-pepper noise plus with Gaussian noise is utilized to test the performances of the suggested model DVAE. In the initial configuration, $\mu_1 = 0, \sigma_1 = 30$ control the noise level of images. Each retinal image is corrupted by uniform impulse noise $p = 25\%$ and Gaussian noise $\sigma_1 = 30$, which are shown in Figs. 2b, and 3b for example.

In the training phase, we have randomly selected 400 normal (cleaned) and corresponding additive noise fundus pairs. Patch size p is set to 128 × 128 pixels. We have evaluated many parameter accumulation for the reconstruction of retinal image, and therefore Gaussian distribution is approximated to $\mathbb{N}(0, \sigma * I)$ in cost function Eq. (5), where σ is the hyper-parameter and I denote the identity matrix. With the increase in hidden layers and neurons, duration of training increases without noticeable performance gain, therefore the number of layers L are set to 16 for all cases. All the experiments are performed on Python 3.6 using Keras and Opencv-python on an Ubuntu 18.04 LTS (Intel *i*58400 CPU and 16 GB RAM).

In the test phase, we have imposed noise in patches of test images are managed by the trained proposed DVAE, so we have used the resultant patches to regenerate the reconstructed image. Three metrics have been adapted for statistical quantification. The primary one is the root mean square error (**MSE**) [30], and later are picture quality score(**PQS**) [24] universal quality index (**UQI**) [31] and structural similarity index metric (**SSIM**) [24, 30], respectively. The highest value among them indicates minimum distortion and better reconstruction except MSE.

Three different baseline methods are compared for the performance evaluation, such as "An adaptive bilateral filter based framework for image denoising" abbre-

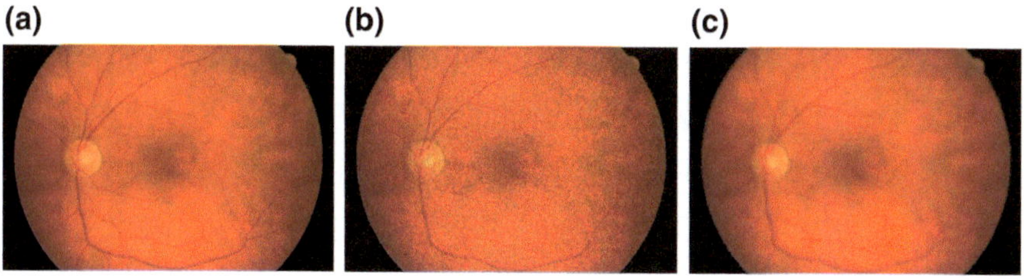

Fig. 2 Simulation with the proposed scheme. **a** The original image. **b** Noised image. **c** Reconstructed image (epoch = 50, batch-size = 64)

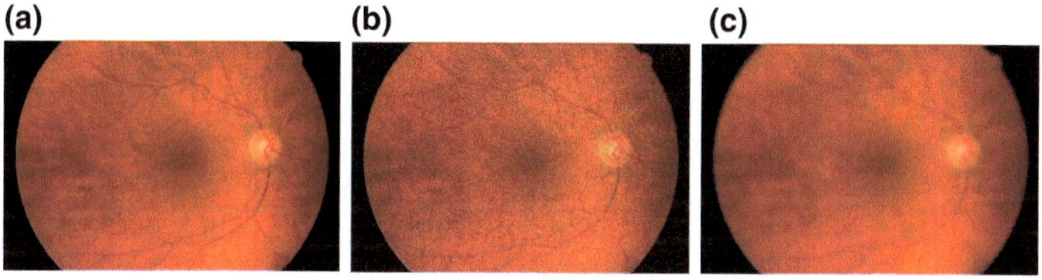

Fig. 3 Simulation with the proposed scheme. **a** The original image. **b** Noised image **c** Reconstructed image (epoch = 50, batch-size = 64)

viated for convention as (ABFD) [5], "Artifact suppressed dictionary learning for low-dose CT image processing" (SDLA) [14], and "Medical image denoising using convolutional denoising auto-encoders" (CDA) [18]. ABFD is the most fashionable image-domain-based denoising strategy which use bilateral filter recently. CDA is projected low-dose CT reconstruction ways supported by neural networks. All the connected parameters within the baseline techniques are fixed as per the recommendation of the primary investigation.

4.3 Analysis of Different Parameters

The efficiency of the proposed DVAE method depends on different parameters of the algorithm. Here, we have examined the two key factors: the amount of training instance and noise level introduced in the databases that have been discussed in this section.

The training data: We have emphasized on the effect of the size of training data from three databases, viz., "DRIVE," "STARE," and "DIARETDB1." In the first phase, 400 image pairs have been randomly chosen from the databases as training data. In the next phase, data augmentation has been taken to magnify the training data supporting the initial 400 image pairs using Keras API. The transformations enclosed rotation (by 2–3), horizontal and vertical flipping, and scaling (varying scale factors

Table 1 Quantitative evaluation of various methods

Image	Method	MSE	PQS	UQI	SSIM
Figure 4	ABFD [5]	0.0276	26.87dB	0.846	0.824
	SDLA [14]	0.0227	27.52dB	0.833	0.841
	CDA [18]	0.0195	28.62dB	0.875	0.887
	DVAE	**0.0187**	**29.12dB**	**0.884**	**0.892**
Figure 5	ABFD [5]	0.0269	25.95dB	0.861	0.830
	SDLA [14]	0.0224	26.62dB	0.873	0.835
	CDA [18]	0.0202	28.14dB	0.884	0.878
	DVAE	**0.0182**	**29.52dB**	**0.902**	**0.921**
Figure 6	ABFD [5]	0.0308	27.53dB	0.894	0.872
	SDLA [14]	0.0238	28.45dB	0.803	0.847
	CDA [18]	0.0214	28.86dB	0.876	0.852
	DVAE	**0.0206**	**29.68dB**	**0.895**	**0.927**
Figure 7	ABFD [5]	0.0263	26.17dB	0.784	0.802
	SDLA [14]	0.0207	26.52dB	0.804	0.785
	CDA [18]	0.0187	27.35dB	0.878	0.825
	DVAE	**0.0176**	**28.72dB**	**0.882**	**0.853**

0.2–0.5). Lastly, similar operations for data augmentation have been executed on the new elite where a hundred image pairs images are achieved.

Noise level: In practice, consistent to noise levels from the training and test sets should not be similar; consequently, it is essential to justify the validity of the suggested method completely for various datasets with different noise levels. However, one testing set with a Gaussian noise level, $\mu_1 = 0$, $\sigma_1 = 30$ and salt-paper noise with $p = 25\%$ area unit is employed in the analysis. The training set preserved a similar noise level CDA [18]. The objective evaluations are shown in Table 1. It is determined that for a given noise level with large epochs the CDA and DVAE have higher MSE, PQS, and UQI values than different baseline methods. Figure 2 and Fig. 3 demonstrate the reconstruction of the retinal image by the proposed DVAE where (b) is a noisy version of (a).

4.4 *Qualitative Analysis*

In test phase, one sequence frame from the test set has been selected to assess the performance of the proposed DAVE. Figures 4, 5, 6, and 7 demonstrated the results achieved from a retinal image by proposed DVAE along with all baseline methods. Figures 4, 5, 6, and 7b show the noise retinal images and all vessels (especially thin vessels) disappeared in the images. Each one of the techniques can productively stifle the noise; however, in Figs. 4, 5, 6, and 7c, ABFD can be affected by conspicuous

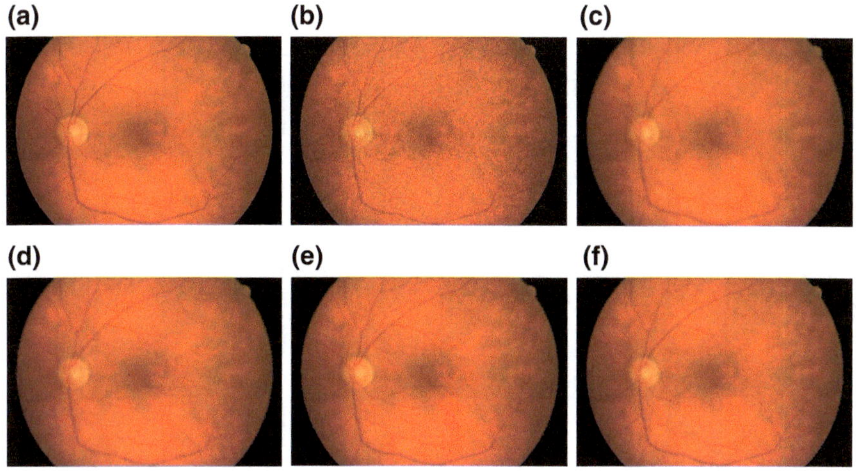

Fig. 4 Results of a Retina (in Fig. 2b) image for comparison with baseline methods. **a** Cleaned image; **b** noised image; **c** ABFD; **d** SDLA; **e** CDA; and **f** DVAE

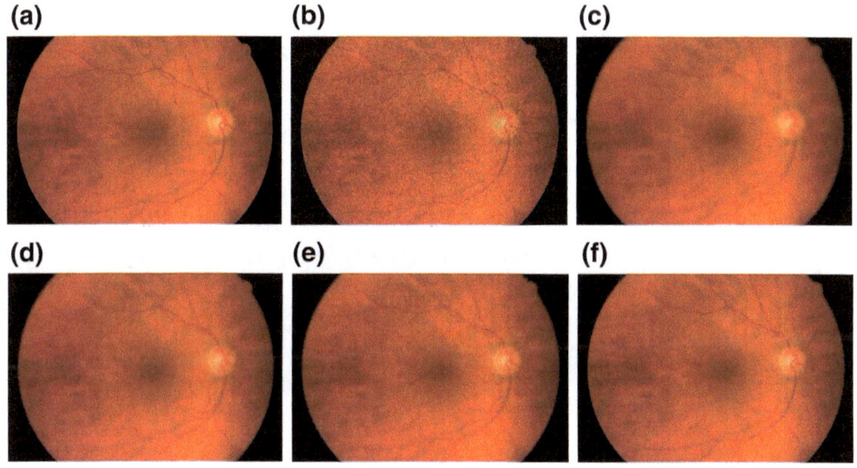

Fig. 5 Results of a Retina (in Fig. 3b) image for comparison with baseline methods. **a** Cleaned image; **b** noised image; **c** ABFD; **d** SDLA; **e** CDA; and **f** DVAE

square effect and oversmoothed a few tiny vessels in the retinal border region. CDA retains major details than SDLA, but the tiny spot in the center region of retinal images still has the noisy appearance with the low contrast levels. Deep-learning-based techniques CDA can protect more subtle elements but the noise and artifacts. Nonetheless, CDA oversmoothed few low contrast domains that are indicated in Figs. 4, 5, 6, and 7e.

Three baseline strategies (ABFD, SDLA, and CDA) have been put in place. It is detected in Figs. 4, 5, 6, and 7c–e that the three strategies preserve the tiny vessels

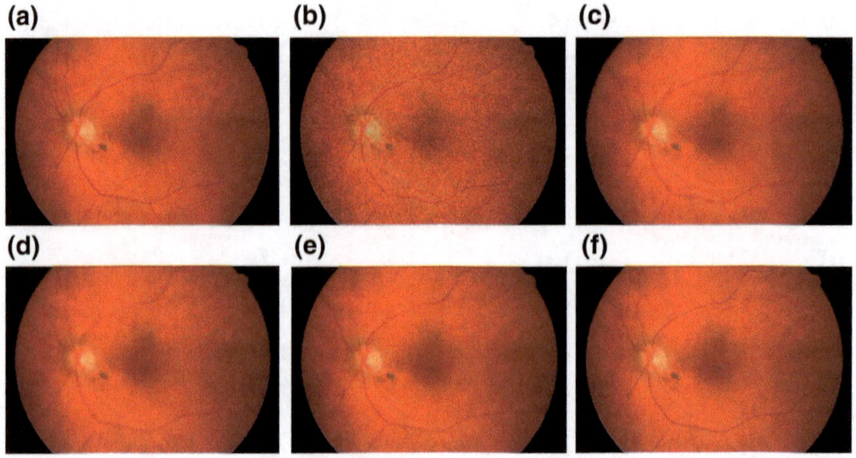

Fig. 6 Results of a Retinal image for comparison with baseline methods. **a** Cleaned image; **b** noised image; **c** ABFD; **d** SDLA; **e** CDA; and **f** DVAE

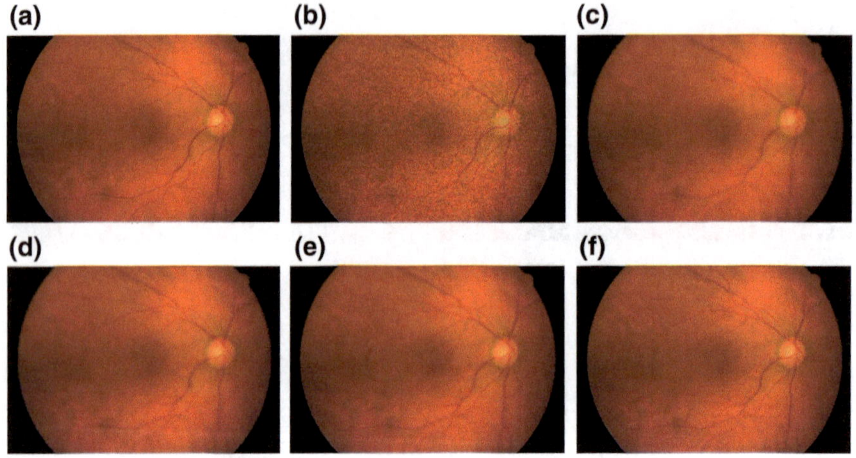

Fig. 7 Results of a Retinal image for comparison with baseline methods. **a** Cleaned image; **b** noised image; **c** ABFD; **d** SDLA; **e** CDA; and **f** DVAE

much distinct than different methods. DVAE additionally would have obtained the most effective performance within the distinctive low region if we had circumstantially determined all resultant data. Additional methods, mainly ABFD and SDLA failed to eliminate impulse noise completely. The proposed methodology DVAE can be preserved additionally by keeping noise-free pixels unchanged on multi-loss utilization and hence we can determine that the details of the restored image by the proposed methodology DVAE has produced the best visual effect compared to other methodologies.

4.5 Quantitative Analysis

The objective performance of the proposed method DVAE is compared with three existing methods as mentioned above shown in Figs. 4, 5, 6, and 7. The statistical results for the four metrics (MSE, PQS, UQI, and SSIM) are shown in Table 1 with bold font with the highest score of DVAE. For all the images, DVAE achieved better scores of PSNR and PQS, UQI and SSIM and interestingly, CDA also achieved better performance with respect to MSE, PQS, UQI, and SSIM, which is the same as DVAE, and it indicates that both are able to sustain more details in image restoration task. Table 1 shows the statistical validation of Figs. 4, 5, 6, and 7c–f in the test phase. Table 1 proves that the suggested DVAE method obtained ambitious achievement with CDA and excellent reconstruction than other baseline methods.

The performances of proposed techniques DVAE are compared with three baseline methods such as ABFD [5], SDLA [14], and CDA [18] in Table 1. The UQI obtained by baseline methods such as ABFD and SDLA are in the range of 0.78 to 0.88, whereas deep-learning-based techniques CDA in UQI are close to the proposed algorithm. SSIM for proposed technique DVAE is more important than ABFD, S-DLA, and CDA and close is to 0.9 in all cases. So, we realize that edges of vessels are not well sustained using ABFD and SDLA, whereas the CDA and suggested method DVAE are able to do so. So, we can state that Table 1 reveals all image quality metrics with improved CDA and proposed technique DVAE than ABFD and SDLA methods. Metric indices of performance metric in Table 1 show superior noise reduction capabilities of the proposed DVAE scheme. In Fig. 8a–b shown that the test and validation loss of the DVAE with a certain number of epochs, respectively. The proposed model DVAE model has taken very less number of epoch (approximate 50) in training phase.

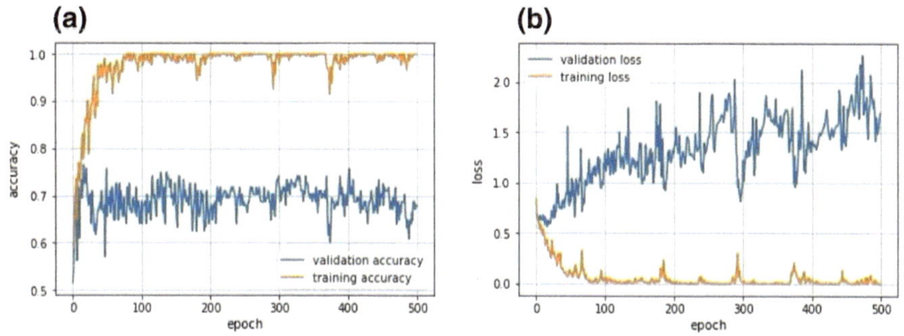

Fig. 8 Performance evaluation of the proposed method. **a** Curve for accuracy, and b curve for validation

5 Conclusion

In this work, deep learning that is based on variational auto-encoder has been recommended for retinal image restoration. The proposed method can increase the picture quality proficiently and rapidly as well as preserve image details more accurately since variational lower bound yield a sensible corruption function which is tighter than the standard lower bound on noisy image. The extracted features map into latent space using latent variable with Gaussian distribution which reduce the dimension of the data and converge for few iterations compared to other deep learning model. Both the subjective and quantitative results have exhibited and investigated its capacity to decrease noise and maintain the structure details. Proposed algorithm has reduced the computational cost because of its fast convergence for the diagnosis of several malfunctions from retinal image such as diabetic retinopathy, glaucoma, and exudates at the first screening of an ophthalmology. It is more robust and flexible than DAE since it uses approximate posterior distribution on input data. Some times variational model fails to restore images due to the presence of high amount of complex noises in the images. In future work, other issues the retinal images for diabetic retinopathy detection, optic disk detection, or exudates detection via deep learning will be exploited.

References

1. Sonka, M., Hlavac, V., Boyle, R.: Image processing analysis and machine vision. Thomson Learn (2001)
2. McBee, M.P., et al.: Deep Learning in Radiology. Elsevier, Academic Radiology (2018)
3. Moraru, L., Obreja, C.D., Dey, N., Ashour, A.S.: Dempster-shafer fusion for effective retinal vessels diameter measurement. Soft Comput. Based Med. Image Anal. 149-160 (2018)
4. Chakraborty, S., Mukherjee, A., Chatterjee, D., Maji, P., Acharjee, S., Dey, N.: A semi-automated system for optic nerve head segmentation in digital retinal images. In: 2014 International Conference on Information Technology (ICIT), IEEE (2014)
5. Zhang, Y., Tian, X., Ren, P.: An adaptive bilateral filter based framework for image denoising. Neurocomputing (2014)
6. Zhang, Y., Zhang, W.H., Chen, H., Yang, N.L., Li, T.Y., Zhou, J.L.: Few-view image reconstruction combining total variation and a high-order norm. Int. J. Imaging Syst. Technol. **23**(3), 249–255 (2013)
7. Hang, Z., Gallo, O., Frosio, I., Kautz, J.: Loss Functions for Image Restoration with Neural Networks. IEEE Transactions on Computational Imaging, ACM, arxiv.org/abs/1511.08861 (2016)
8. Abramoff, M.D., Garvin, M.K., Sonka, M.: Retinal imaging and image analysis. IEEE Rev. Biomed. Eng. **3**, 169–208 (2010)
9. Ghosh, S.K., Ghosh, A., Chakrabarti, A.: VEA: Vessels extraction algorithm and a novel wavelet analyser for diabetic retinopathy detection. Int. Journal of Image and Graphics-Worldscientific **18**(2) (2018)
10. Singer, D.E., Nathan, D.M., Fogel, H.A., Schachat, A.P.: Screening for diabetic retinopathy. Ann. Intern. Med. **116**, 660–671 (1992)
11. Kanski JJ, Bowling B (2012) Synopsis of Clinical Ophthalmology. Elsevier Health Sci

12. Schmidhuber, J.: Deep learning in neural networks: an overview. Neural Netw. **61**, 85–117 (2015)
13. Goodfellow, I., Bengio, Y., et al.: Deep Learning. MIT Press (2016)
14. Chen, Y., Shi, L., Feng, O., Yang, J., Shu, H., et al.: Artifact suppressed dictionary learning for low-dose CT image processing. IEEE Trans. Med. Imaging **33**(12), 2271–2292 (2014)
15. Zhang, Y., Wang, Y., et al.: Statistical iterative reconstruction using adaptive fractional order regularization. Biomed. Opt. Exp. **7**(3), 1015–1029 (2016)
16. Li, Z., et al.: Adaptive nonlocal means filtering based on local noise level for CT denoising. Med. Phys. **41**(1) (2014)
17. Huang, Z.K., Li, Z.H.: Comparison of different image denoising algorithms for Chinese calligraphy images. Neurocomputing **188**, 102–112 (2016)
18. Gondara, L.: Medical image denoising using convolutional denoising autoencoders. Pro. of the IEEE Int. Con. Data Mining Workshops (ICDMW 2016) pp. 241–246 (2016)
19. Doersch, C.: Tutorial on variational autoencoders. Mach. Learn. arxiv.org/abs/1606.05908 (2016)
20. Im, D.J., Ahn, S., Memisevic, R., Bengio, Y.: Denoising Criterion for Variational auto-encoding framework. In: Proceedings of the Thirty-First AAAI Conference on Artificial Intelligence (AAAI-17) arXiv:1511.06406 (2017)
21. Pu, Y., Gan, Z., Henao, R., Yuan, X., Li, C., Stevens, A.: Variational autoencoder for deep learning of images, Labels and Captions. In: 30th Conference on Neural Information Processing Systems (NIPS 2016), arXiv:1609.08976 (2016)
22. Vincent, P.: A connection between score matching and denoising autoencoders. Neural Comput. **23**(7), 1661–1674 (2011)
23. Mao, X.J., Shen, C., Yang, Y.B.: Image restoration using convolutional autoencoders with symmetric skip connections. In: Proceeding of NIPS (2016)
24. Wang, Z., Bovik, A. C., Sheikh, H. R., Simoncelli, E. P.: Image quality assessment: from error visibility to structural similarity. IEEE Tran. of Image Processing **13**(4) (2004)
25. Niemeijer, M., Staal, J.J., Ginneken, B., Loog, M., Abramoff, M.D.: Comparative study of retinal vessel segmentation methods on a new publicly available database In: Fitzpatrick, J.M., Sonka M. (eds.), SPIE Medical Imaging, SPIE 5370, pp. 648-656 (2004)
26. Available from: https://www.isi.uu.nl/Research/Databases/DRIVE/
27. Hoover, A., Kouznetsova, V., Goldbaum, M.: Locating blood vessels in retinal images by piecewise threshold probing of a matched filter response. IEEE Trans. Med. Imaging **19**(3), 203–210 (2000)
28. Available from: http://cecas.clemson.edu/~ahoover/stare/
29. Kalesnykiene, V., Kamarainen, J. K., Voutilainen, R., Pietil, J., Klviinen, H., Uusitalo, H.: DIARETDB1 diabetic retinopathy database and evaluation protocol (2007)
30. Gu, K., Zhai, G., Yang, X., Zhang, W.: Using free energy principle for blind image quality assessment. IEEE Trans. Multimedia **17**(1), 50–63 (2015)
31. Wang, Z., Bovik, A.C.: A universal image quality index. IEEE Signal Process. Lett. **9**(3) (2002)

Optical Marker- and Vision-Based Human Gait Biomechanical Analysis

Ganesh Roy, Thomas Jacob, Dinesh Bhatia and Subhasis Bhaumik

Abstract Human gait is an important area of study in the field of physical therapy, medical diagnostics and biomedical engineering. Using multi-camera system associated with necessary software tools, the 3D human gait analysis is prepared. The images captured from each of the camera are assembled together to develop a 3D model with complete motion of human gait pattern. The passive optical markers are attached to the different locations of the lower limb of a human subject. The static position and the dynamic movement of the markers are obtained, and a 3D model is initialized. By taking the raw data from each of the camera, 3D tracking is performed. Also, a force plate placed in the walking platform assists in procuring kinetics data involved in movement and locomotion. Finally, the six joints of the lower limb are tracked and an inverse kinematics along with inverse dynamics library for the human gait is developed and validated with analytical geometrical results.

Keywords Camera · Gait · 3D tracking · Inverse kinematics and dynamics

1 Introduction

Human locomotion is possible for the bipedal support of the lower limbs and the necessary movements in the required direction which is well known as gait. The gait analysis finds in various fields of applications such as clinical biomechanics, medical rehabilitation, prosthetics, and sports science etc. [1]. Patient's condition, treatment and required rehabilitation can be decided by the physiotherapist, orthopaedist and

G. Roy · S. Bhaumik
Indian Institute of Engineering Science and Technology, Shibpur, Howrah 711103, West Bengal, India

T. Jacob · D. Bhatia (✉)
North-Eastern Hill University, Shilong 793022, Meghalaya, India
e-mail: bhatiadinesh@rediffmail.com

G. Roy
Central Institute of Technology, Kokrajhar 783370, Assam, India

S. Bhattacharyya et al. (eds.), *Hybrid Machine Intelligence for Medical Image Analysis*, Studies in Computational Intelligence 841,
https://doi.org/10.1007/978-981-13-8930-6_11

neurologist utilizing the proper analysis of the gait motion. In case of medical rehabilitation problem when a person is unable to walk after affected by SCI (spinal cord injury) or stroke, an external electromechanical supporting device called lower limb exoskeleton is provided to regenerate his mobility. The lower limb exoskeleton device is governed by a suitable power backup. And the human-like gait motion trajectory has been provided externally through the essential arrangement. In the light of these issues, the authors get the motivation of the present work. Therefore, proper detection of the gait pattern is quite important. A survey report to guide the rehabilitation technique for tracking human motion is available in [2]. Mainly there are two kinds of procedures available to record the gait motion [3]. The first one is contact-based direct measurement technique which includes accelerometer and goniometer sensors to find the necessary movement. Another one is the vision-based technique using multi-camera system which is popular and accurate for the three-dimensional (3D) tracking of the human movement. The second method is further classified into two types either marker-less or with active/passive markers. Globally extensive research work is being carried out to improve the human gait library with the help of these techniques to analyse the biomechanics of human gait [4–6]. The optical system based on passive markers has been reported for analysis the kinematics of gait by Prakash et al. [7]. In another application, the marker-less Kinect camera-based gait analysis has been presented [3, 8]. In the present work, we are employing Simi Motion 9.0 system from Simi Reality Motion Systems GmbH, Germany, which uses light reflective passive markers [9]. The main aim of the software system is image processing for analysing motion. By the change in coordinates in a certain time span, the motion of a subject can be created physically where coordinates and camera calibration are the main focus for the image processing. For selecting an impeccable image for processing, two things should be considered. Firstly, the motion plane must fill as much of the picture or image area as possible. Secondly, there should be a considerable distance between the motion plane and the focal plane [9].

There are a number of parameters which are useful for studying the human gait theoretically, clinically and biomechanically [10]. Out of them inverse kinematics and inverse dynamics analysis are the most important choices [11]. However, the parameters recorded from a vision-based system and the validation of those values analytically have never been published elsewhere as far as the author's knowledge goes. Hence, in the current work, the 3D human gait motion has been tracked by the passive marker-based system. Totally eight number of cameras are used for image recording purpose. A group of 23 markers have been connected in the different specified parts of the lower body to get the exact kinematics of the six different joints of the lower body. The joints are namely hip, knee and ankle for the both legs. All the inverse kinematics parameter data has been generated from the captured video file from those cameras. Additionally, one force plate having dimension of 40 cm × 60 cm is placed on the way of the walking path. The reaction forces generated from the force plate are utilized to calculate the corresponding joint moments. Therefore, the inverse dynamics parameters can be determined from the video recording. The

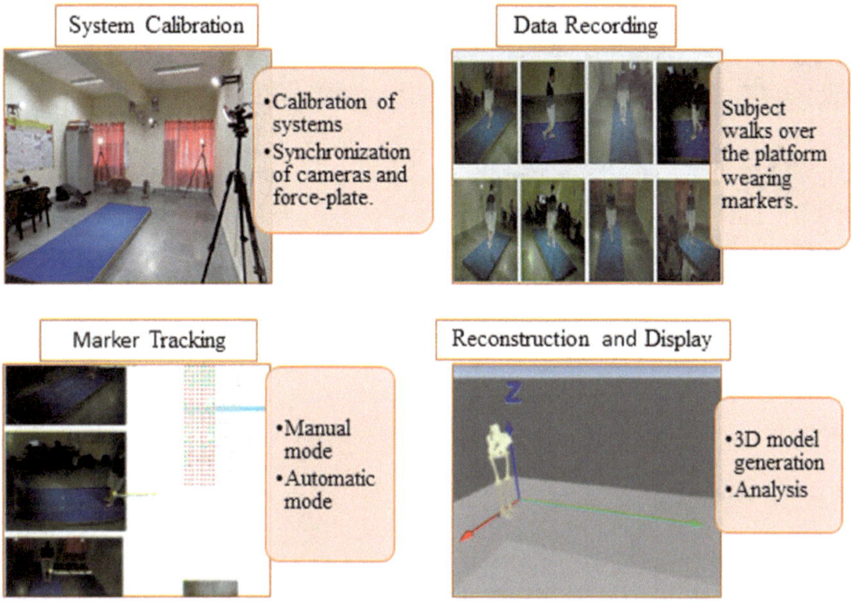

Fig. 1 Overall block diagram

overall block diagram has been depicted in Fig. 1. In the subsequent sections of this paper, all the steps shown in Fig. 1 have been elaborated.

2 Methodology of 3D Gait Analysis System

2.1 Camera Calibration

To get the error-free reading, the calibration for all the cameras is highly recommended. For obtaining 3D video recordings, a rate of frame of 25 fps or 50 fields per second (PAL) or 30/60 (NTSC) is required. We have used Basler acA1300-75gc GigE camera which captures videos at a frame rate of 50 fps with a resolution of 1.3 mega pixels (MP). To create a 3D reconstruction, at least two cameras are required. The alignment of the optical axes of the two cameras should be in between 60° and 120°. And also the images should be recorded concurrently.

Two types of calibrations are done: static (2D) and dynamic (3D). Static calibration refers to the removal of distortion in the image caused by the camera lens. Most common is the barrel distortion where the vertical and horizontal straight lines appear as curved lines, more curved towards the periphery of the visual field of the camera causing a fisheye effect. If x and y are the coordinates for ideal image projection, the actual coordinate (x_d, y_d) is given as in Eq. (1).

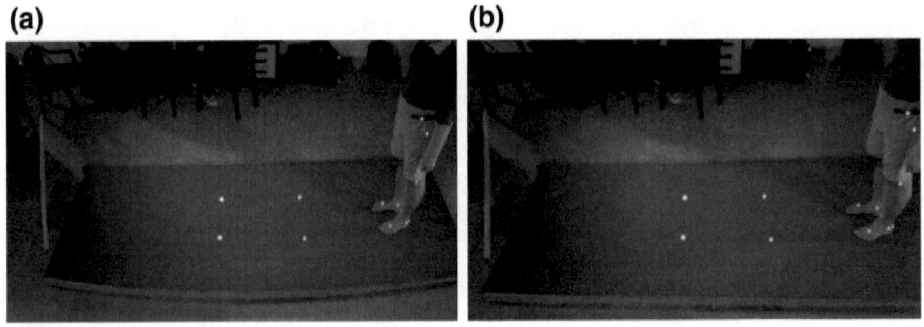

Fig. 2 Distorted (**a**) and undistorted (**b**) images

$$\begin{bmatrix} x_d \\ y_d \end{bmatrix} = \begin{bmatrix} x + \Delta x \\ y + \Delta y \end{bmatrix} = D(r)\begin{bmatrix} x \\ y \end{bmatrix} \quad (1)$$

where $(\Delta x, \Delta y)$ are the deviation of actual points from ideal projection points along with the *X*- and *Y*-axis. $D(r)$ is the radial distortion factor which is a function of radius, r [12]. This distortion is corrected by calibrating the camera using a calibrating device (checkerboard or wand). The system identifies the predefined calibration of object and straightens its image by creating straight line connecting the detected points. As an example, the distorted and undistorted images have been depicted in Fig. 2.

The motion system transforms a 3D world point (X_W) to a 2D image point (X_I) according to the principle of central projection as noted in Eq. (2).

$$X_I = P X_W \quad (2)$$

The camera projection matrix (*P*) describes the parameters that map X_W to X_I. Calibration helps to obtain the parameters that define camera projection matrix (*P*).

2.2 Data Recording

The point's coordinates are recorded as a raw data during motion capture. There are two methods available for data recording. The first method called manual tracking that needs to operate by selecting the respective points or frames separately. The automatic tracking is the second method where the image processing algorithm is used to track the points. Before starting of the data recording, 23 markers are to be placed for tracking. The light reflector markers are used in different standard position of the human body so that the lower limb joints can be determined after tracking those marks. In manual tracking, the required anatomical points are selected in each camera by clicking and progressing to next frame to repeat the whole process. The process of manual tracking has been shown in Fig. 3. In automatic tracking, the first frame is manually tracked and based on pattern recognition; the consequent movements

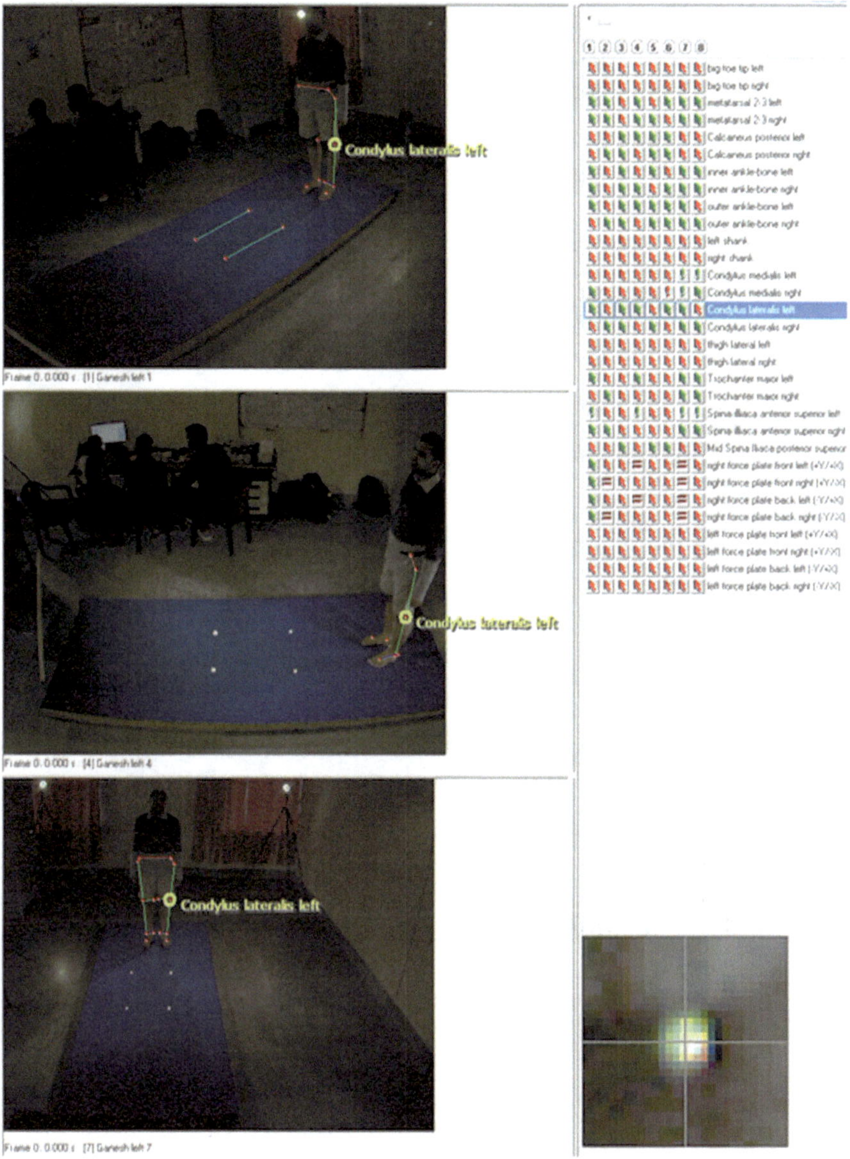

Fig. 3 Manual 3D tracking

of the selected points are tracked. Any error in the automatic tracking needs to be manually checked and corrected. When all the video frames and points in the required sequence have been processed, then motion capture is now completed and the data is available for the 3D calculations.

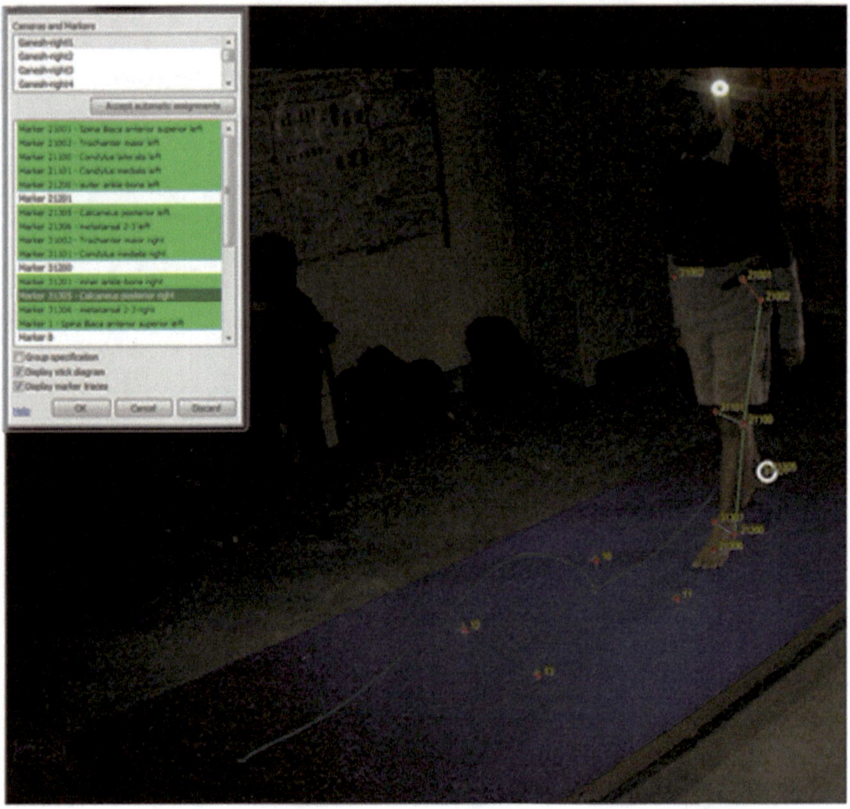

Fig. 4 Automatic marker tracking

2.3 *Marker Tracking*

The marker visibility and contrast with the background are the major concerns for the automatic tracking method. At first, all the marker points in a single frame are required to track manually to avail a better and easier analysis from the automatic tracking method. 2D coordinates for each point are obtained by each camera. The set of points are projected onto a 3D space to obtain the 3D reconstructed image using Eq. (2). The calibration parameters are employed to produce the 3D model. The pictorial view of the automatic tracking has been depicted in Fig. 4.

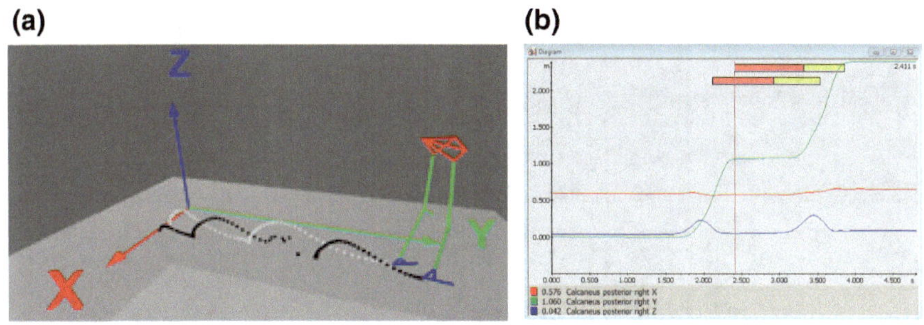

Fig. 5 **a** Line drawing and point tracing. **b** Graphical representation of data

2.4 Calculating the Scaled 3D Coordinates

To calculate the 3D coordinates, every camera is not required to be used. In theory, only two cameras are necessary to carry out a 3D calculation. However, more than two cameras are used to compensate for the hidden markers in some frames, resulting in more accurate calculations with fewer problems for model development. If a point is missed in 2–3 frames, it can be obtained by interpolation. Incorrect markings or momentary transition of markers can be corrected by moving average filter with a reasonable window length.

2.5 Reconstruction and Displaying the Data

All the recorded data can be displayed by initializing the model. Trajectory of any point could be traced, and line joining any two points can be obtained from the model. The data can, alternatively, be obtained as graphs, or the values can be exported in tabular form. The data includes the coordinates of each point, the joint angles or any new data that can be created by mathematical or statistical operation on the results of the tracking. Figure 5a shows the line model of the subject in 3D space and a trace of posterior calcaneus of both legs. The collected data is displayed in Fig. 5b.

3 Experimental Setup

The experimental setup consists of eight digital video cameras, a walking platform with a force-plate integrated to it and a personal computer with necessary software and data acquisition system. All the cameras having model number of Basler acA1300-75gc GigE which capture videos at a rate of 50 fps with a resolution of 1.3 mega pixels (MP) are used for capturing movement data. The cameras are initialized in

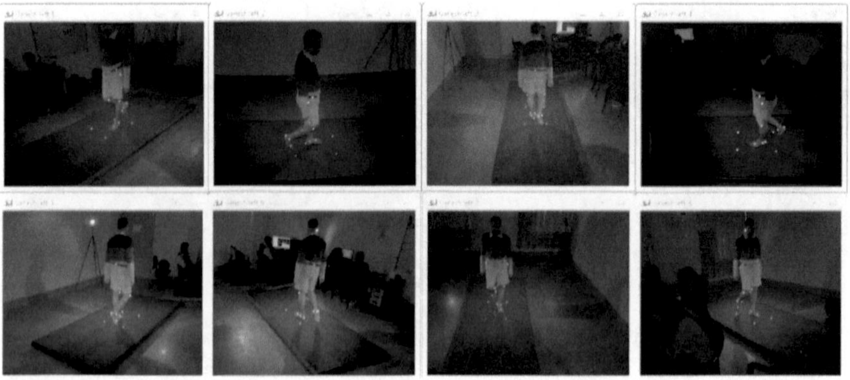

Fig. 6 Walking motion of subject on the force platform

the Simi Motion software along with the force plate for synchronous data recording. The length and width of the walking platform are 3.096 m and 1.260 m, respectively. Bertec Force plate 4060, with dimensions 600 mm × 400 mm, is mounted on the platform. When the subject walks on the platform, the eight cameras record the video simultaneously. One frame has been depicted with respect to eight cameras in Fig. 6 when the left foot is on the force plate. The force plate is marked by placing four markers over the walking platform.

A total number of 23 different optical passive markers are attached to the body in the different positions of the lower limb. Figure 7 shows the positions of the markers. Here only the right leg positions are labelled. The left leg markers can also be labelled in the similar way.

3.1 Force-Plate Coordinate System

The moments generated at each joint of the human gait cycle are associated with the ground reaction force (GRF) which is measured by using a force plate in the gait laboratory. A local coordinate scheme is considered for the measurement purpose. The axes X_f, Y_f and Z_f have shown in Fig. 8 for the ground reaction force in the middle of the plate for that reason. The fixed coordinate system placed in the corner of the force plate having the axis of X_L, Y_L and Z_L. The position of local coordinate system is measured with respect to the fixed coordinate system in any motion analysis system. It is notable from Fig. 8 that the *X* and *Z* axes for both the coordinate systems are placed opposite to each other, whereas the *Y* axes are in same direction. The position of the origin is located as 20 and 30 cm from the fixed coordinate along *X* and *Y* directions respectively. This position is represented by a vector, $\bar{r}$ by Eq. (3).

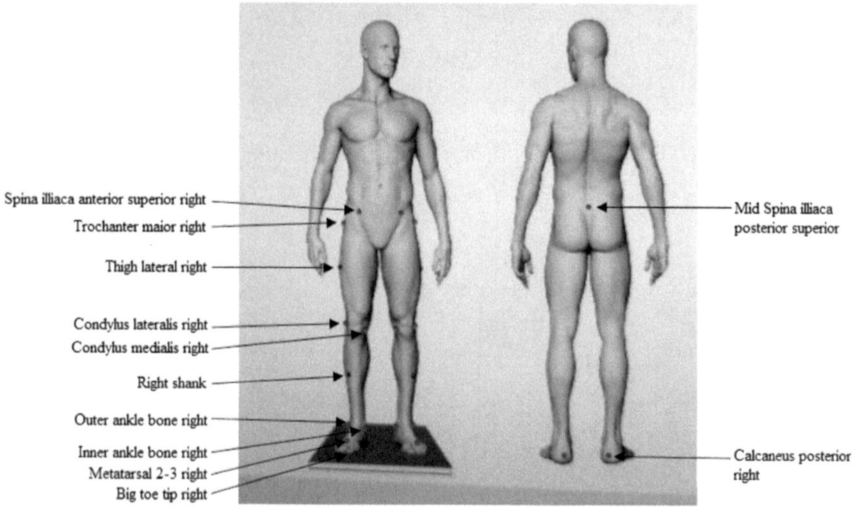

Fig. 7 Different positions of markers [9]

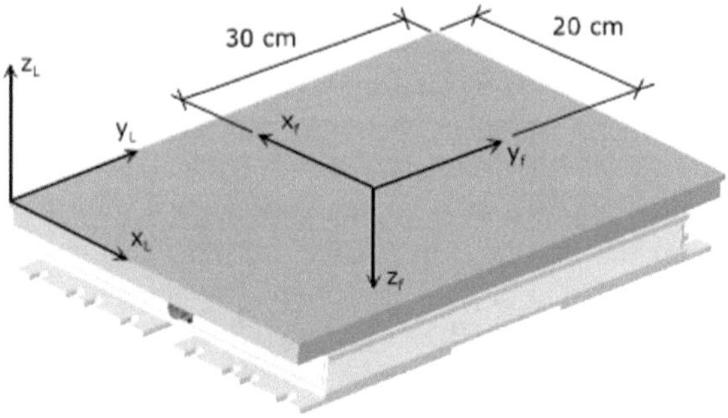

Fig. 8 Ground reaction force [13]

$$\bar{r} = \begin{bmatrix} 20 \\ 30 \\ 0 \end{bmatrix} \quad (3)$$

It is already have been shown that the *X*- and *Z*-axis direction between the two coordinates are opposite to each other and the *Y* axes are in same direction. Now, the measured plate angle (θ) would be $\theta_{11} = \theta_{33} = 180°$ for *X* and *Z* axes, respectively, and the $\theta_{22} = 0°$ for *Y* axes. Except these three angles all the angles are 90° or – 90°. The transformation matrix calculated in this way is given in Eq. (4).

$$T = \begin{bmatrix} -1 & 0 & 0 \\ 0 & 1 & 0 \\ 0 & 0 & -1 \end{bmatrix} \tag{4}$$

The new force (F') and moment (M') values are obtained as given in Eqs. (5) and (6), respectively.

$$\begin{bmatrix} F'_x \\ F'_y \\ F'_z \end{bmatrix} = [T] \cdot \begin{bmatrix} F_x \\ F_y \\ F_z \end{bmatrix} \tag{5}$$

$$\begin{bmatrix} M'_x \\ M'_y \\ M'_z \end{bmatrix} = [T] \cdot \begin{bmatrix} M_x \\ M_y \\ M_z \end{bmatrix} + \bar{r} \times \begin{bmatrix} F'_x \\ F'_y \\ F'_z \end{bmatrix} \tag{6}$$

4 Experimental Results

After recording the videos and processing them, the inverse kinematics and inverse dynamics data library has been prepared from 11 healthy subjects. In this paper, only a 30-year-old male subject's data has been provided. The subject having weight and height of 74 kg and 170 cm, respectively, with normal walking pattern is considered. The walking speed is recorded as 0.90 m/s. Although the results are available in the three different planes, for the convenience and most significant use in [14, 15], the sagittal plane is presented in this description. The ankle inverse kinematics and kinetics are obtained from the Simi Reality Motion Systems which has been shown in Fig. 9a, b, respectively. The images describe the ankle angle, i.e. the dorsiflexion and plantar-flexion variation and the necessary torque produced at the ankle joint in a full gait cycle. The knee joint and the hip joint flexion/extension are plotted in Figs. 10a and 11a, respectively. Figures 10b and 11b represent the torque profile for knee and hip joint. For the kinetics only the right leg data is provided. A comparative study between the experimentally recorded data in the gait laboratory and the analytically calculated data is reported in Table 1.

5 Discussions

The kinetics of the human motion is quite difficult to access without a force platform. Mathematically using a link segment model of the lower limb, the joints moments can also be determined [16]. This process is well known as inverse dynamics. In this process, the muscle activity, i.e. the resultant joint moments, can be calculated from the kinematics as in Eq. (7). The equation is better known as dynamic equation of motion [17].

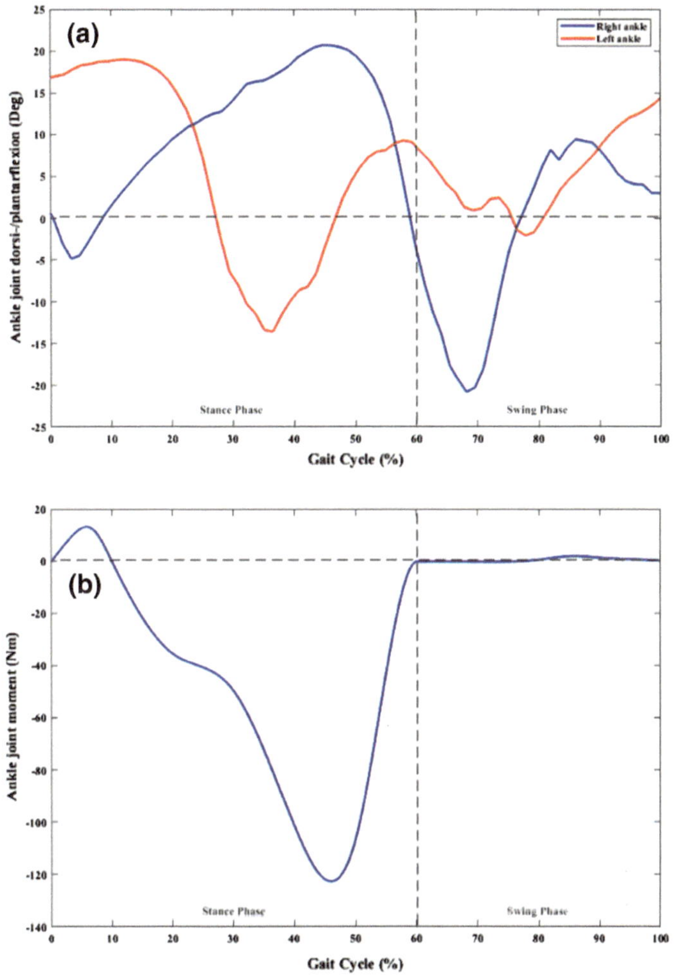

Fig. 9 **a** Ankle joint kinematics. **b** Ankle joint kinetics

$$M[q]\ddot{q} + C[q]\dot{q} + G[q] = \tau \tag{7}$$

where q be the position vector, M be the system mass matrix, C be the Coriolis and centrifugal force, G be the gravitational force, and τ be the joint torque.

The total lower body can be assumed as a model of link segment where all the segments are treated as rigid body as shown in Fig. 12a. The FBD (free body diagram) is drawn to determine the necessary moments for the joints.

As an example, the ankle joint moment can be determined from its FBD as shown in Fig. 12b. About the centre of mass (COM), the total muscle moment of any segment is described by Eq. (8).

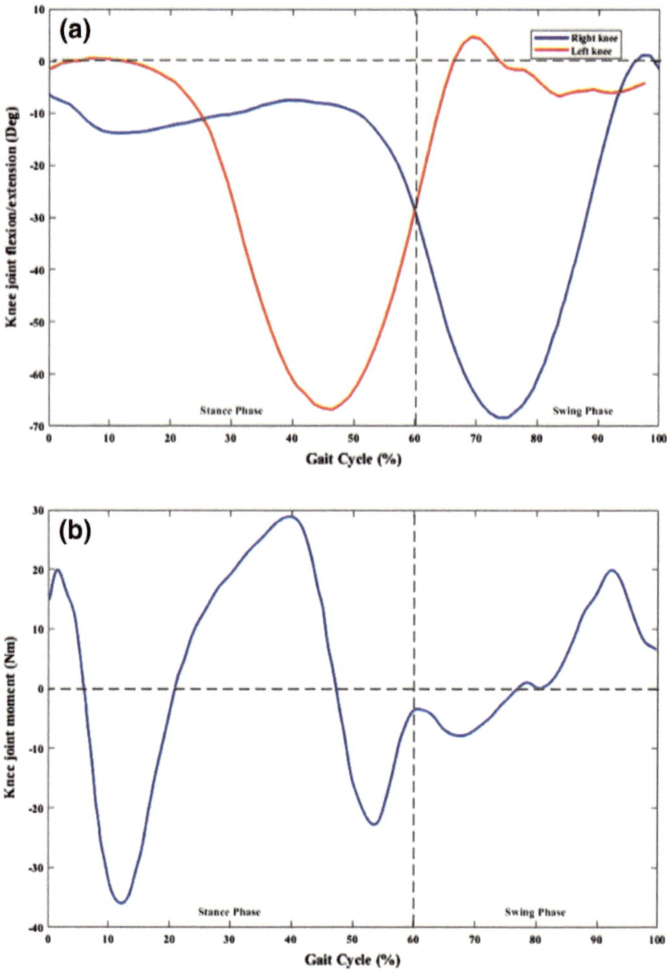

Fig. 10 **a** Knee joint kinematics. **b** Knee joint kinetics

$$M = I\alpha \tag{8}$$

where M = Total moment, I = Segment inertia, and α = Segment angular acceleration.

Three gait events have been considered in the present work to calculate the inverse dynamics geometrically from the kinematics and the force-plate data as given in Table 1.

Firstly, when the subject is standing in a static position by keeping his right foot on the force plate (20% gait cycle), the ground reaction force (GRF) is acting at a distance of d_1 from the ankle joint. This position is also known as centre of pressure (COP) as shown in Fig. 12b. The two reaction forces on the two ends of the ankle

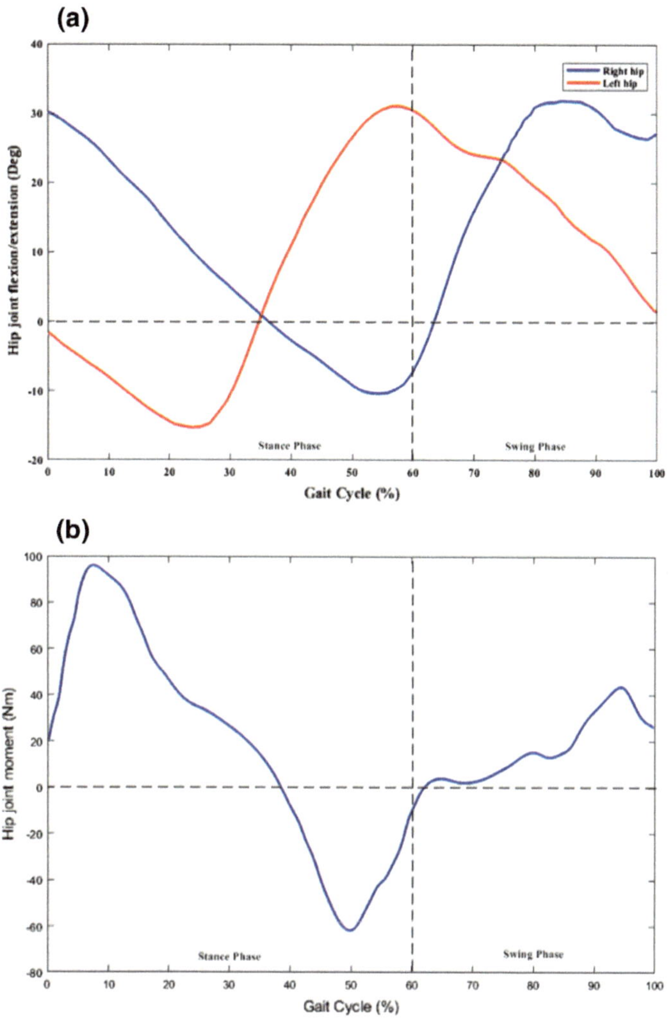

Fig. 11 **a** Hip joint kinematics. **b** Hip joint kinetics

link are R_1 and R_2. The y-axis component of R_1 is R_{1y} is actually the force due to gravitation acting in the upward direction. $R_{1y} = 74$ kg × 9.8 m/s^2 = 725.2 N. The net x- and y-directional forces are given by Eqs. (9a) and (9b), respectively.

$$R_{2x} + R_{1x} = ma_x \tag{9a}$$

$$R_{2y} + R_{1y} - m \times g = ma_y \tag{9b}$$

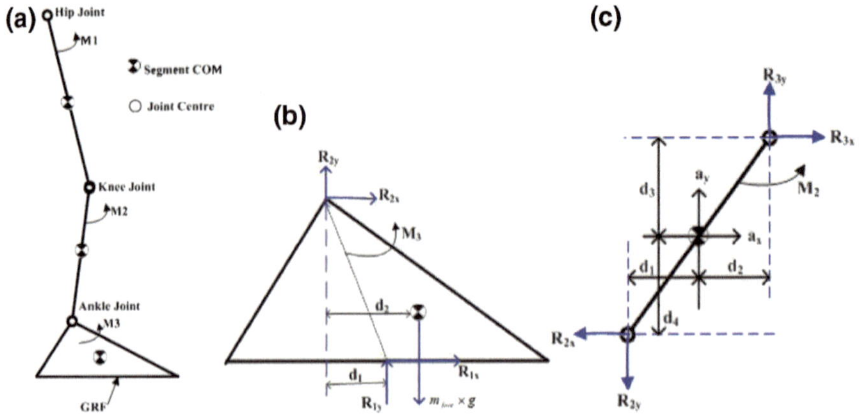

Fig. 12 **a** Link segment model of lower limb during stance phase, **b** FBD of foot and **c** FBD of shank

where m is the body segment weight (here as a foot) and a_x and a_y are the accelerations of the segment along x and y directions respectively. For static position, the $a_x = 0 = a_y$. If the mass of the foot, $m = 1$ kg, then the $R_{2y} = -715.4$ N.

Then, the resultant moment about the COM of the foot is given in Eq. (10).

$$M_3 - R_{1y} \times (d_2 - d_1) - R_{2y} \times d_2 = I\alpha \tag{10}$$

For the static position, the angular acceleration (α) becomes zero. Hence, the net ankle joint moment becomes $M_3 = -28.42$ N m, considering $d_1 = 4$ cm and $d_2 = 6$ cm. Here, the total muscle moments having negative sign indicate the clockwise rotation of ankle joint and beginning of plantar-flexion motion. The ankle joint

Table 1 Comparative study between analytical and experimental results

Gait event	Joint name	Joint moments (N m)	
		Analytical	Experimental
Midstance (20% of the gait cycle)	Ankle	−28.42	−35.48
	Knee	−9.57	−4.68
	Hip	11.37	46.12
Terminal stance (45% of the gait cycle)	Ankle	−120.43	−121.43
	Knee	−27.83	−15.58
	Hip	−24.75	−39.84
Midswing (80% of the gait cycle)	Ankle	1.59	1.01
	Knee	1.38	0.143
	Hip	1.51	15.04

moments shown in Fig. 9b clearly resemble the event. The inertial data and segment mass have been determined from the anthropometric relation [1]. The angular and linear velocities have been determined from the inverse kinematics data.

A second event is considered as the push-off condition (at 45% gait cycle) when the ankle joint realizes the maximum torque. Using Eqs. (9a) and (9b), the reaction forces are calculated as

$$\begin{aligned}
R_{1x} &= 91.17\ \text{N (From force plate)},\\
R_{2x} &= ma_x - R_{1x} = 1 \times 2.86 - 91.17 = -88.31\ \text{N}\\
R_{1y} &= 761.54\ \text{N and}\\
R_{2y} &= ma_y + m \times g - R_{1y} = 1 \times 1.70 + 1 \times 9.8 - 761.54 = -750.04\ \text{N}.
\end{aligned}$$

The total torque at ankle joint by considering $\alpha = -250.32\ \text{rad/s}^2$ is therefore

$$\begin{aligned}
M_3 + R_{1x} \times d_4 + R_{1y} \times (d_2 - d_1) - R_{2x} \times d_3 - R_{2y} \times d_2 &= I\alpha;\\
M_3 = 0.0139 \times (-250.32) - 91.17 \times 0.082 - 761.54 \times 0.077&\\
- 750.04 \times 0.054 - 88.31 \times 0.074&\\
M_3 = -120.43\ \text{N m}&
\end{aligned}$$

Lastly, the swing phase (80% gait cycle) is considered, where there is no effect of the GRF. At this time, the reaction forces are

$$\begin{aligned}
R_{1x} &= 0,\ R_{2x} = ma_x = 1 \times 7.6 = 7.6\ \text{N},\ R_{1y} = 0\ \text{and}\\
R_{2y} &= ma_y + m \times g = 1 \times (-3.32) + 1 \times 9.8 = 6.48\ \text{N}
\end{aligned}$$

Hence, the total torque attained by the ankle joint is

$$M_3 - R_{2x} \times d_3 - R_{2y} \times d_2 = I\alpha; \quad M_3 = 1.59\ \text{N m}.$$

In the same way, the knee joint moments have been determined. The FBD of the shank segment of the lower limb is depicted in Fig. 12c. According to the FBD, the knee joint moment is shown in Eq. (11).

$$M_2 - M_3 - R_{2x} \times d_4 + R_{2y} \times d_1 - R_{3x} \times d_3 + R_{3y} \times d_2 = I\alpha \tag{11}$$

For the hip joint moment, the FBD should be same as Fig. 12c. Only differences are the lower-end and upper-end reaction forces R_3 and R_4, respectively. Hence, Eq. (12) represents the final equation of motion for the hip joint.

$$M_1 - M_2 - R_{3x} \times d_4 + R_{3y} \times d_1 - R_{4x} \times d_3 + R_{4y} \times d_2 = I\alpha \tag{12}$$

All the analytically calculated values for the three gait events have been given in Table 1. When the torque is attending a positive value, it indicates the counterclockwise rotation for the corresponding link segment and vice versa. The movement of the leg is caused by the combination of two reaction forces at each end of the leg segment.

6 Conclusion

Throughout this paper, the generation of normal human gait library and the different biomechanics parameters of the human gait are investigated. The acquired 3D inverse kinematics and dynamics are validated with an open-source software OpenSim3.3 [18]. Thereafter, the joint moments have also been calculated by following the geometrical load distribution method using link segment model. All the analytically calculated values of the joint moments closely resemble the optical marker vision-based system. Hence, it has been observed and verified that the optical marker-based multi-camera system is quite efficient and accurate method to find out appropriate human gait patterns for biomechanics and biomedical studies. The gait library generated in this present work can easily be transferred to create a human-like motion of a lower limb exoskeleton robotic device as a medical rehabilitation research application.

References

1. Winter, D.A.: Biomechanics and Motor Control of Human Movement, 3rd edn. Wiley, New York (2009)
2. Zhou, H., Hu, H.: Human motion tracking for rehabilitation—a survey. Biomed. Signal Process. Control **3**(2008), 1–18 (2008)
3. Roy, G., Bhuiya, A., Mukherjee, A., Bhaumik, S.: Kinect camera based gait data recording and analysis for assistive robotics-an alternative to goniometer based measurement technique. Procedia Comput. Sci. **133**, 763–771 (2018)
4. Moeslund, T.B., Hilton, A., Krüger, V.: A survey of advances in vision-based human motion capture and analysis. Comput. Vis. Image Underst. **104**, 90–126 (2006)
5. Colyer, S.L., Evans, M., Cosker, D.P., Salo, A.I.: A review of the evolution of vision-based motion analysis and the integration of advanced computer vision methods towards developing a markerless system. Sports Med. Open **4**, 24 (2008)
6. Mehrizi, R., Peng, X., Xu, X., Zhang, S., Metaxas, D., Li, K.: A computer vision based method for 3D posture estimation of symmetrical lifting. J. Biomech. **69** (2018). https://doi.org/10.1016/j.jbiomech.2018.01.012
7. Prakash, C., Gupta, K., Mittal, A., Kumar, R., Laxmi, V.: Passive marker based optical system for gait kinematics for lower extremity. Procedia Comput. Sci. **45**, 176–185 (2015)
8. Destelle, F., Ahmadi, A., O'Connor, N.E., Moran, K., Chatzitofis, A., Zarpalas, D., Daras, P.: Low-cost accurate skeleton tracking based on fusion of kinect and wearable inertial sensors. In: 2014 22nd European Signal Processing Conference (EUSIPCO), Lisbon, pp. 371–375 (2014)
9. www.simi.com

10. Vaughan, C.L., Davis, B.L., O'Connor, J.C.: Dynamics of Human Gait, 2nd edn. Kiboho Publisher, Cape Town (1992)
11. Davy, D.T., Audu, M.L.: A dynamic optimization technique for predicting muscle forces in the swing phase of gait. J. Biomech. **20**(2), 187–201 (1987)
12. Bader, J., Senner, V.: Validation of a dynamic calibration method for video supported movement analysis. Master thesis, Technische Universität München (2011)
13. Operating Manual: Bertec Force plates, BertecCoopereation, Version 1.0.0 (March 2012)
14. Zoss, A.B., Kazerooni, H., Member, IEEE, Chu, A.: Biomechanical design of the Berkeley lower extremity exoskeleton (BLEEX). IEEE/ASME Trans. Mech. **11**(2) (2006)
15. Latif, N., Shaari, A., Isa, I.M., Jun, T.C.: Torque analysis of the lower limb exoskeleton robot design. ARPN J. Eng. Appl. Sci. **10**(19) (2015)
16. Kirtley, C.: Clinical Gait Analysis. Theory and Practice. Churchill Livingstone (2006)
17. Pizzolato, C., Reggiani, M., Modenese, L., Lloyd, D.G.: Real-time inverse kinematics and inverse dynamics for lower limb applications using OpenSim. Comput. Methods Biomech. Biomed. Eng. **20**(4), 436–445 (2017). https://doi.org/10.1080/10255842.2016.1240789
18. Delp, S.L., Anderson, F.C., Arnold, A.S., Loan, P., Habib, A., John, C.T., Guendelman, E., Thelen, D.G.: OpenSim: open-source software to create and analyze dynamic simulations of movement. IEEE Trans. Biomed. Eng. **54**(11), 1940–1950 (2007)

Author Index

S. Bhattacharyya et al. (eds.), *Hybrid Machine Intelligence for Medical Image Analysis*, Studies in Computational Intelligence 841,
https://doi.org/10.1007/978-981-13-8930-6